Switch-Mode Power Converters

Switch-Mode Power Converters

Design and Analysis

Keng C. Wu

ELSEVIER
ACADEMIC
PRESS

Amsterdam • Boston • Heidelberg • London • New York • Oxford
Paris • San Diego • San Francisco • Singapore • Sydney • Tokyo

Elsevier Academic Press
30 Corporate Drive, Suite 400, Burlington, MA 01803, USA
525 B Street, Suite 1900, San Diego, California 92101-4495, USA
84 Theobald's Road, London WC1X 8RR, UK

This book is printed on acid-free paper. ∞

Library of Congress Cataloging-in-Publication Data
Wu, Keng C., 1948-
Switch-mode power converters / Keng C. Wu.
p. cm.
Includes bibliographical references and index.
ISBN 0-12-088795-9 (alk. paper)
1. Electric current converters. 2. Switching power supplies. I. Title.
TK7872.C8W853 2006
621.31'7–dc22

2005014319

British Library Cataloguing in Publication Data
A catalogue record for this book is available from the British Library

For information on all Academic Press publications visit our Web site at www.academicpress.com

Transferred to Digital Printing 2009

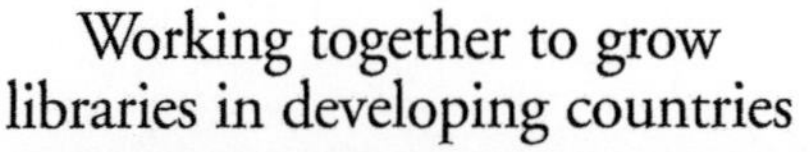

To
My wife, Shwu
My daughter, Stephanie
My mother, Tai

Table of Contents

Preface

This is not a cookbook, for switch-mode power converter design is a serious topic that must be treated with the utmost care. Therefore, the book makes a major departure from most existing texts covering the same subjects. It uses mathematics extensively, employing, for example, symbolic closed-form solutions for conduction times of a loaded full-wave-rectifier with a capacitor filter. At the first sight, readers may feel discouraged, but there is no shortcut. I sincerely urge readers to be patient, for the reward is profound.

The book covers in depth the three basic topologies: step-down (buck, forward), step-up (boost), step-down/up (flyback); push–pull; current-fed; resonant converters and their derivatives; AC–DC power factor correction. Depending on the operating conditions, switch-mode power converters may operate either in continuous conduction mode (CCM) or discontinuous conduction mode (DCM). Under transient conditions, the operation of power converters may slide in and out of both modes. For closed-loop control of converters, two fundamental mechanisms, voltage-mode control or current-mode control, are generally employed. Current-mode control has been understood to offer superior performance. Current mode control is further subdivided into average-current control and peak-current control. While most switch-mode converters utilize pulse-width modulation, resonant converters use frequency modulation. In addition to the main operation mechanism, many supporting circuits are also needed to make power converters viable. These include switch drivers, error amplifiers, and feedback isolators.

The presentation follows a fairly consistent pattern. The relationship between steady-state output and control variables (duty cycle, in the case of PWM, or frequency, in the case of resonance) is established first for both the CCM and the DCM operation. By examining the cyclical current waveforms of CCM, geometrical properties of the waveforms are extracted. These lead to the identification of critical inductance, which marks the boundary distinguishing CCM and DCM operation.

Under each operation mode and given a selected control mechanism, steady-state closed-loop output formulation that includes feedback ration, error amplifier, PWM gain (or frequency-modulation gain), and power stage is then established. In some simplified cases that exclude losses, the output formulation may be placed in the explicit form. When losses are included, the desire to obtain an explicit form is prohibitively impractical and abandoned. Instead, implicit functions and Jacobian determinants are employed to study output sensitivity and regulation.

With the steady state firmly established, the small-signal AC stability issues are examined for both control modes. Loop stability with voltage-mode control based on the average model (Dr. R. Middlebrook) is formulated and validated. Current-mode control necessitates the addition of current-loop gains surrounding the original average mode. In effect, the Middlebrook average model is extended to current-mode control and remains as valid.

This book also introduces accelerated steady-state analysis in the time domain. The technique connects the concept of the continuity of state and the periodic, steady-state output of converters. The analysis uses two approaches: Laplace transformation and state transitions. The latter calls on eigenvalues, eigenvectors, and matrix exponentials, the core of matrix theory associated with system theory.

Nowadays, simulations always play some role in almost all fields of studies. For power converters, there is no exception. This book, however, approaches it from a more fundamental way, which is quite distinctive from the graphic-based simulations available commercially. The latter suffers convergence issues frequently. Our approach avoids such nagging difficulties.

The book is written for those already exposed to the basics of switch-mode power converters and seek higher dimensions. It is suitable for graduate students and professionals majoring in electrical engineering. In particular, readers with training in linear algebra will find the techniques of state transition being applied very inspiring.

Acknowledgment

Finally and most importantly, profound gratitude is extended to Charles B. Glaser, senior acquisition editor and his staff at Elsevier Inc., Burlington, MA ; Annie Martin, production director, Elsevier Ltd., England; and Sheryl Avruch, copyeditors, typesetters, and staff at SPI Publisher Services.

Chapter 1

Isolated Step-Down (Buck) Converter

The power stage of an isolated buck converter in its simplest form is presented in Figure 1.1. Depending on the output loading and the value of filter inductor L, the power stage can be operated in two distinctive modes: continuous conduction mode (CCM) and discontinuous conduction mode (DCM). In the CCM, the inductor current, i, always stays above zero. In the DCM, the current, for a certain duration, stays at zero. It is also understood that, in the CCM, the power stage alternates between two topologies while, on the contrary, it experiences three in the DCM.

1.1 CCM Open-Loop Output and Duty Cycle Determination

If ideal rectifiers are assumed and series losses are ignored, the requirement of flux conservation, that is, the volt-second balance, across the inductor gives

$$\left(\frac{N_s}{N_p} V_{in} - V_o\right) D \cdot T_s + (-V_o)(1 - D)T_s = 0 \tag{1.1}$$

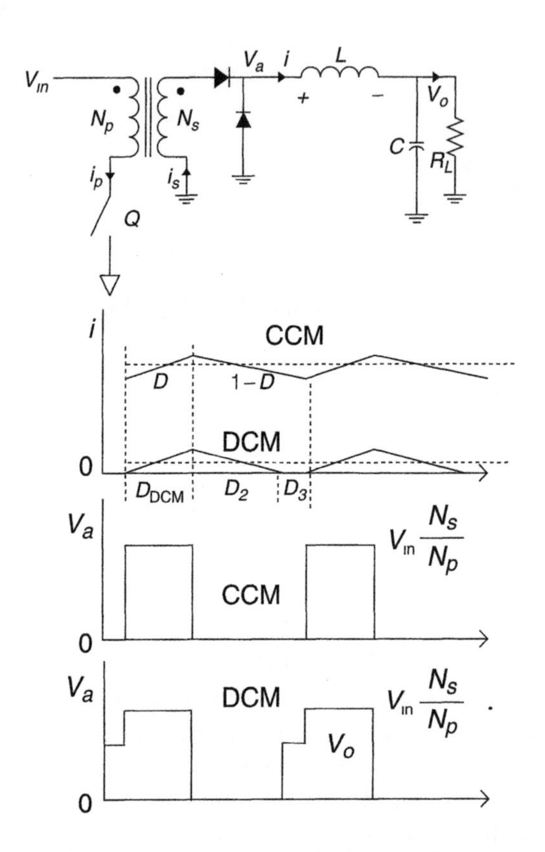

Figure 1.1: Power stage of an ideal forward converter

when the circuit alternates between two topologies under the steady state with a defined switch(Q)-on duty cycle, D, and a given clock rate T_s. Obviously, (1.1) results in

$$D = \frac{V_o}{\frac{N_s}{N_p} V_{\text{in}}} \tag{1.2}$$

As a matter of fact, (1.2) can also be given a different interpretation. That is, the rectangular wave, V_a, driving the loaded LC filter contains a DC component:

$$V_o = \frac{N_s}{N_p} V_{\text{in}} \cdot D \tag{1.3}$$

This latter view aligns well with the ultimate goal of the converter operation, extracting the average voltage embedded in the transformed input drive and regulating the output voltage by fine-tuning the turn ratio with variable duty cycle, D.

However, in reaching (1.1)–(1.3), we made an expedient, but unrealistic, assumption, which is the zero forward voltage a rectifier diode offers when it is conducting. We shall make the necessary corrections by first forgoing the assumption of the ideal diode. Rather, the rectifier's forward voltage is given a nonzero value, V_D. With it, and referring to Figure 1.2, (1.1)–(1.3) are modified and become

$$\left(\frac{N_s}{N_p}V_{\text{in}} - V_D - V_o\right)D \bullet T_s + (-V_D - V_o)(1 - D)T_s = 0 \tag{1.4}$$

$$D = \frac{V_o + V_D}{\frac{N_s}{N_p}V_{\text{in}}} \tag{1.5}$$

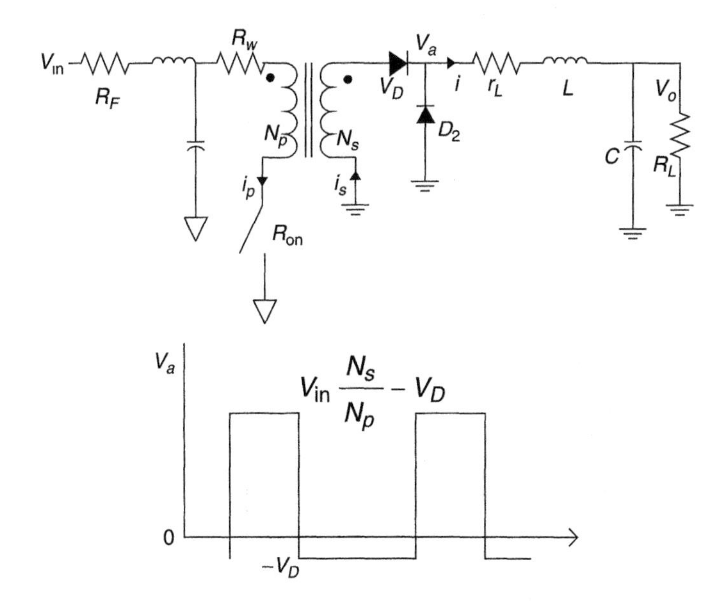

Figure 1.2: Nonideal power stage

$$V_o = \frac{N_s}{N_p} V_{\text{in}} \bullet D - V_D \tag{1.6}$$

Next, we consider series losses—first the secondary side losses then the primary side losses. We include the secondary side losses by examining (1.6). What (1.6) offers is the voltage presented by an ideal source that has zero source resistance. If a nonzero source resistance, r_L, exists, (1.6) evolves into

$$V_o = \frac{N_s}{N_p} V_{\text{in}} \bullet D - V_D - r_L \bullet \frac{V_o}{R_L} \tag{1.7}$$

or

$$V_o = \frac{\frac{N_s}{N_p} V_{\text{in}} \bullet D - V_D}{1 + \frac{r_L}{R_L}} \tag{1.8}$$

We also note the presence of primary side resistance, including the input filter series resistance, the transformer primary winding resistance, and the switch-on resistance. It is also understood that the input filter resistance experiences a DC current while the transformer's primary winding resistance registers a pulsating current. In other words, (1.8) is modified as

$$V_o = \frac{\frac{N_s}{N_p}\left[\begin{array}{l} V_{in} - \frac{N_s}{N_p} \bullet \frac{V_o}{R_L}(R_w + R_{\text{on}}) - \\ \frac{N_s}{N_p} \bullet \frac{V_o}{R_L} \bullet D \bullet R_F \end{array}\right] D - V_D}{1 + \frac{r_L}{R_L}} \tag{1.9}$$

Readers are cautioned in applying (1.9), for it is an implicit function in V_o and a quadratic equation for D. With a little patience, (1.9) yields

$$V_o = \frac{(n \bullet D \bullet V_{\text{in}} - V_D)R_L}{[D^2 \bullet R_F + (R_w + R_{\text{on}})D]n^2 + R_L + r_L}, \quad n = \frac{N_s}{N_p} \tag{1.10}$$

It can also be reformulated as

$$n^2 \cdot R_F \cdot V_o \cdot D^2 + [n^2(R_w + R_{\text{on}})V_o - n \cdot R_L \cdot V_{\text{in}}]D + (R_L + r_L)V_o + R_L \cdot V_D = 0 \quad (1.11)$$

1.2 DCM Open-Loop Output and Duty Cycle Determination

The DCM operation is rarely used for actual design. However, it does have educational merits from an analytical point of view. We consider only diode losses for demonstration purposes. As shown in Figure 1.1, there are three distinctive operation intervals for the DCM. It is no longer a simple task identifying the duty cycle using the concept of volt–second balance alone. Yes, the concept is still, and always, applicable, but we need more than that. Again, we first apply Faraday's law of flux conservation:

$$(n \cdot V_{\text{in}} - V_D - V_o)D_{\text{DCM}} \cdot T_s + (-V_D - V_o)D_2 \cdot T_s = 0 \quad (1.12)$$

Equation (1.12), however, has two unknowns, D_{DCM} and D_2. We need one more equation. This need can be met by examining the inductor current form given in Figure 1.3.

The key is the fact that the load current, I_o, equals the average value contained in the current waveform. That is,

$$\frac{n \cdot V_{\text{in}} - V_D - V_o}{2 \cdot L} D_{\text{DCM}} \cdot T_s \cdot (D_{\text{DCM}} + D_2) \cdot T_s \cdot \frac{1}{T_s} = \frac{V_o}{R_L} \quad (1.13)$$

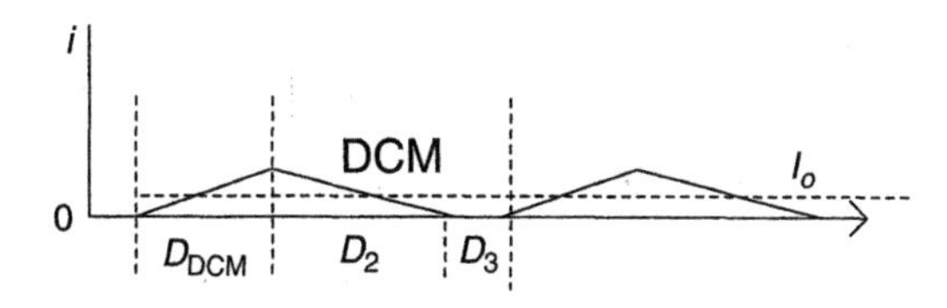

Figure 1.3: Inductor current for DCM

Equations (1.12) and (1.13) can then be solved together and the symbolic solutions are

$$D_{\mathrm{DCM}} = \sqrt{\frac{2 \cdot L \cdot f_s \cdot \dfrac{V_o}{R_L}}{(n \cdot V_{\mathrm{in}} - V_D - V_o)\dfrac{n \cdot V_{\mathrm{in}}}{V_D + V_o}}} \tag{1.14}$$

$$D_2 = \sqrt{\frac{(n \cdot V_{in} - V_D - V_o)2 \cdot L \cdot f_s \cdot \dfrac{V_o}{R_L}}{(V_D + V_o)n \cdot V_{\mathrm{in}}}} \tag{1.15}$$

1.3 CCM to DCM Transition, Critical Inductance

It is very interesting to compare (1.5) and (1.14). The obvious difference is in the form of equation (1.5), which is very simple, while (1.14) looks formidable with all circuit components and switching frequency, f_s, involved in setting the duty cycle. Readers may then ask, What critical part does a designer control to determine the mode of operation? The answer is the inductor. Given the required input, output, loading, and selected switching frequency, there is a critical inductor value that marks the boundary of CCM to DCM transition. How do we obtain that value? There is more than one way to determine the critical value. We will present two approaches.

The first approach recognizes that, when the operating condition changes to a point, the DCM duty cycle equals that of the CCM:

$$D_{\mathrm{DCM}} = \sqrt{\frac{2 \cdot L \cdot f_s \cdot \dfrac{V_o}{R_L}}{(n \cdot V_{\mathrm{in}} - V_D - V_o)\dfrac{n \cdot V_{\mathrm{in}}}{V_D + V_o}}} = \frac{V_o + V_D}{n \cdot V_{\mathrm{in}}} \tag{1.16}$$

Equation (1.16) yields the critical inductance:

$$L_{\mathrm{cri}} = \frac{(n \cdot V_{\mathrm{in}} - V_D - V_o)(V_o + V_D)}{2 \cdot f_s \cdot \dfrac{V_o}{R_L} \cdot n \cdot V_{\mathrm{in}}} \tag{1.17}$$

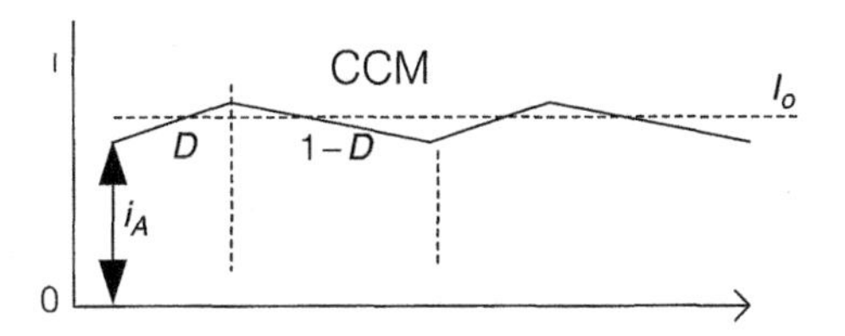

Figure 1.4: Inductor current for CCM

The other approach takes a little extra effort but gives additional insight. This time, the inductor current under the CCM operation is reexamined in Figure 1.4. An AC ripple current is superimposed on top of the DC load current, I_o.

The ripple current has a magnitude of

$$\Delta i = \frac{(n \bullet V_{\text{in}} - V_D - V_o)}{L} D \bullet T_s = \frac{(n \bullet V_{\text{in}} - V_D - V_o)(V_o + V_D)}{L \bullet f_s \bullet n \bullet V_{\text{in}}} \tag{1.18}$$

The trough magnitude is therefore

$$i_A = \frac{V_o}{R_L} - \frac{(n \bullet V_{\text{in}} - V_D - V_o)(V_o + V_D)}{2 \bullet L \bullet f_s \bullet n \bullet V_{\text{in}}} \tag{1.19}$$

It is easy to see that the power stage enters the DCM operation when the trough current equals zero. In other words, the condition $i_A = 0$ gives the critical inductance, and it is the same as (1.17).

1.4 Gain Formula for Nonideal Operational Amplifiers

In most existing electronics textbooks dealing with operational amplifiers, the concept of virtual ground, Figure 1.5, is often invoked. The concept emerges from the assumption that both the noninverting, V_1, and inverting, V_2, inputs track each other and that one of the inputs is generally at a fixed DC voltage. As a result, both inputs can be treated as zero potential for signal analysis purposes. However, both the logic and the concept suffer unnecessarily from many deficiencies. The first,

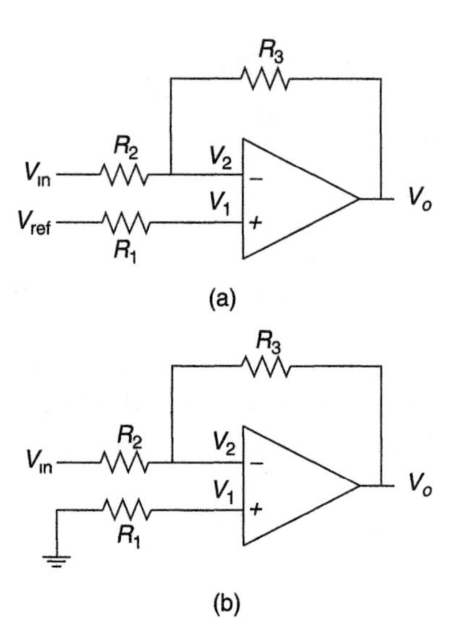

Figure 1.5: (a) Typical op-amp circuit, (b) inverting configuration

and perhaps the worst of all, is the assumption of infinite gain and bandwidth. Second, and no worse, is the missing information about the DC operating state. Third, the nonideal open-loop gain is not accounted for.

The situation can be improved significantly by getting rid of the virtual ground concept and using the voltage view and the superposition principle. Referring to Figure 1.6, the noninverting node gives $V_p = V_{\text{ref}}$, while the inverting node gives

$$V_n = \frac{Z_f(s) \cdot V_i + Z_i(s) \cdot V_o}{Z_f(s) + Z_i(s)} \tag{1.20}$$

The output is therefore given as

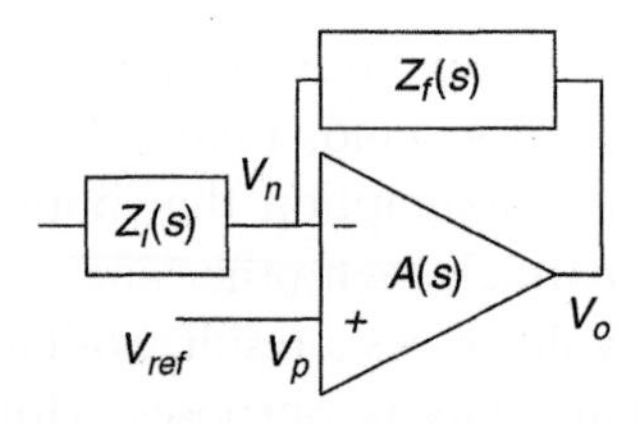

Figure 1.6: General op-amp circuit

$$V_o = A(s)(V_p - V_n) = A(s)\left[V_{\text{ref}} - \frac{Z_f(s) \cdot V_i + Z_i(s) \cdot V_o}{Z_f(s) + Z_i(s)}\right] \tag{1.21}$$

With further manipulation, (1.21) gives

$$V_o = \frac{V_{\text{ref}} - \dfrac{Z_f(s)}{Z_f(s) + Z_i(s)} \cdot V_i}{\dfrac{1}{A(s)} + \dfrac{Z_i(s)}{Z_f(s) + Z_i(s)}} \tag{1.22}$$

If $A(s) \gg 0$, (1.22) degenerates into

$$V_o = \left[1 + \frac{Z_f(s)}{Z_i(s)}\right] V_{\text{ref}} - \frac{Z_f(s)}{Z_i(s)} V_i \tag{1.23}$$

This is the form given in many books with the assumption of infinite open-loop gain and bandwidth. However, we stick with (1.22) from here on, since it accounts for the nonideal gain $A(s)$. As a matter of fact, the nonideal gain can also be approximated by a single, first-order pole

$$A(s) = \frac{A_0}{\dfrac{s}{2 \cdot \pi \cdot f_p} + 1} \tag{1.24}$$

where A_0 stands for the open-loop low frequency gain of op-amp (integrated circuit) and f_p is the 3-db roll-off frequency. These figures are always given in manufacturers' data sheets.

1.5 Feedback under Voltage-Mode Control

To obtain a precise and well-regulated output voltage against input or load changes, a feedback technique is always used in modern switch-mode power converters. Early converters, in the 1960s, tended to use voltage-mode control alone. By the late 1970s, the concept of current-mode control began to show up. A typical voltage-mode control scheme is shown in Figure 1.7.

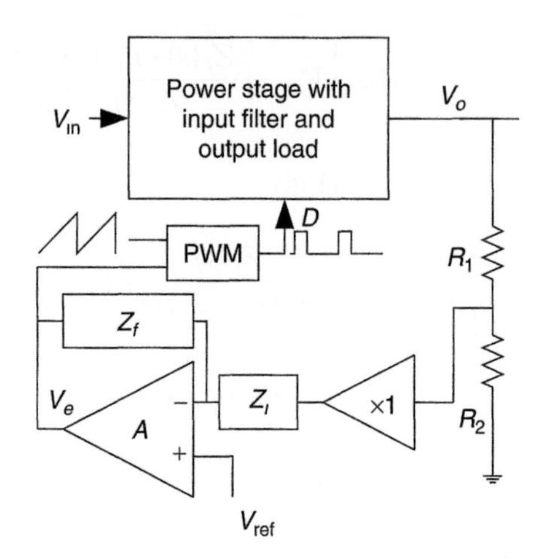

Figure 1.7: Voltage-mode control scheme

In this scheme, the output to be regulated is fed back through a resistive voltage divider and compared with a precision reference voltage, V_{ref}. The error voltage, V_e, is then feeding a pulsewidth modulator, which has an embedded sawtooth reference. We develop the feedback path gains based on this typical schematic. First, the error amplifier and its associated circuit is depicted in its steady-state form (Figure 1.8).

It is fairly straightforward to show that the error voltage is

$$V_e = A\left(R_3 \cdot I_b + V_{ref} - \frac{R_2}{R_1 + R_2} V_o - V_{os} - \frac{R_1 \cdot R_2}{R_1 + R_2} I_b\right) \tag{1.25}$$

where op-amp offset voltage and bias current are also accounted for.

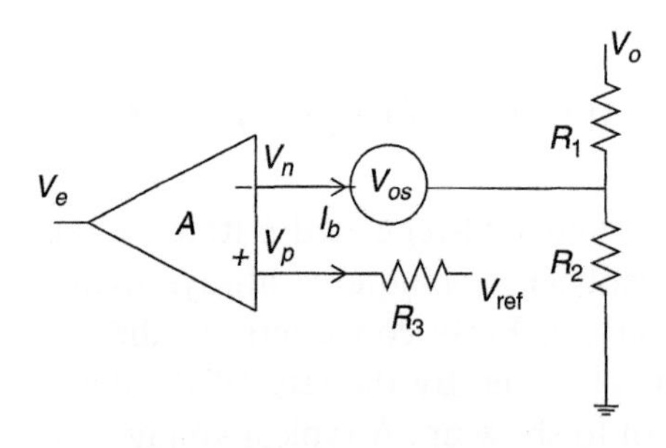

Figure 1.8: Error amplifier

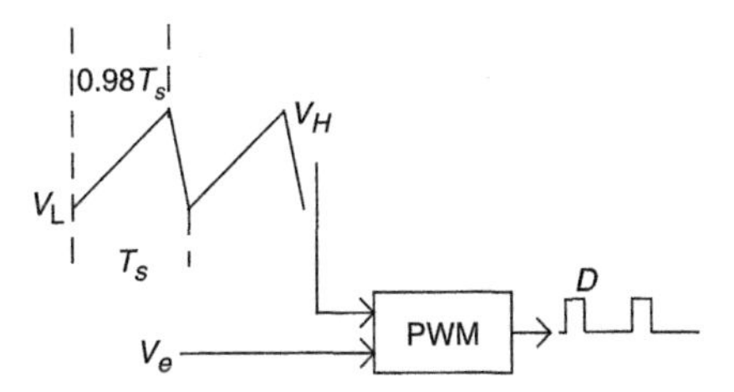

Figure 1.9: PWM block and external sawtooth

Next the PWM gain is derived. In the case of voltage-mode control, it is again quite simple. The circuit surrounding the PWM is given a little more detail as shown in Figure 1.9.

The sawtooth reference is a periodic clock consisting of two parts, the active up ramp and the dead-time down ramp. In general, the up ramp takes up 98% of the clock cycle and the down ramp the remaining 2%. The reference swings between two values, a nonzero V_L and V_H. Given that, the up ramp can be described as

$$V_L + \frac{V_H - V_L}{0.98} f_s \bullet t \tag{1.26}$$

At steady-state, the cyclic open-loop duty cycle is determined when the up ramp intercepts the error voltage:

$$A(V_p - V_n) = V_L + \frac{V_H - V_L}{0.98} f_s \bullet D \bullet T_s = V_L + \frac{V_H - V_L}{0.98} D \tag{1.27}$$

That is,

$$D = \frac{0.98(V_e - V_L)}{V_H - V_L} = \frac{0.98[A(V_p - V_n) - V_L]}{V_H - V_L} \tag{1.28}$$

1.6 Voltage-Mode CCM Closed Loop

By this time, we basically have developed all the essential blocks for step-down power converters in CCM operation. We are ready to close the loop, which can be represented in block diagram form (Figure 1.10).

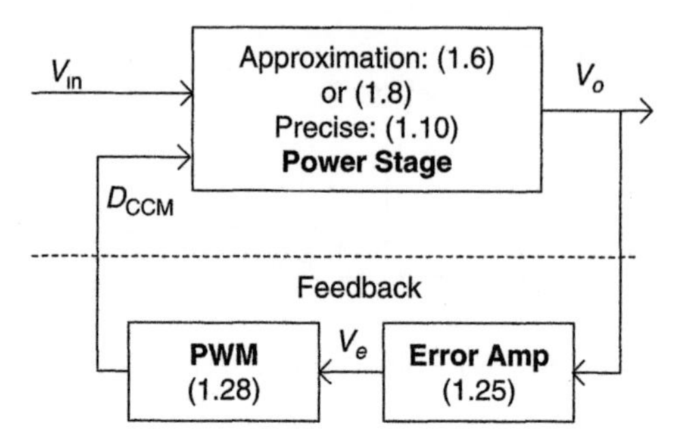

Figure 1.10: Buck converter CCM in a closed loop

The block diagram is intentionally partitioned (dashed line) into two parts—the feedback path and the power stage (plant, in control system terminology). Inside each block, the relevant equation governing the block function is given in parentheses. Equations (1.25) and (1.28) can be combined to give the open-loop duty cycle in terms of circuit components and open-loop output:

$$D = \frac{0.98\left\{A\left[R_3 \bullet I_b + V_{\text{ref}} - \dfrac{R_2}{R_1 + R_2}V_o - V_{os} - \dfrac{R_1 \bullet R_2}{R_1 + R_2}I_b\right] - V_L\right\}}{V_H - V_L} \tag{1.29}$$

In theory, (1.29) can be further combined with (1.10) to yield the closed-loop output. But anyone attempting to do so soon realizes that it is a mission impossible, for (1.10) contains a D^2 term and squaring (1.29) is not a simple matter. Furthermore, even after plugging in the D^2 and D terms, (1.10) does not give the closed-loop output explicitly, because V_o appears on both sides of the equation. One can use the approximation (1.8) instead but at the expense of accuracy. Do we have a way out? Yes, we can handle the situation using an implicit function. We first define two implicit functions from (1.29) and (1.10):

$$p(D, V_o, R_x, \ldots) =$$

$$D - \frac{0.98\left[A(R_3 \bullet I_b + V_{\text{ref}} - \dfrac{R_2}{R_1+R_2}V_o - V_{os} - \dfrac{R_1 \bullet R_2}{R_1+R_2}I_b - V_L)\right]}{V_H - V_L} = 0 \tag{1.30}$$

$$q(D, V_o, V_{\text{in}}, \ldots) = V_o - \frac{(n \cdot D \cdot V_{\text{in}} - V_D)R_L}{[D^2 \cdot R_F + (R_w + R_{\text{on}})D]n^2 + R_L + r_L} = 0 \quad (1.31)$$

Given the two functions, and using the Jacobian determinant, the output sensitivity against all circuit components and variables can be easily obtained. For instance, the load sensitivity is given as

$$\frac{\partial V_o}{\partial R_L} = -\frac{\begin{vmatrix} \dfrac{\partial p}{\partial D} & \dfrac{\partial p}{\partial R_L} \\ \dfrac{\partial q}{\partial D} & \dfrac{\partial q}{\partial R_L} \end{vmatrix}}{\begin{vmatrix} \dfrac{\partial p}{\partial D} & \dfrac{\partial p}{\partial V_o} \\ \dfrac{\partial q}{\partial D} & \dfrac{\partial q}{\partial V_o} \end{vmatrix}} \quad (1.32)$$

Of course, the steady-state closed-loop output, and duty cycle can both be solved simultaneously by solving (1.30) and (1.31) numerically using mathematical software MathCAD, MATLAB, MAPLE, or Mathematica.

1.7 Voltage-Mode DCM Closed Loop

As mentioned before, buck converters operating in the DCM are not desirable. But, for academic completeness, the closed-loop formulation for this operation mode is also given. We first consolidate (1.12) and (1.15):

$$V_o = \frac{\left[(n \cdot V_{\text{in}} - V_D)D_{\text{DCM}} - V_D \cdot \sqrt{\dfrac{(n \cdot V_{\text{in}} - V_D - V_o)2 \cdot L \cdot f_s \cdot \dfrac{V_o}{R_L}}{(V_D + V_o)n \cdot V_{\text{in}}}} \right]}{D_{\text{DCM}} + \sqrt{\dfrac{(n \cdot V_{\text{in}} - V_D - V_o)2 \cdot L \cdot f_s \cdot \dfrac{V_o}{R_L}}{(V_D + V_o)n \cdot V_{\text{in}}}}} \quad (1.33)$$

We then replace the power stage of Figure 1.10 with one for the DCM. This step leads to Figure 1.11 for the DCM.

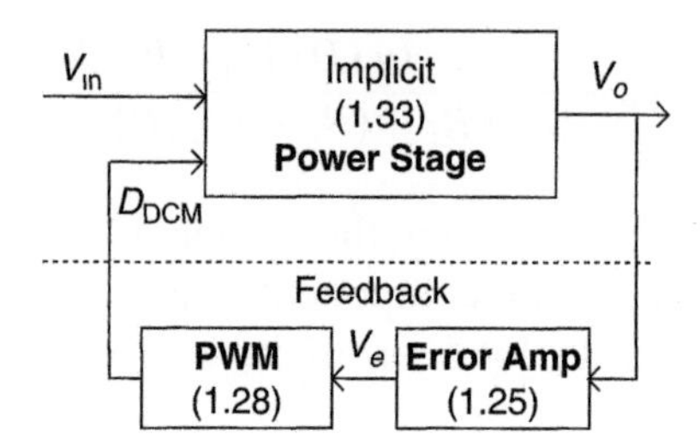

Figure 1.11: Buck converter DCM in a closed loop

By defining a new implicit function based on (1.33), we certainly can perform the same sensitivity study as outlined in the previous section. It is not repeated here.

1.8 Voltage-Mode CCM Small-Signal Stability

By nature, switch-mode power converters with feedback control are nonlinear control systems. Nonlinear control systems certainly are not easily subjected to the conventional linear system analysis, in which the superposition principle applies and the classical system stability theory is also applicable. Stated differently, switch-mode converters without the support of a grand vision cannot enjoy the vast amount of analytical benefits maturely developed in 1950–1980 for linear systems. Fortunately, that grand vision came in the mid-1970s. Dr. R. Middlebrook and his then graduate student Slobodan Cuk at the California Institute of Technology conceived the concept of state-space averaging. Based on the concept, nonlinear power converters, power stages in particular, are given equivalent linear models. Once that hurdle was surmounted, switch-mode power converters have been well investigated, employing those tools originally developed for linear systems. Since then, streams of in-depth studies and insightful results have been generated and reported. We utilize many models developed by those two visionary figures without proof but with great appreciation. Readers interested in the topics should refer to [1] for details.

Based on the state-space averaging technique, the nonisolated buck converter power stage in the CCM can be represented by Figure 1.12 for

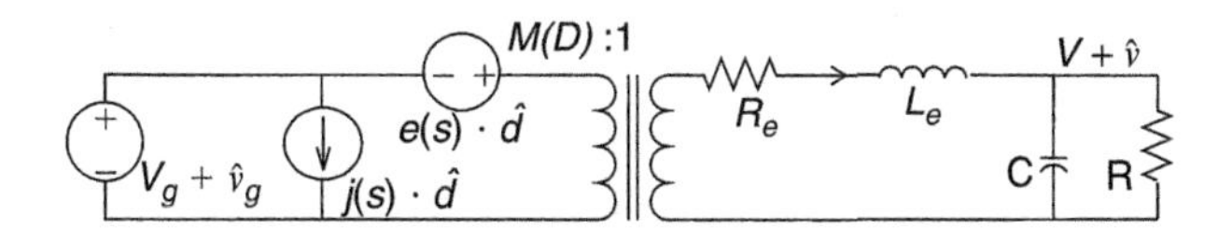

Figure 1.12: Small-signal model for a nonisolated converter in the CCM

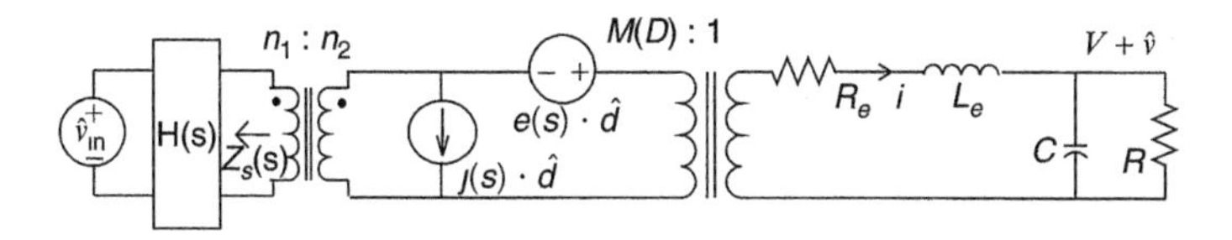

Figure 1.13: Small-signal model for an isolated converter in the CCM

small-signal studies. To fit our application in which isolation transformer and line filter are included, the model must be modified as Figure 1.13.

Figure 1.13 with transfer function $H(s)$ and source impedance $Z_s(s)$ representing the line filter first is simplified to make it easier for equation formulation. This can be done by reflecting the two dependent sources, $e(s)$ and $j(s)$, to the transformer's primary side, converting the input circuit to its Thevenin equivalent and consolidating two transformers into a single one. These steps lead to Figure 1.14.

On the input circuit side, two equations, a node and a loop, can be written. The input node gives

$$\text{source current} = \frac{n_2}{n_1} j(s) \bullet \hat{d} + M' \bullet \hat{i} = \frac{n_2}{n_1} j(s) \bullet \hat{d} + M' \bullet \frac{\hat{v}_{se}}{Z_{ei}} \tag{1.34}$$

The input loop gives

$$\left[\frac{n_2}{n_1} j(s) \bullet \hat{d} + M' \bullet \frac{\hat{v}_{se}}{Z_{ei}}\right] Z_s(s) - \frac{n_1}{n_2} \bullet e(s) \bullet \hat{d} + \hat{v}_p = H(s) \bullet \hat{v}_{in} \tag{1.35}$$

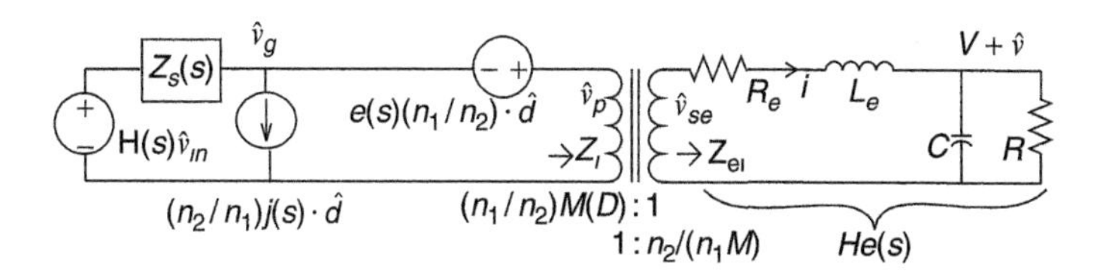

Figure 1.14: Consolidated AC model for an isolated converter in the CCM

By reflecting the secondary voltage to the primary one, (1.35) gives

$$\left[\frac{n_2}{n_1} j(s) \cdot \hat{d} + (M')^2 \cdot \frac{\hat{v}_p}{Z_{ei}}\right] Z_s(s) - \frac{n_1}{n_2} \cdot e(s) \cdot \hat{d} + \hat{v}_p = H(s) \cdot \hat{v}_{\text{in}} \tag{1.36}$$

where $M' = n_2/(n_1 \cdot M)$. Equation (1.36) ultimately gives

$$\hat{v}_p = \frac{H(s) \cdot \hat{v}_{\text{in}} + \frac{e(s)}{\frac{n_2}{n_1}}\left[1 - \frac{j(s)}{e(s)}\left(\frac{n_2}{n_1}\right)^2 Z_s(s)\right]\hat{d}}{1 + \frac{Z_s(s)}{\left[\frac{n_1}{n_2} M\right]^2 Z_{ei}(s)}} \tag{1.37}$$

$$\hat{v}_{se} = M' \cdot \hat{v}_p$$

Equation (1.37) hints that we can translate Figure 1.14 to a block diagram form by rewriting the equation as

$$\hat{v}_{se} = M'[G_{V_p g}(s) \cdot \hat{v}_g + G_{V_p d}(s) \cdot \hat{d}] \tag{1.38}$$

where

$$G_{V_p g}(s) = \frac{1}{1 + \frac{Z_s(s)}{\left[\frac{n_1}{n_2} M\right]^2 Z_{ei}(s)}} \tag{1.39}$$

and

$$G_{V_p d}(s) = \frac{\frac{e(s)}{\frac{n_2}{n_1}}\left[1 - \frac{j(s)}{e(s)}\left(\frac{n_2}{n_1}\right)^2 Z_s(s)\right]}{1 + \frac{Z_s(s)}{\left[\frac{n_1}{n_2} M\right]^2 Z_{ei}(s)}} \tag{1.40}$$

$$Z_{ei}(s) = R_e + L_e \cdot s + \left(\frac{1}{R} + C \cdot s\right)^{-1} \tag{1.41}$$

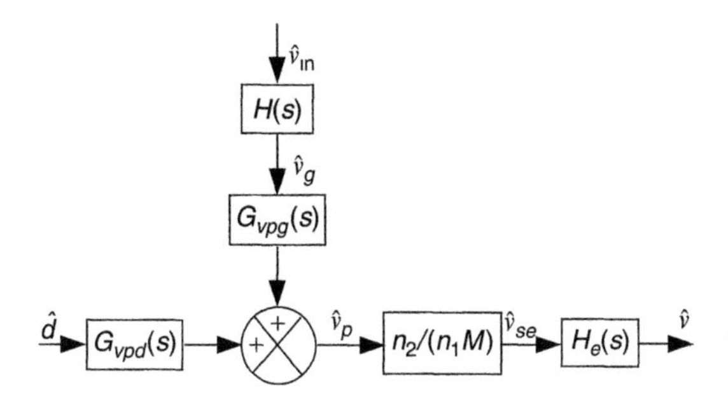

Figure 1.15: Small-signal block diagram for an isolated buck converter

In block diagram form, Figure 1.14 becomes Figure 1.15, where $H_e(s)$ is the effective, loaded output filter transfer function:

$$H_e(s) = \frac{\left(\frac{1}{R} + C \bullet s\right)^{-1}}{R_e + L_e \bullet s + \left(\frac{1}{R} + C \bullet s\right)^{-1}} \tag{1.42}$$

Again, readers are reminded that R_e, L_e, $e(s)$, $j(s)$, and several other model parameters and variables are given in [1].

At this point, we are almost ready again to close the loop for AC small-signal studies. However, we need three more blocks to finish the job: the feedback ratio, the error amplifier, and the voltage-mode PWM gain. The feedback ratio is quite simple. It is the voltage division ratio given in Figure 1.7:

$$K_f = \frac{R_2}{R_1 + R_2} \tag{1.43}$$

The error amplifier transfer function is the inverting part of (1.23) if the approximation is invoked. The sign is taken care of later:

$$EA(s) = \frac{Z_f(s)}{Z_i(s)} \tag{1.44}$$

If a more accurate form is desired, the error amplifier gain is the inverting part of (1.22):

$$EA(s) = \frac{A(s) \bullet Z_f(s)}{Z_f(s) + [1 + A(s)]Z_i(s)} \tag{1.45}$$

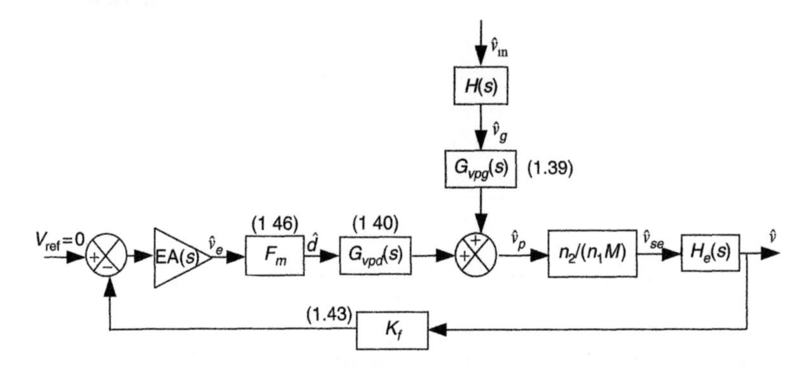

Figure 1.16: Closed-loop AC block diagram for an isolated buck converter

The last block, PWM gain, can be derived from (1.28) with the understanding that we are dealing with a small signal; that is, the gain of the PWM is the partial derivative of (1.28) against the error voltage:

$$F_m = \frac{\partial D}{\partial V_e} = \frac{0.98}{V_H - V_L} \tag{1.46}$$

Given Figure 1.15 and equations (1.42) through (1.46), we close the complete loop in Figure 1.16.

Generally speaking, the AC loop stability studies are conducted under the condition of constant input voltage. This further simplifies Figure 1.16 and leads to Figure 1.17, where the input effects are essentially zero.

Clearly, Figure 1.17 gives the overall loop gain as

$$T(s) = -K_f \bullet EA(s) \bullet F_m \bullet G_{V_pd}(s) \bullet \frac{n_2}{n_1 \bullet M} \bullet H_e(s) \tag{1.47}$$

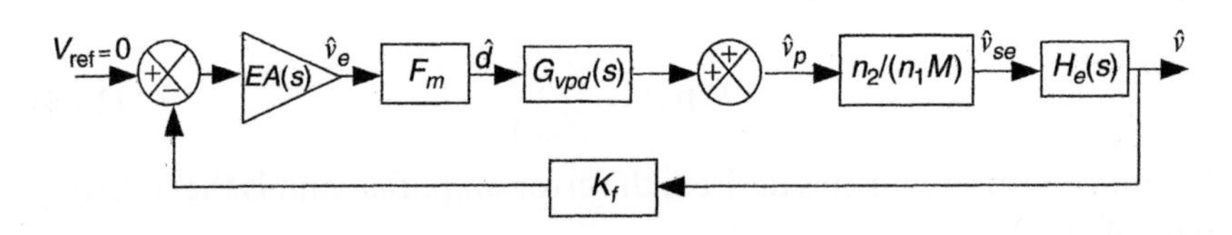

Figure 1.17: Closed-loop AC block diagram for an isolated buck converter with zero input effects

Furthermore, Figure 1.16 also gives the conducted susceptibility, which is a measure of how much input perturbation ends up at the output:

$$C_s(s) = \frac{\hat{v}}{\hat{v}_{\text{in}}} = H(s) \cdot G_{V_{pg}}(s) \frac{\dfrac{n_2}{n_1 \cdot M} \cdot H_e(s)}{1 - T(s)} \tag{1.48}$$

1.9 Current-Mode Control

By nature, signals in current forms have advantages over those in voltage form, since voltage is an accumulation of electron flux and, therefore, slow in time as far as control mechanism is concerned. In the early 1980s, this understanding spawned a new tide in switch-mode power supply design, namely, the current-mode control. In this control mode, the averaged or peak current of magnetic origin is employed in the feedback loop of switch-mode power converters. However, by adding a current loop, the conventional concept of loop gain is blurred, since multiple loops exist and make it difficult to identify the main loop. The current-mode control techniques, in addition to introducing difficulties in loop identification, also create new territories for analysis. In this and following sections, we give this superior technique an in-depth study. We again cover both CCM and DCM operations using the peak current. The average-current current-mode control is set aside and covered in a separate chapter because of its mathematical complexity. But, let us first look at the general feature of a current-mode control scheme, Figure 1.18.

Evidently, the sole and key difference between Figures 1.7 and 1.18 is the way the reference ramp is generated. In the case of voltage-mode control, the ramp is external from the viewpoint of the power plant, whereas for current-mode control, it is internal. However, the superficial view alone does not unveil the superior nature of peak current-mode control. We definitely need to probe further into the dynamics of the technique. For that, we examine how the current sensing ramp is produced. A typical example of that is given in Figure 1.19.

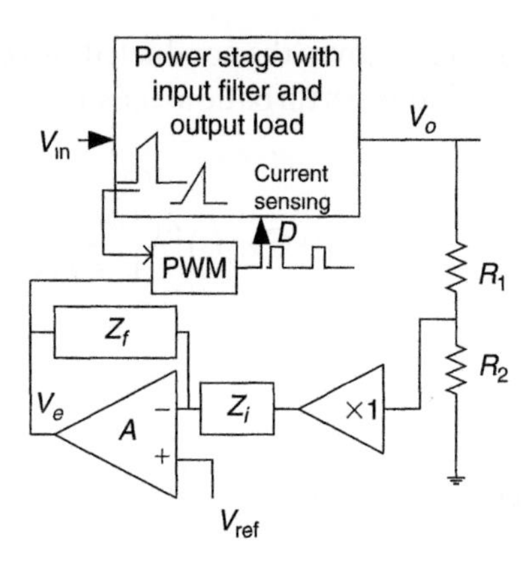

Figure 1.18: Peak-current current-mode control scheme

The figure shows that the instantaneous switch current is sensed by a current transformer that provides isolation and current scaling. The switch is turned on at a clock edge. It is turned off when the sensed current in voltage form intercepts the error voltage, V_e (Figure 1.20).

The performance merits of current-mode control over voltage-mode control can be appreciated more by looking at the transient response when the converter is subjected to a step-load disturbance or a step-line change. Figure 1.19 shows that the sensed signal has a ramp-up part. This part is attributed to a magnetic device with a ferrous core. When subjected to a volt–second (flux) drive, the magnetic device develops a current. The time rate of such a current is easily expressed as

$$\frac{di}{dt} \propto \frac{\text{voltage across}}{\text{inductance}} = \frac{V_{\text{input}} - V_o}{L} \tag{1.49}$$

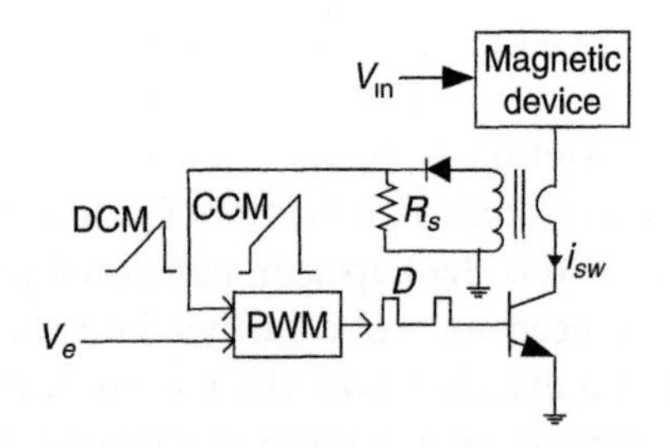

Figure 1.19: Peak current sensing

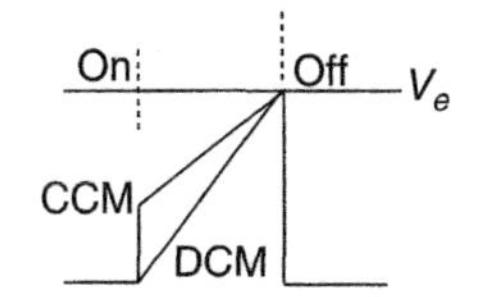

Figure 1.20: Determination of the peak current-mode duty cycle

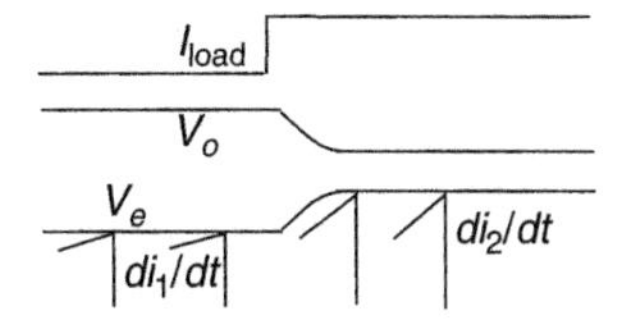

Figure 1.21: Current-mode control mechanism

The equation tells us loud and clear that the magnetic device's current rate of change is in phase with either the input or the output, voltage changes that, in turn, are reflected in error voltage. In pictorial form, Figure 1.21 shows how a step load initiates a chain of events that is the hallmark of current-mode control.

As the figure shows, $di_2/dt > di_1/dt$ when a step load commences, given constant input. In essence, the average value of the magnetic device's current rapidly tracks the step load and minimizes the error voltage. Another way of praising the mechanism is to say that the phase delay property of a magnetic device is removed. In the jargon of control system theory, a lagging pole is eliminated and system response speed is improved. That is the key merit of current-mode control.

1.10 CCM Current-Mode Control in a Closed-Loop Steady State

In the previous section, the general feature of current-mode operation and its advantages were briefly reviewed. In this section, we give a thorough treatment of CCM. For that, we refer to Figure 1.1 (or 1.2). The primary winding current (switch current) is understood to consist of three components: the reflected load, the reflected ripple of the output

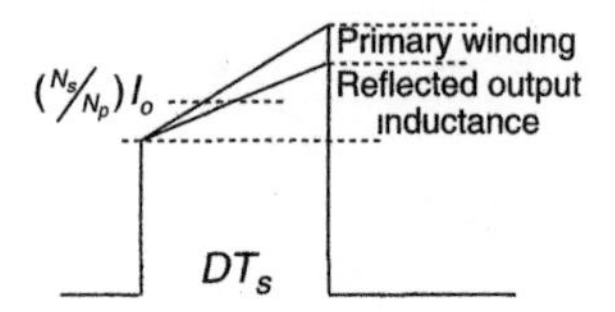

Figure 1.22: Main switch current composition

inductor, and the primary magnetization, L_p. Given in Figure 1.22, these will help us formulate and perform the analysis to follow.

The total ramp-up current profile can be written as

$$i_p(t) = \frac{N_s}{N_p}\left[I_o - \frac{\left(\frac{N_s}{N_p}V_{\text{in}} - V_D - V_o\right)D}{2L \cdot f_s} + \frac{\frac{N_s}{N_p}V_{\text{in}} - V_D - V_o}{L} \cdot t \right] + \frac{V_{\text{in}}}{L_p} \cdot t \quad (1.50)$$

Referring also to Figure 1.19 and assuming a 1-to-n_i current transformer, the steady-state open-loop duty cycle is therefore decided when the ramp-up signal meets the error voltage; that is,

$$\frac{i_p(D \cdot T_s)}{n_i} \cdot R_s = V_e \quad (1.51)$$

or

$$D(V_e, V_{\text{in}}, V_o) = \frac{\frac{n_i \cdot V_e}{R_s} - \frac{N_s}{N_p}\frac{V_o}{R_L}}{\frac{N_s}{N_p}\frac{\frac{N_s}{N_p}V_{\text{in}} - V_D - V_o}{2 \cdot L \cdot f_s} + \frac{V_{\text{in}}}{L_p \cdot f_s}} \quad (1.52)$$

Compared with (1.28) for voltage-mode control, the intricacy of current-mode control is simply amazing. It seems to have built-in intelligence by incorporating all the essential variables. Moreover, we can easily modify Figure 1.10 and infuse it with the sophistication of current-mode control. This leads to Figure 1.23, in which only the mechanism of the PWM is modified.

For those of you interested in the closed-loop output, we do the following procedure. But, considering that (1.10) is too prohibitively

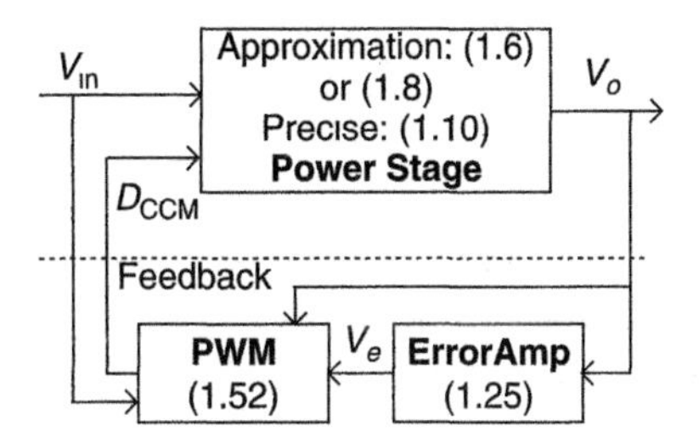

Figure 1.23: Current-mode buck converter in a CCM closed loop

complicated to use, we invoke only (1.6), the approximation. Starting from (1.6), we replace the variable D with (1.52):

$$V_o = \frac{N_s}{N_p} V_{\text{in}} \bullet \frac{\dfrac{n_i \bullet V_e}{R_s} - \dfrac{N_s}{N_p}\dfrac{V_o}{R_L}}{\dfrac{N_s}{N_p}\dfrac{\dfrac{N_s}{N_p}V_{\text{in}} - V_D - V_o}{2 \bullet L \bullet f_s} + \dfrac{V_{\text{in}}}{L_p \bullet f_s}} - V_D \tag{1.53}$$

We then close the loop by plugging in (1.25), replacing V_e:

$$V_o = \frac{\dfrac{N_s}{N_p} V_{\text{in}} \left[\dfrac{n_i \bullet A \left(\begin{array}{l} R_3 \bullet I_b + V_{\text{ref}} - \\ \dfrac{R_2}{R_1 + R_2} V_o - V_{os} - \dfrac{R_1 \bullet R_2}{R_1 + R_2} I_b \end{array} \right)}{R_s} - \dfrac{N_s}{N_p}\dfrac{V_o}{R_L} \right]}{\dfrac{\dfrac{N_s}{N_p}\left(\dfrac{N_s}{N_p} V_{\text{in}} - V_D - V_o\right)}{2 \bullet L \bullet f_s} + \dfrac{V_{\text{in}}}{L_p \bullet f_s}} - V_D \tag{1.54}$$

Unfortunately, even by using only the approximation, the output voltage is still not in an explicit form. Although it is not impossible to solve V_o symbolically, given modern software, we shall not attempt to do so, since we would not gain more than what we have so far.

1.11 CCM Current-Mode Control Small-Signal Stability

In the previous section, it was mentioned that the sole difference between Figures 1.10 and 1.23 is the PWM block. This statement holds true for AC small-signal studies. We then modify whatever surrounds the PWM, F_m, in Figure 1.16. This is done by expressing the total derivative of (1.52) in terms of three partial derivatives against error voltage perturbation, input disturbance, and output deviation:

$$dD = \frac{\partial D}{\partial V_e}\delta V_e + \frac{\partial D}{\partial V_{\text{in}}}\delta V_{\text{in}} + \frac{\partial D}{\partial V_o}\delta V_o$$
$$= F_m \bullet \delta V_e + F_{vb} \bullet \delta V_{\text{in}} + F_v \bullet \delta V_o \tag{1.55}$$

where, for instance,

$$F_m = \frac{\partial D(V_e, V_{\text{in}}, V_o)}{\partial V_e} = \frac{1}{\dfrac{N_s}{N_p}\dfrac{\dfrac{N_s}{N_p}V_{\text{in}} - V_D - V_o}{2 \bullet L \bullet f_s} + \dfrac{V_{\text{in}}}{L_p \bullet f_s}} \bullet \frac{n_I}{R_s} \tag{1.56}$$

The other two gain coefficients, $F_{vb}(= \partial D/\partial V_{\text{in}})$ and $F_v(= \partial D/\partial V_o)$, in symbolic forms are extremely burdensome to write and omitted in print with the understanding that they are readily computable given modern software. Anyway, (1.55) needs a new block description, Figure 1.24, for a current-mode PWM.

Then, using Figure 1.24, we replace the F_m block in Figure 1.16. The complete block diagram, Figure 1.25, for CCM operation is done. It

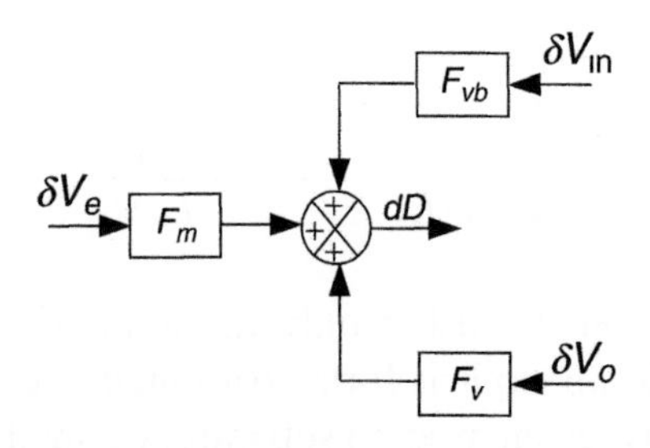

Figure 1.24: A current-mode CCM PWM

clearly reflects the added complexity of the current-mode control mechanism.

Again, for loop gain evaluation, Figure 1.25 is simplified by assuming constant input. This simplification leads to Figure 1.26.

Figure 1.26 shows two loops, an inner current loop and an outer voltage loop. We first absorb the current loop, and Figure 1.26 is further simplified to Figure 1.27.

The loop gain for current-mode control is, by inspection,

$$T(s) = -K_f \bullet EA(s) \bullet F_m \bullet \frac{G_{vpd}(s) \bullet \dfrac{n_2}{n_1 \bullet M(D)} \bullet H_e(s)}{1 - F_v \bullet G_{vpd}(s) \bullet \dfrac{n_2}{n_1 \bullet M(D)} \bullet H_e(s)} \tag{1.57}$$

Figure 1.25 also gives the current-mode CCM conducted susceptibility

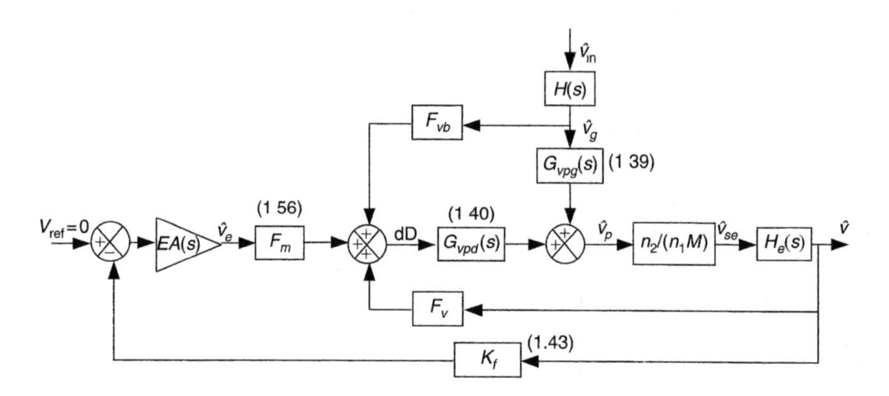

Figure 1.25: Closed-loop AC block diagram for the CCM current mode

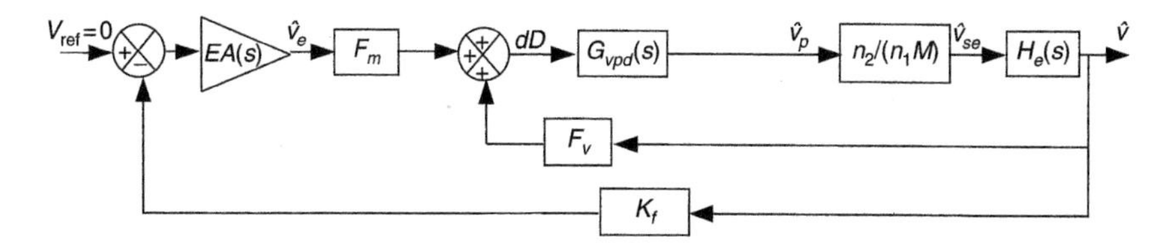

Figure 1.26: CCM current-mode closed-loop diagram for loop gain

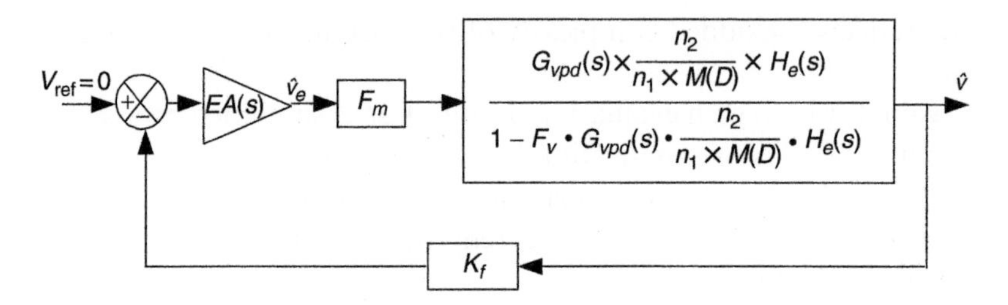

Figure 1.27: CCM closed-loop diagram with the current loop absorbed

$$C_s(s) = H(s) \cdot \left\{ \frac{G_{vpg}(s) \cdot M' \cdot H_e(s) + F_{vb} \cdot G_{vpd}(s) \cdot M' \cdot H_e(s)}{1 - [F_v + K_f \cdot EA(s) \cdot F_m] G_{vpd}(s) \cdot M' \cdot H_e(s)} \right\} \tag{1.58}$$

1.12 Output Capacitor Size and Accelerated Steady-State Analysis

The output filter capacitor, C in Figure 1.1, plays a major role in setting the output ripple voltage amplitude, which always is considered a very important part of power supply specification. As such, a reliable technique for selecting the appropriate capacitor value tied to a given ripple requirement is highly sought after. We study two techniques for the CCM: one graphic based and one time-domain based. In the first approach, we reexamine the output inductor current in detail (Figure 1.28).

It is understood that the capacitor passes only AC current. Therefore, the triangular area above I_o and its equivalent charge, δQ, can be placed in an equation connecting the ripple requirement δv and the capacitor value required:

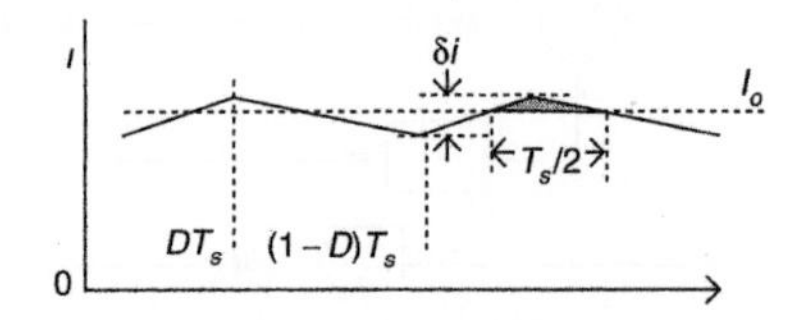

Figure 1.28: CCM output inductor current

$$C = \frac{\delta Q}{\delta v} = \frac{\frac{1}{2} \cdot \frac{\delta i}{2} \cdot \frac{T}{2}}{\delta v} = \frac{\delta i}{8 \cdot \delta v \cdot f_s} = \frac{\frac{V_o + V_D}{L}(1 - D)T_s}{8 \cdot \delta v \cdot f_s}$$

$$= \frac{(V_o + V_D)\left(1 - \dfrac{V_o + V_D}{\dfrac{N_s}{N_p} V_{\text{in}}}\right)}{8 \cdot L \cdot \delta v \cdot f_s^2} \tag{1.59}$$

This first approach, however, has some deficiencies. We all understand that an LC filter without damping exhibits peaking that can easily destabilize a converter loop. Although the load resistance also provides damping, load-dependent damping is not desirable. A well-designed output filter generally incorporates critical damping that comes in the form of Figure 1.29.

With damping included, (1.59) is no longer applicable because of unknown current division between two capacitor branches. Wu [2] presented an interesting way of choosing values for both capacitors. It was based on the equation form and the requirement of critical damping. Readers are encouraged to refer to the book. Here, we offer another way, based on the time-domain analysis and the concept of continuity of state. The driving source is identified as v_g with two alternating states and a duty cycle D at switching frequency f_s. The input loop gives a voltage equation:

$$\frac{di}{dt} + \frac{r_L}{L} \cdot i + \frac{1}{L} \cdot v = \frac{v_g}{L} \tag{1.60}$$

The damping capacitor node gives a current equation:

$$\frac{dv_d}{dt} + \frac{1}{r_d \cdot C_d} \cdot v_d - \frac{1}{r_d \cdot C_d} \cdot v = 0 \tag{1.61}$$

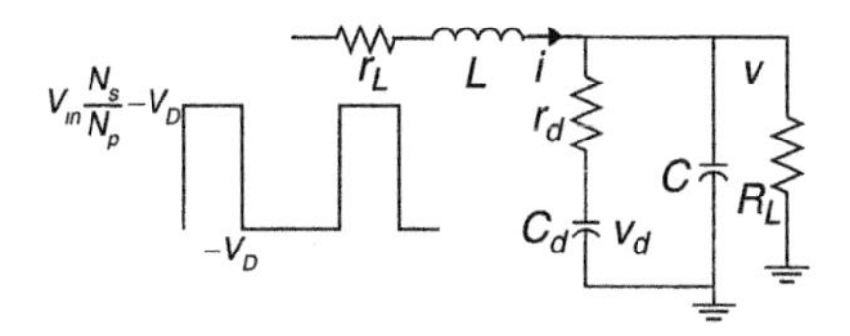

Figure 1.29: Output filter with damping

The output node yields another current equation:

$$-\frac{1}{C}\bullet i-\frac{1}{r_d\bullet C}\bullet v_d+\frac{dv}{dt}+\left(\frac{1}{r_d}+\frac{1}{R_L}\right)\frac{1}{C}\bullet v=0 \tag{1.62}$$

By taking a Laplace transformation with unknown starting conditions, I_0, V_{d0}, and V_0, (1.60)–(1.62) are transformed to

$$\left(s+\frac{r_L}{L}\right)I(s)+\frac{1}{L}\bullet V(s)=I_0+\frac{V_g(s)}{L} \tag{1.63}$$

$$\left(s+\frac{1}{r_d\bullet C_d}\right)V_d(s)-\frac{1}{r_d\bullet C_d}\bullet V(s)=V_{d0} \tag{1.64}$$

$$-\frac{1}{C}\bullet I(s)-\frac{1}{r_d\bullet C}\bullet V_d(s)+\left[s+\left(\frac{1}{r_d}+\frac{1}{R_L}\right)\frac{1}{C}\right]V(s)=V_0 \tag{1.65}$$

The transformed equation set gives

$$I(s)=\frac{\begin{vmatrix} I_0+\frac{V_g(s)}{L} & 0 & \frac{1}{L} \\ V_{d0} & \left(s+\frac{1}{r_d\bullet C_d}\right) & -\frac{1}{r_d\bullet C_d} \\ V_0 & -\frac{1}{r_d\bullet C} & \left[s+\left(\frac{1}{r_d}+\frac{1}{R_L}\right)\frac{1}{C}\right] \end{vmatrix}}{D_e(s)} \tag{1.66}$$

$$V_d(s)=\frac{\begin{vmatrix} \left(s+\frac{r_L}{L}\right) & I_0+\frac{V_g(s)}{L} & \frac{1}{L} \\ 0 & V_{d0} & -\frac{1}{r_d\bullet C_d} \\ -\frac{1}{C} & V_0 & \left[s+\left(\frac{1}{r_d}+\frac{1}{R_L}\right)\frac{1}{C}\right] \end{vmatrix}}{D_e(s)} \tag{1.67}$$

$$V(s)=\frac{\begin{vmatrix} \left(s+\frac{r_L}{L}\right) & 0 & I_0+\frac{V_g(s)}{L} \\ 0 & \left(s+\frac{1}{r_d\bullet C_d}\right) & V_{d0} \\ -\frac{1}{C} & -\frac{1}{r_d\bullet C} & V_0 \end{vmatrix}}{D_e(s)} \tag{1.68}$$

where

$$D_e(s) = \begin{vmatrix} \left(s + \frac{r_L}{L}\right) & 0 & \frac{1}{L} \\ 0 & \left(s + \frac{1}{r_d \cdot C_d}\right) & -\frac{1}{r_d \cdot C_d} \\ -\frac{1}{C} & -\frac{1}{r_d \cdot C} & \left[s + \left(\frac{1}{r_d} + \frac{1}{R_L}\right)\frac{1}{C}\right] \end{vmatrix} \tag{1.69}$$

The numerator of the inductor current transfer function can be expanded and grouped. The transfer function is then expressed as

$$I(s) = \frac{\begin{vmatrix} \left(s + \frac{1}{r_d \cdot C_d}\right) & -\frac{1}{r_d \cdot C_d} \\ -\frac{1}{r_d \cdot C} & \left[s + \left(\frac{1}{r_d} + \frac{1}{R_L}\right)\frac{1}{C}\right] \end{vmatrix}}{D_e(s)} I_0 + \frac{-\begin{vmatrix} 0 & \frac{1}{L} \\ -\frac{1}{r_d \cdot C} & \left[s + \left(\frac{1}{r_d} + \frac{1}{R_L}\right)\frac{1}{C}\right] \end{vmatrix}}{D_e(s)} V_{d0} + \frac{\begin{vmatrix} 0 & \frac{1}{L} \\ \left(s + \frac{1}{r_d \cdot C_d}\right) & -\frac{1}{r_d \cdot C_d} \end{vmatrix}}{D_e(s)} V_0 + \frac{\begin{vmatrix} \left(s + \frac{1}{r_d \cdot C_d}\right) & -\frac{1}{r_d \cdot C_d} \\ -\frac{1}{r_d \cdot C} & \left[s + \left(\frac{1}{r_d} + \frac{1}{R_L}\right)\frac{1}{C}\right] \end{vmatrix}}{L \cdot D_e(s)} V_g(s) \tag{1.70}$$

With a little patience, we can do the same thing for $V_d(s)$ and $V(s)$. We also understand that the corresponding $V_g(s)$ is

$$V_g(s) = \begin{cases} \left(\frac{N_s}{N_p} V_{\text{in}} - V_D\right)\frac{1}{s} = V_{ga}(s) & 0 < \text{t} < \text{D} \cdot \text{Ts} \\ -\frac{V_D}{s} = V_{gb}(s) & \text{D} \cdot \text{Ts} < \text{t} < \text{Ts} \end{cases} \tag{1.71}$$

if we designate the time interval $0 < t < D \cdot T_s$ as a and $D \cdot T_s < t < T_s$ as b. Then, during the a interval, the inductor current transfer function (1.70) can be placed in (1.72) with unknown starting states designated as I_{0a}, V_{d0a}, and V_{0a}:

$$
\begin{aligned}
I_a(s) &= F_1(s) \cdot I_{0a} + F_2(s) \cdot V_{d0a} + F_3(s) \cdot V_{0a} \\
&\quad + \frac{\begin{vmatrix} \left(s + \dfrac{1}{r_d \cdot C_d}\right) & -\dfrac{1}{r_d \cdot C_d} \\ -\dfrac{1}{r_d \cdot C} & \left[s + \left(\dfrac{1}{r_d} + \dfrac{1}{R_L}\right)\dfrac{1}{C}\right] \end{vmatrix}}{L \cdot D_e(s)} V_{ga}(s) \\
&= F_1(s) \cdot I_{0a} + F_2(s) \cdot V_{d0a} + F_3(s) \cdot V_{0a} + F_4(s)
\end{aligned}
\tag{1.72}
$$

We can do the same for the damping capacitor voltage and the output voltage:

$$V_{da}(s) = G_1(s) \cdot I_{0a} + G_2(s) \cdot V_{d0a} + G_3(s) \cdot V_{0a} + G_4(s) \tag{1.73}$$

$$V_a(s) = H_1(s) \cdot I_{0a} + H_2(s) \cdot V_{d0a} + H_3(s) \cdot V_{0a} + H_4(s) \tag{1.74}$$

By the same token, during interval b with unknown starting states designated as I_{0b}, V_{d0b}, and V_{0b} and considering time shift (delay) and driving source change, the three transfer functions become

$$
\begin{aligned}
I_b(s) &= \left\{ \begin{array}{l} F_1(s) \cdot I_{0b} + F_2(s) \cdot V_{d0b} + F_3(s) \cdot V_{0b} + \\ \dfrac{\begin{vmatrix} \left(s + \dfrac{1}{r_d \cdot C_d}\right) & -\dfrac{1}{r_d \cdot C_d} \\ -\dfrac{1}{r_d \cdot C} & \left[s + \left(\dfrac{1}{r_d} + \dfrac{1}{R_L}\right)\dfrac{1}{C}\right] \end{vmatrix}}{L \cdot D_e(s)} V_{gb}(s) \end{array} \right\} e^{-D \cdot T_s \cdot s} \\
&= [F_1(s) \cdot I_{0b} + F_2(s) \cdot V_{d0b} + F_3(s) \cdot V_{0b} + F_5(s)] e^{-D \cdot T_s \cdot s}
\end{aligned}
\tag{1.75}
$$

$$V_{db}(s) = [G_1(s) \cdot I_{0b} + G_2(s) \cdot V_{d0b} + G_3(s) \cdot V_{0b} + G_5(s)] e^{-D \cdot T_s \cdot s} \tag{1.76}$$

$$V_b(s) = [H_1(s) \cdot I_{0b} + H_2(s) \cdot V_{d0b} + H_3(s) \cdot V_{0b} + H_5(s)] e^{-D \cdot T_s \cdot s} \tag{1.77}$$

Next we perform inverse Laplace transformation of (1.72)–(1.74) and obtain

$$i_a(t) = f_1(t) \cdot I_{0a} + f_2(t) \cdot V_{d0a} + f_3(t) \cdot V_{0a} + f_4(t) \tag{1.78}$$

$$v_{da}(t) = g_1(t) \cdot I_{0a} + g_2(t) \cdot V_{d0a} + g_3(t) \cdot V_{0a} + g_4(t) \tag{1.79}$$

$$v_a(s) = h_1(t) \cdot I_{0a} + h_2(t) \cdot V_{d0a} + h_3(t) \cdot V_{0a} + h_4(h) \tag{1.80}$$

Equations (1.78)–(1.80) certainly can be placed in a matrix form:

$$\begin{bmatrix} i_a(t) \\ v_{da}(t) \\ v_a(t) \end{bmatrix} = \begin{bmatrix} f_1(t) & f_2(t) & f_3(t) \\ g_1(t) & g_2(t) & g_3(t) \\ h_1(t) & h_2(t) & h_3(t) \end{bmatrix} \begin{bmatrix} I_{0a} \\ V_{d0a} \\ V_{0a} \end{bmatrix} + \begin{bmatrix} f_4(t) \\ g_4(t) \\ h_4(t) \end{bmatrix} \tag{1.81}$$

Then, at $t = D \cdot T_s$, (1.81) results in

$$\begin{bmatrix} i_a(D \cdot T_s) \\ v_{da}(D \cdot T_s) \\ v_a(D \cdot T_s) \end{bmatrix} = \begin{bmatrix} f_1(D \cdot T_s) & f_2(D \cdot T_s) & f_3(D \cdot T_s) \\ g_1(D \cdot T_s) & g_2(D \cdot T_s) & g_3(D \cdot T_s) \\ h_1(D \cdot T_s) & h_2(D \cdot T_s) & h_3(D \cdot T_s) \end{bmatrix} \begin{bmatrix} I_{0a} \\ V_{d0a} \\ V_{0a} \end{bmatrix} + \begin{bmatrix} f_4(D \cdot T_s) \\ g_4(D \cdot T_s) \\ h_4(D \cdot T_s) \end{bmatrix} \tag{1.82}$$

We place (1.82) in compact, closed form:

$$A_1 \cdot X_a + B_1 = X_b \tag{1.83}$$

where

$$A_1 = \begin{bmatrix} f_1(D \cdot T_s) & f_2(D \cdot T_s) & f_3(D \cdot T_s) \\ g_1(D \cdot T_s) & g_2(D \cdot T_s) & g_3(D \cdot T_s) \\ h_1(D \cdot T_s) & h_2(D \cdot T_s) & h_3(D \cdot T_s) \end{bmatrix},$$

$$B_1 = \begin{bmatrix} f_4(D \cdot T_s) \\ g_4(D \cdot T_s) \\ h_4(D \cdot T_s) \end{bmatrix}, \quad X_a = \begin{bmatrix} I_{0a} \\ V_{d0a} \\ V_{0a} \end{bmatrix}, \quad X_b = \begin{bmatrix} I_{0b} \\ V_{d0b} \\ V_{0b} \end{bmatrix} \tag{1.84}$$

What (1.83) means is that the yet unknown starting states, X_a, propagates to the end states, X_b, during interval a. We repeat a similar procedure for (1.75)–(1.77) using X_b as the starting state for interval b:

$$i_b(t) = f_1(t - D \cdot T_s) \cdot I_{0b} + f_2(t - D \cdot T_s) \cdot V_{d0b} + f_3(t - D \cdot T_s) \cdot V_{0b} + f_5(t - D \cdot T_s) \tag{1.85}$$

$$v_{db}(t) = g_1(t - D \cdot T_s) \cdot I_{0b} + g_2(t - D \cdot T_s) \cdot V_{d0b} + g_3(t - D \cdot T_s) \cdot V_{0b} + g_5(t - D \cdot T_s) \tag{1.86}$$

$$v_b(t) = h_1(t - D \cdot T_s) \cdot I_{0b} + h_2(t - D \cdot T_s) \cdot V_{d0b} + h_3(t - D \cdot T_s) \cdot V_{0b} + h_5(t - D \cdot T_s) \tag{1.87}$$

Equations (1.85)–(1.87) can be placed in matrix form, too:

$$\begin{bmatrix} i_b(t) \\ v_{db}(t) \\ v_b(t) \end{bmatrix} = \begin{bmatrix} f_1(t - D \cdot T_s) & f_2(t - D \cdot T_s) & f_3(t - D \cdot T_s) \\ g_1(t - D \cdot T_s) & g_2(t - D \cdot T_s) & g_3(t - D \cdot T_s) \\ h_1(t - D \cdot T_s) & h_2(t - D \cdot T_s) & h_3(t - D \cdot T_s) \end{bmatrix} \begin{bmatrix} I_{0b} \\ V_{d0b} \\ V_{0b} \end{bmatrix} + \begin{bmatrix} f_5(t - D \cdot T_s) \\ g_5(t - D \cdot T_s) \\ h_5(t - D \cdot T_s) \end{bmatrix} \tag{1.88}$$

Then, at $t = T_s$ and considering that the end states of interval b must equal the starting states of interval a under a steady state, the following is derived from (1.88) (if the wraparound is not met, it is not a steady state):

$$\begin{bmatrix} f_1[(1 - D)T_s] & f_2[(1 - D)T_s] & f_3[(1 - D)T_s] \\ g_1[(1 - D)T_s] & g_2[(1 - D)T_s] & g_3[(1 - D)T_s] \\ h_1[(1 - D)T_s] & h_2[(1 - D)T_s] & h_3[(1 - D)T_s] \end{bmatrix} \begin{bmatrix} I_{0b} \\ V_{d0b} \\ V_{0b} \end{bmatrix} + \begin{bmatrix} f_5[(1 - D)T_s] \\ g_5[(1 - D)T_s] \\ h_5[(1 - D)T_s] \end{bmatrix} = \begin{bmatrix} I_{0a} \\ V_{d0a} \\ V_{0a} \end{bmatrix} \tag{1.89}$$

In matrix form, (1.89) gives

$$A_2 \cdot X_b + B_2 = X_a \tag{1.90}$$

where matrices A_2 and B_2 are self-evident.

Equations (1.83) and (1.90) together give

$$X_a = (I - A_2 \bullet A_1)^{-1}(A_2 \bullet B_1 + B_2) \tag{1.91}$$

In other words, under a steady state, the unknown starting state vector actually is not unknown at all. Given X_a, X_b (1.83) follows. And the complete, cyclic steady-state solution of the circuit is done. For instance, the inductor current is

$$\begin{aligned} i(t) = {} & \{[f_1(t)\, f_2(t)\, f_3(t)]X_a + f_4(t)\}[u(t) - u(t - D \bullet T_s)] \\ & + \left\{ \begin{matrix} [f_1(t - D \bullet T_s) \quad f_2(t - D \bullet T_s) \quad f_3(t - D \bullet T_s)]X_b \\ + f_5(t - D \bullet T_s) \end{matrix} \right\} \\ & \bullet [u(t - D \bullet T_s) - u(t - T_s)] \end{aligned} \tag{1.92}$$

As for the damping capacitor and the output voltages, they have similar forms, of course, and are not repeated here. Anyway, given $v_d(t)$ and $v(t)$, the damping resistor power can also be described analytically:

$$p_r(t) = \frac{[v(t) - v_d(t)]^2}{r} \tag{1.93}$$

This is to say the preceding technique can easily identify a filter capacitor's esr power dissipation and RMS current, if so desired. This latter benefit is not readily obtainable by other means.

1.13 A Complete Example

A. Closed-Loop Output Equation

Refer to Figure 1.30, showing the schematic of a forward converter with peak current-mode control. The operation of the converter can be briefly described as follows. At the initiation of an internal clock, f_s, residing in the PWM integrated circuit, SG1843 (silicon general semiconductor), the

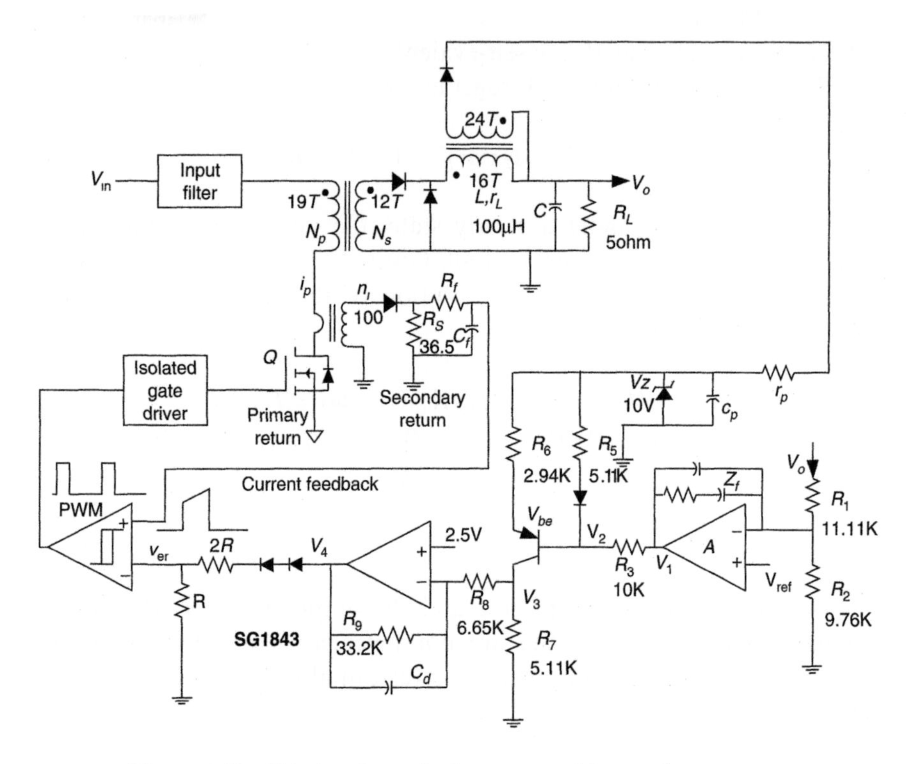

Figure 1.30: Circuit schematic for an actual forward converter

power switch, Q, is turned on. This action generates a ramping up current, i_p, in the primary winding. Meanwhile, the output voltage, V_o, is processed by the feedback loop and eventually transformed into a control (error) voltage, v_{er}. Both the control voltage and the scaled version of the ramp current are fed to a hysteretic comparator. At the instant the feedback current in voltage form intercepts the control voltage, the power switch ceases conduction and remains off until the next clock-on cycle. The converter switching frequency equals 125 KHz.

At a steady state, that is, under constant load and constant line input, the collector current of the bipolar transistor can be expressed as

$$I_C = K_1 \cdot V_o + K_2 \tag{1.94}$$

where

$$K_1 = \frac{A \cdot R_5 \cdot \dfrac{R_2}{R_1 + R_2} \cdot h_{FE}}{R_5 \cdot R_3 + (1 + h_{FE}) R_6 \cdot (R_3 + R_5)} \tag{1.95}$$

$$K_2 = \frac{-A \cdot R_5 \cdot V_{\text{ref}} + R_3 \cdot V_D + R_5 \cdot V_z - (R_3 + R_5) V_{be}}{R_5 \cdot R_3 + (1 + h_{FE}) R_6 \cdot (R_3 + R_5)} \cdot h_{FE} \tag{1.96}$$

Succeeding control voltage v_{er}, equivalent to V_e of Figure 1.18, can also be expressed as

$$v_{\text{er}} = K_3 \cdot V_o + K_4 \tag{1.97}$$

where

$$K_3 = -\frac{2}{3} \cdot \frac{R_7 \cdot R_9}{R_7 + R_8} \cdot K_1$$

$$K_4 = \frac{2}{3}\left[2.5\left(1 + \frac{R_9}{R_8}\right) - \frac{R_7 \cdot R_9}{R_7 + R_8} \cdot K_2 - \frac{2.5 R_7 \cdot R_9}{(R_7 + R_8) R_8} - 2 V_D\right] \tag{1.98}$$

and V_D stands for either signal diode forward drop or rectifier diode forward voltage.

Equations (1.50)–(1.52) give the open-loop, steady-state duty cycle, while (1.53) gives the open-loop output. For closed-loop output, we replace the control error voltage V_e in (1.53) with (1.97):

$$V_o = \frac{N_s}{N_p} V_{\text{in}} \cdot \frac{\dfrac{n_i (K_3 \cdot V_o + K_4)}{R_S} - \dfrac{N_S}{N_p} \dfrac{V_o}{R_L}}{\dfrac{N_S}{N_p} \dfrac{\dfrac{N_S}{N_p} V_{\text{in}} - V_D - V_o}{2 \cdot L \cdot f_S} + \dfrac{V_{\text{in}}}{L_p \cdot f_S}} - V_D \tag{1.99}$$

In theory, one can plug in K_4, K_3, K_2, and K_1 in succession and express the closed-loop output as an implicit function. But such an expression is so unruly we have to leave it as it is and use numerical computation instead. However, it is not a total waste, because we can use the Jacobian technique outlined in (1.30)–(1.32) to perform very meaningful studies.

B. Closed-Loop AC Studies

For AC small-signal (designated in lower case letters) loop gain studies, we need more preparation. We begin with the error amplifier. Figure 1.31 gives the corresponding small-signal equivalent circuit of the error amplifier. Based on the equivalent circuit and by also considering the non-ideal gain, $A(s)$, of the operational amplifier, the input-to-output transfer function is obtained.

$$\frac{v_1}{v_o} = -K_f \bullet EA(s), \quad EA(s) = \frac{A(s)}{R_p\left[\frac{1}{R_p} + \frac{1 + A(s)}{Z_f(s)}\right]}, \quad K_f = \frac{R_2}{R_1 + R_2} \tag{1.100}$$

where $R_p = R_1//R_2$ and

$$A(s) = \frac{A}{\left(\frac{s}{2 \bullet \pi \bullet f_p} + 1\right)} \tag{1.101}$$

represents the open-loop gain of the operational amplifier with single-pole gain-rolloff at frequency f_p and DC gain A.

For the transistor voltage-to-current converter, more steps are involved. Figure 1.32 gives both the DC and AC circuits. Three nodal equations at v_1, v_2, and v_3 can be written for Figure 1.32(b).

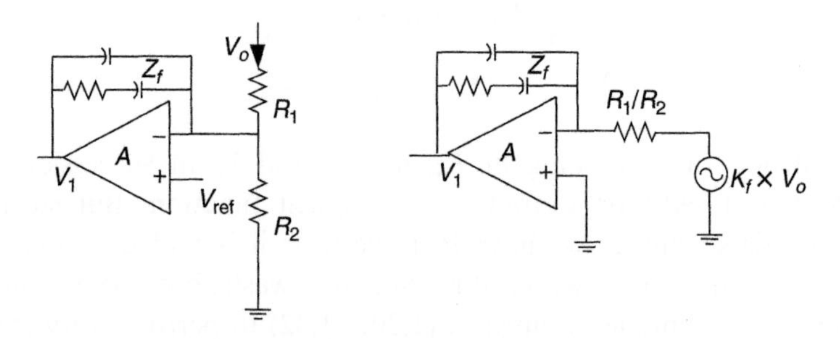

Figure 1.31: Error amplifier

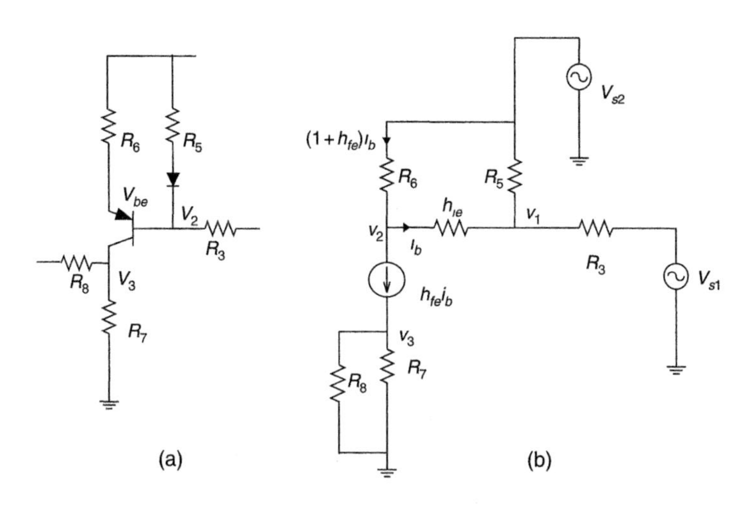

Figure 1.32: Transistor voltage-to-current converter: (a) DC, (b) AC

$$
\left(\frac{1}{R_3}+\frac{1}{R_5}+\frac{1}{h_{ie}}\right)v_1-\frac{1}{h_{ie}}v_2=\frac{v_{s2}}{R_5}+\frac{v_{s1}}{R_3},
$$

$$
\frac{1+h_{fe}}{h_{ie}}v_1-\left(\frac{1}{R_6}+\frac{1+h_{fe}}{h_{ie}}\right)v_2=-\frac{v_{s2}}{R_6}, \tag{1.102}
$$

$$
\frac{h_{fe}}{h_{ie}}v_1-\frac{h_{fe}}{h_{ie}}v_2+\frac{1}{R_{cp}}v_3=0, \qquad R_{cp}=\frac{R_7R_8}{R_7+R_8}
$$

The equation set yields, at node v_3,

$$
v_3=\frac{\begin{vmatrix} \frac{1}{R_3}+\frac{1}{R_5}+\frac{1}{h_{ie}} & \frac{-1}{h_{ie}} & \frac{v_{s2}}{R_5}+\frac{v_{s1}}{R_3} \\ \frac{1+h_{fe}}{h_{ie}} & -\left(\frac{1}{R_6}+\frac{1+h_{fe}}{h_{ie}}\right) & \frac{-v_{s2}}{R_6} \\ \frac{h_{fe}}{h_{ie}} & \frac{-h_{fe}}{h_{ie}} & 0 \end{vmatrix}}{D_e} \tag{1.103}
$$

and the transistor stage gains

$$G_{t1}(s) = \frac{\partial v_3}{\partial v_{s2}} = \frac{\begin{vmatrix} \frac{1+h_{fe}}{h_{ie}} & -\left(\frac{1}{R_6} + \frac{1+h_{fe}}{h_{ie}}\right) \\ \frac{h_{fe}}{h_{ie}} & -\frac{h_{fe}}{h_{ie}} \end{vmatrix} \frac{1}{R_5} + \begin{vmatrix} \frac{1}{R_3} + \frac{1}{R_5} + \frac{1}{h_{ie}} & \frac{-1}{h_{ie}} \\ \frac{h_{fe}}{h_{ie}} & -\frac{h_{fe}}{h_{ie}} \end{vmatrix} \frac{1}{R_6}}{D_e} \tag{1.104}$$

$$G_{t2}(s) = \frac{\partial v_3}{\partial v_{s1}} = \frac{\begin{vmatrix} \frac{1+h_{fe}}{h_{ie}} & -\left(\frac{1}{R_6} + \frac{1+h_{fe}}{h_{ie}}\right) \\ \frac{h_{fe}}{h_{ie}} & -\frac{h_{fe}}{h_{ie}} \end{vmatrix} \frac{1}{R_3}}{D_e}$$

where

$$D_e = \begin{vmatrix} \frac{1}{R_3} + \frac{1}{R_5} + \frac{1}{h_i} & \frac{-1}{h_i} & 0 \\ \frac{1+h_f}{h_i} & -\left(\frac{1}{R_6} + \frac{1+h_f}{h_i}\right) & 0 \\ \frac{h_f}{h_i} & \frac{-h_f}{h_i} & \frac{1}{R_{cp}} \end{vmatrix} \tag{1.105}$$

Following the voltage-to-current converter, the internal error amplifier of the PWM IC yields one more transfer function, based on Figure 1.33:

$$G_s(s) = \frac{v_{er}}{v_3} = -\frac{1}{3} \bullet \frac{\left(\frac{1}{R_9} + C_d \bullet s\right)^{-1}}{R_8} \tag{1.106}$$

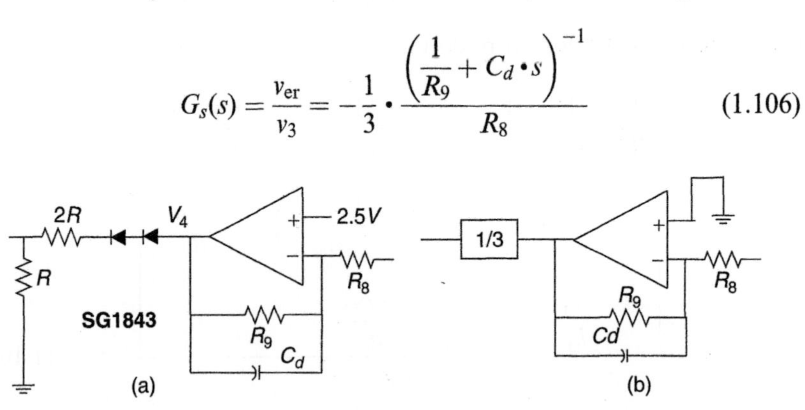

Figure 1.33: PWM internal error amplifier: (a) DC, (b) AC

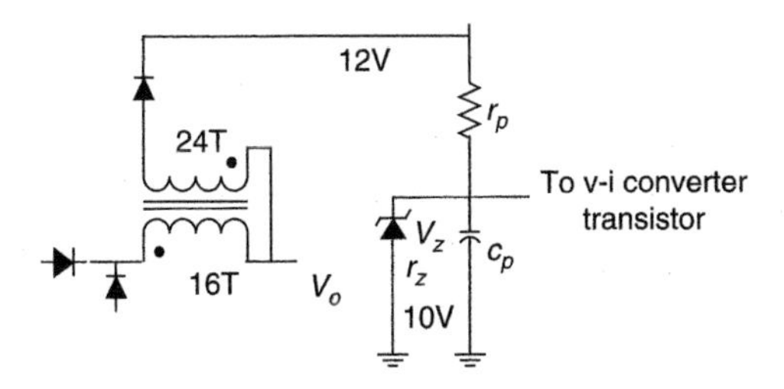

Figure 1.34: Housekeeping supply

The local supply supporting the voltage-to-current converter also presents a sneaky transfer function (Figure 1.34):

$$G_c(s) = -\frac{24}{16} \bullet \frac{\left(\frac{1}{r_z} + c_p \bullet s\right)^{-1}}{r_p + \left(\frac{1}{r_z} + c_p \bullet s\right)^{-1}} \tag{1.107}$$

With both the effective error voltage and the current feedback available at the input terminals of the PWM comparator, the steady-state duty cycle is established, given by (1.52). Based on (1.52), PWM gain factors are given by (1.55) and (1.56) and so forth. The next block to be treated is the output filter, including the main load and any housekeeping load. This block is shown in Figure 1.35, in which Z_h represents the internal housekeeping circuit, Z_L the main load, and other parasitic elements.

Its transfer function, transformer secondary to output, is

$$H_e(s) = \frac{Z_h Z_L}{(Z_h + Z_m)\left[\left(\frac{1}{Z_h} + \frac{1}{Z_m}\right)^{-1} + r_w\right]} \tag{1.108}$$

In section 1.8, a state-space averaged model is invoked to derive (1.34)–(1.40) for power stage transfer functions. A slightly simpler and just as effective approach can also be taken for power stage. We use (1.8) and do the following:

$$dV_o = \frac{n \bullet V_{\text{in}}}{1 + \frac{r_L}{R_L}} \bullet \delta D + \frac{n \bullet D}{1 + \frac{r_L}{R_L}} \bullet \delta V_{\text{in}} = G_{pd} \bullet \delta D + G_{pv} \bullet \delta V_{\text{in}} \tag{1.109}$$

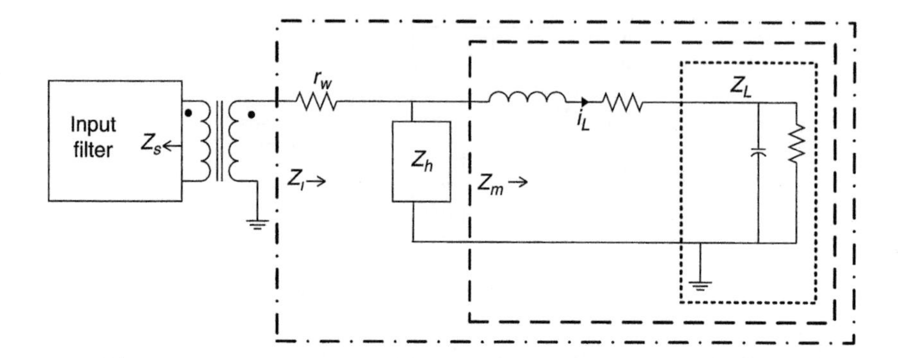

Figure 1.35: Output filter and source impedance interactions

The form of (1.109) suggests a summation similar to Figure 1.24 for PWM gains. This concludes the development of individual transfer functions for all key blocks. They can all be interconnected in the overall block diagram of Figure 1.36.

For loop gain studies, Figure 1.36 can be further simplified to Figure 1.37, since the input voltage, in general, is held constant for the purpose.

Two more steps will absorb the two inner loops, one due to current mode control and the other due to the transconductance amplifier. Once that is done, the loop gain is easily given as

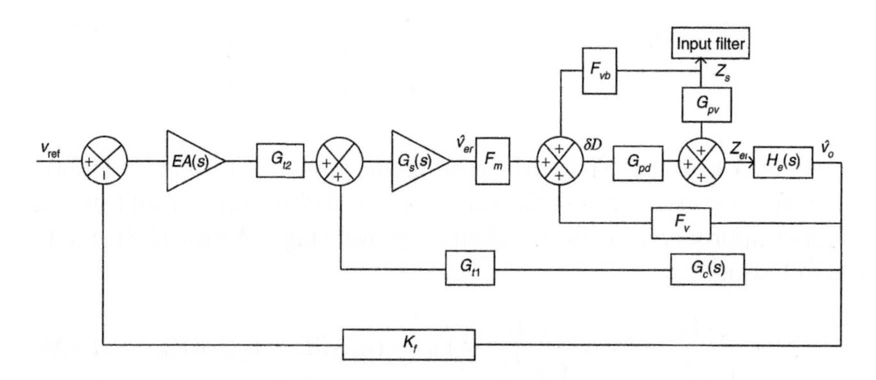

Figure 1.36: Overall block diagram

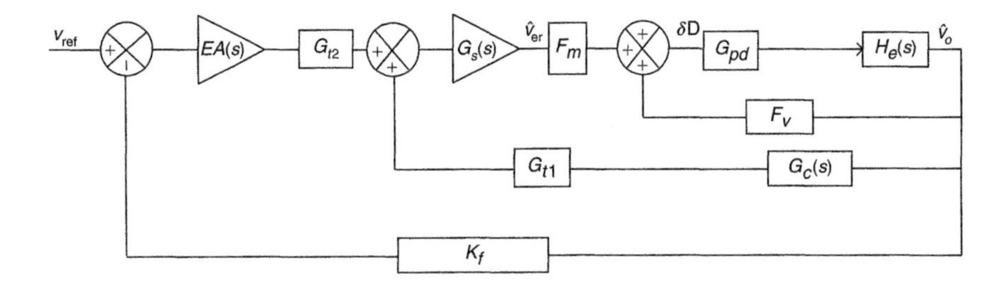

Figure 1.37: Simplified block diagram

$$T(s) = \frac{K_f \cdot EA(s) \cdot G_{t2}(s) \cdot G_s(s) \cdot F_m \cdot \dfrac{G_{pd} \cdot H_e(s)}{1 - F_v \cdot G_{pd} \cdot H_e(s)}}{1 - G_{t1}(s) \cdot G_c(s) \cdot G_s(s) \cdot F_m \cdot \dfrac{G_{pd} \cdot H_e(s)}{1 - F_v \cdot G_{pd} \cdot H_e(s)}} \cdot \frac{1 - \dfrac{Z_s(s)}{R_{el}}}{1 + \dfrac{Z_s(s)}{Z_{el}(s)}} \tag{1.110}$$

Readers can refer to [1] for the last impedance interaction factor in (1.110). Anyway, for the example given, (1.110) gives the theoretical loop gain shown in Figure 1.38. The theoretical prediction compares very well against the actual measurement (Figure 1.39).

C. Power Stage Losses

In the process leading to (1.8), the series resistive losses of the power stage were not included. If more accuracy is required, (1.9) should be used. With (1.9), the power stage gains are modified:

$$G_{pdm} = \frac{\partial V_o(D, V_{\text{in}})}{\partial D}, \quad G_{pvm} = \frac{\partial V_o(D,\ V_{\text{in}})}{\partial V_{\text{in}}} \tag{1.111}$$

In (1.9), R_w stands for the series winding resistance of input filter inductor, R_{on} the MOSFET on-resistance. It is also understood that the primary winding resistance and the MOSFET on-resistance experience the pulsating reflected load current, while the series winding resistance of the input filter sees only the averaged reflected load. However, the

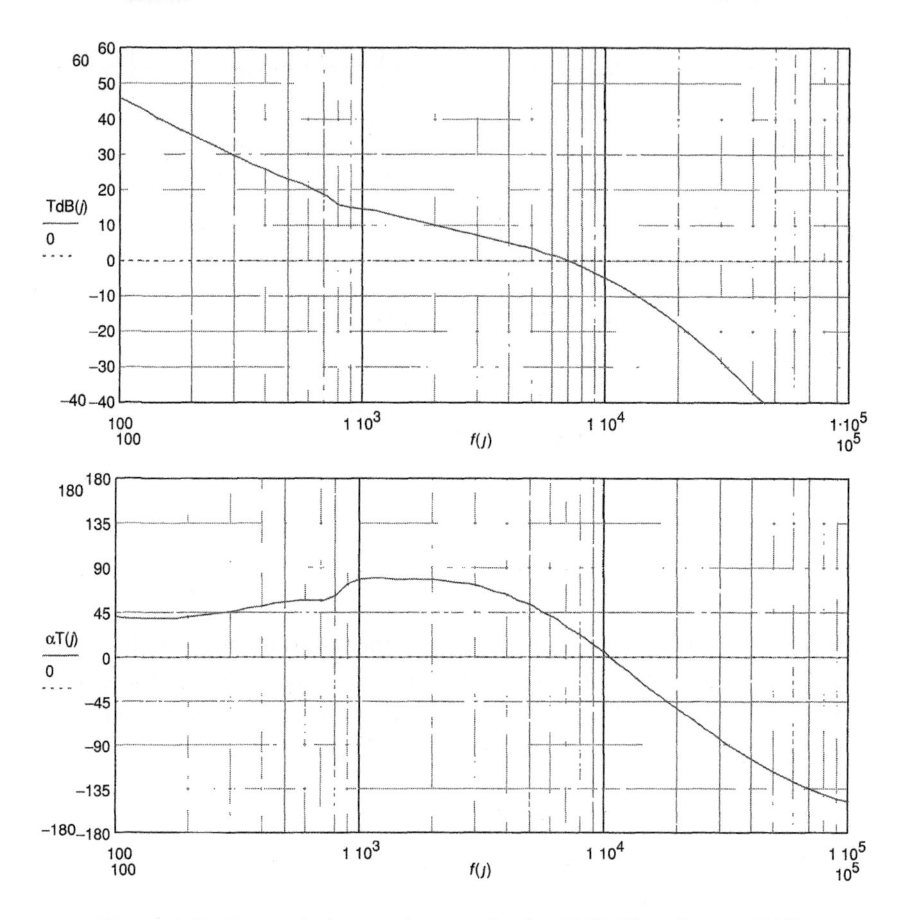

Figure 1.38: Example loop gain magnitude, 10dB/div; phase, 45°/div

magnitude of these additional effects is quite small in comparison. They can be safely ignored without introducing significant error.

D. Additional Filtering

Additional RC-filtering, R_f and C_f in Figure 1.30, is often also added following R_s to reduce high-frequency noise. If this is the case, the steady-state duty cycle can no longer be expressed in the compact, closed form of (1.52). Instead, it is expressed in an implicit form (Appendix 1) as

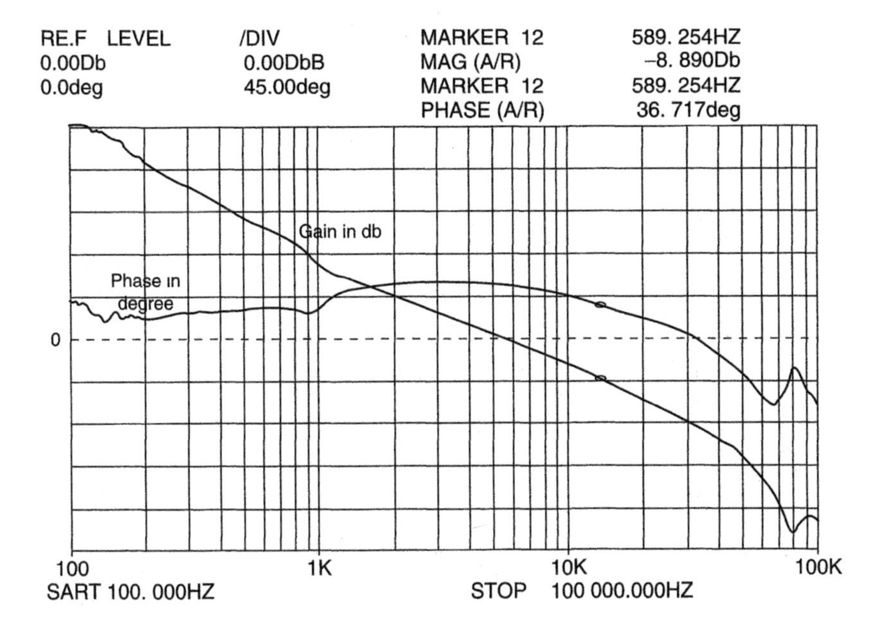

Figure 1.39: Example loop gain measurement, 10dB/div, 45°/div

$$w(V_{\text{in}}, V_o, v_{\text{er}}, D) = V_{\phi 1} e^{-\frac{D \cdot T_s}{\tau}} + k \cdot a \cdot f_1(D \cdot T_s) + k \cdot b \cdot f_2(D \cdot T_s) - v_{\text{er}} = 0 \tag{1.112}$$

This form gives a modified gain F_{mm} (Jacobian)

$$F_{mm} = \frac{\partial D}{\partial v_{\text{er}}} = -\frac{\dfrac{\partial w}{\partial v_{\text{er}}}}{\dfrac{\partial w}{\partial D}} \tag{1.113}$$

With the filter pole included, further improvement is made. This step yields a modified gain $G_{ef}(s)$ that replaces the simple constant gain F_m in Figure 1.37 and equation (1.110):

$$G_{ef}(s) = \frac{F_{mm}}{(R_s + R_f) \cdot C_f \cdot s + 1} \tag{1.114}$$

However, the impact of this additional pole is insignificant.

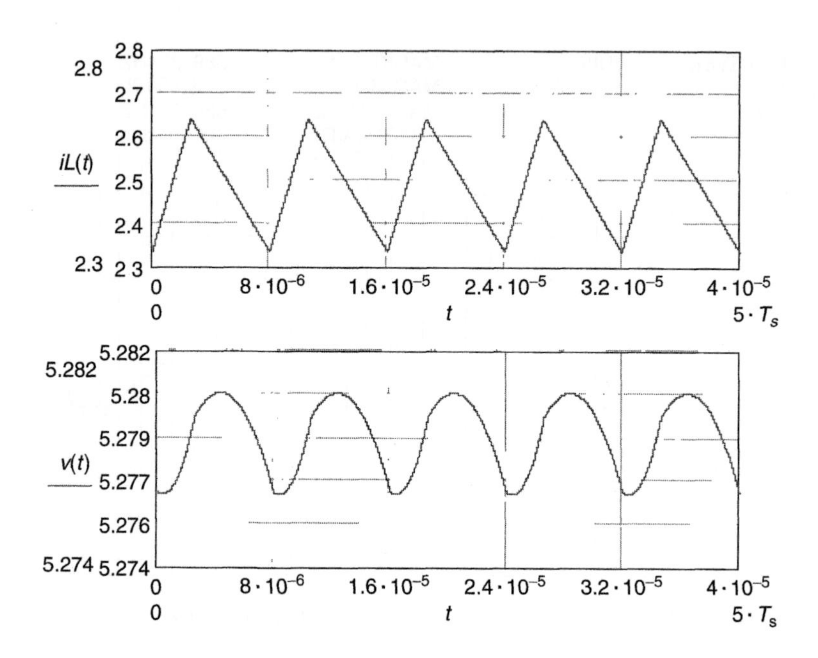

Figure 1.40: Theoretical example of output voltage and inductor current

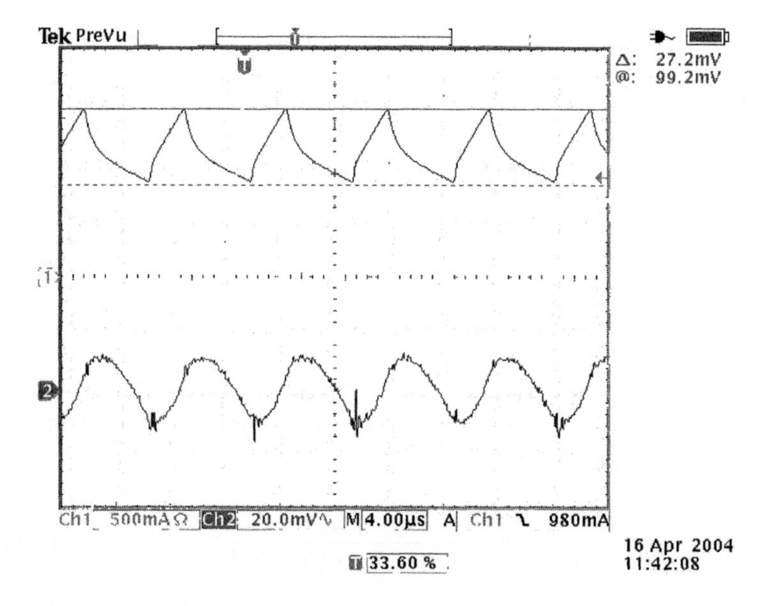

Figure 1.41: Example of inductor current, 0.5 A/div; output voltage, 20mv/div

E. Output Ripple

By applying the accelerated steady-state analysis technique outlined in section 1.12, the output voltage ripple and output inductor current are predicted (Figure 1.40). They match the actual measurement (Figure 1.41).

1.14 State Transition Technique

In section 1.12, a technique called *accelerated steady-state analysis* was invoked to obtain expeditiously the main output including ripple under the steady state. The technique calls for many Laplace transformations and its inverse. The approach can quickly become unmanageable if the order of circuit exceeds three. In this section, we present a different approach that deals with everything directly in the time domain. We use Figure 1.29 as the basis. Since the present goal is to introduce the technique, we use a simplified version of Figure 1.29, deleting r_d and C_d and adding r_C as esr for the main filter capacitor, C. In other words, we are dealing with Figure 1.42.

With the inductor current, i, and the capacitor voltage, v, selected as the two state variables, the following is established:

$$\frac{d}{dt}x = \frac{d}{dt}\begin{bmatrix} i \\ v \end{bmatrix} = \begin{bmatrix} -\dfrac{r_L + R_p}{L} & -\dfrac{k}{L} \\ \dfrac{R_p}{r_C \bullet C} & -\dfrac{1-k}{r_C \bullet C} \end{bmatrix} \begin{bmatrix} i \\ v \end{bmatrix} + \begin{bmatrix} \frac{1}{L} \\ 0 \end{bmatrix} E = A \bullet x + B \bullet E \tag{1.115}$$

where R_p equals r_c in parallel with R_L, while E stands for the rectangular input source driving the filter; A and B are the state matrix; **x** is the state

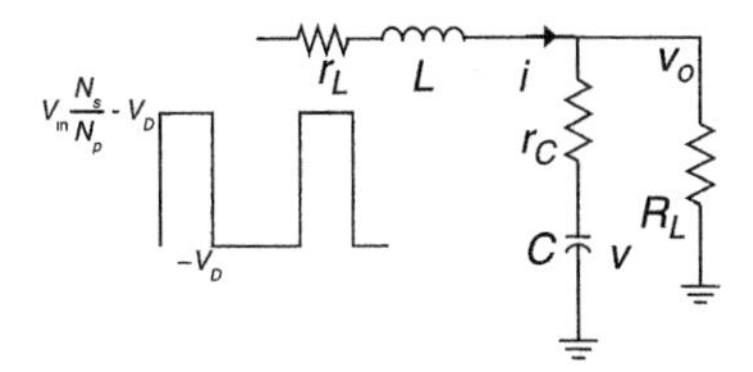

Figure 1.42: Simplified output filter

vector. It is well known in linear system theory, linear algebra, and matrix theory that the solution for (1.115) can be given as

$$x(t) = e^{A \cdot t} \cdot x(0) + \int_0^t e^{A(t-\tau)} \cdot B \cdot E \cdot d\tau \tag{1.116}$$

Furthermore, if the driving source happens to be constant during the time of interest, the solution is reduced to

$$x(t) = e^{A \cdot t} \cdot x(0) - A^{-1}(I - e^{A \cdot t})B \cdot E \tag{1.117}$$

Referring to Figure 1.42 and assuming two unknown starting states—$x(0) = X_1$ for the on-time segment, DT_s, and $x(0) = X_2$ for the off-time segment, $(1 - D)T_s$—we arrive at two solutions for the two segments, DT_s and $(1 - D)T_s$:

$$\begin{aligned} x_1(t) &= e^{A \cdot t} \cdot X_1 - A^{-1}\left(I - e^{A \cdot t}\right)B\left(\frac{N_s}{N_p} V_{\text{in}} - V_D\right) \\ x_2(t) &= e^{A(t-D \cdot T_s)} \cdot X_2 + A^{-1}\left[1 - e^{A(t-D \cdot T_s)}\right]B \cdot V_D \end{aligned} \tag{1.118}$$

Three items, the matrix exponential, X_1, and X_2, need to be evaluated. We first find the starting states. For that, we apply the concept of continuity of states at the two time boundaries, $t = DT_s$ for $x_1(t)$ and $t = T_s$ for $x_2(t)$:

$$\begin{aligned} &e^{A \cdot D \cdot T_s} \cdot X_1 - A^{-1}\left(I - e^{A \cdot D \cdot T}\right)B\left(\frac{N_s}{N_p} V_{\text{in}} - V_D\right) = X_2 \\ &e^{A \cdot (1-D) \cdot T_s} \cdot X_2 + A^{-1}\left[I - e^{A \cdot (1-D) \cdot T}\right]B \cdot V_D = X_1 \end{aligned} \tag{1.119}$$

At this juncture, we have to invoke matrix theory for computing matrix exponentials, $e^{A \cdot D \cdot T_s}$ and $e^{A \cdot (1-D) \cdot T_s}$. (We carry out only one as a demonstration.) It is understood that the 2×2 matrix, $A \cdot D \cdot T_s = A_1$, has two eigenvalues, λ_0 and λ_1, such that both values enable one to express the matrix exponentials in polynomial form:

$$\alpha_0 + \alpha_1 \cdot \lambda_0 = e^{\lambda_0}, \quad \alpha_0 + \alpha_1 \cdot \lambda_1 = e^{\lambda_1} \tag{1.120}$$

We then solve

$$\alpha_0 = \frac{\begin{vmatrix} e^{\lambda_0} & \lambda_0 \\ e^{\lambda_1} & \lambda_1 \end{vmatrix}}{\begin{vmatrix} 1 & \lambda_0 \\ 1 & \lambda_1 \end{vmatrix}}, \quad \alpha_1 = \frac{\begin{vmatrix} 1 & e^{\lambda_0} \\ 1 & e^{\lambda_1} \end{vmatrix}}{\begin{vmatrix} 1 & \lambda_0 \\ 1 & \lambda_1 \end{vmatrix}} \tag{1.121}$$

Based on the Cayley-Hamilton theorem, the matrix exponential is then expressed as

$$e^{A \cdot D \cdot T_s} = e^{A_1} = \alpha_0 \cdot I + \alpha_1 \cdot A_1 \tag{1.122}$$

By the same procedure, the other matrix exponential is written in polynomial form:

$$e^{A \cdot (1-D) \cdot T_s} = e^{A_2} = \mu_0 \cdot I + \mu_1 \cdot A_2 \tag{1.123}$$

Equation (1.119) then allows us to solve the cyclic starting states X_1 and X_2 in sequence:

$$\begin{aligned} X_1 &= [I - (\mu_0 \cdot I + \mu_1 \cdot A_2)(\alpha_0 \cdot I + \alpha_1 \cdot A_1)]^{-1} \cdot \\ &\left\{ \begin{aligned} &-(\mu_0 \cdot I + \mu_1 \cdot A_2) \cdot A^{-1} \cdot [I - (\alpha_0 \cdot I + \alpha_1 \cdot A_1)] \cdot B \cdot \left(\frac{N_s}{N_p} V_{\text{in}} - V_D\right) \\ &\quad + A^{-1} \cdot [I + (\mu_0 \cdot I + \mu_1 \cdot A_2)] \cdot B \cdot V_D \end{aligned} \right\} \\ X_2 &= (\alpha_0 \cdot I + \alpha_1 \cdot A_1) X_1 - A^{-1} \cdot [I - (\alpha_0 \cdot I + \alpha_1 \cdot A_1)] \cdot B \cdot \left(\frac{N_s}{N_p} V_{\text{in}} - V_D\right) \end{aligned} \tag{1.124}$$

Next, the matrix exponential, $e^{A \cdot t}$, is also placed in the polynomial form based on the preceding procedure:

$$\begin{aligned} &e^{A \cdot t} = \beta_0(t) \cdot I + \beta_1(t) \cdot A, \\ &\beta_0(t) = \frac{\begin{vmatrix} e^{\lambda_0 \cdot t} & \lambda_0 \\ e^{\lambda_1 \cdot t} & \lambda_1 \end{vmatrix}}{\begin{vmatrix} 1 & \lambda_0 \\ 1 & \lambda_1 \end{vmatrix}}, \quad \beta_1(t) = \frac{\begin{vmatrix} 1 & e^{\lambda_0 \cdot t} \\ 1 & e^{\lambda_1 \cdot t} \end{vmatrix}}{\begin{vmatrix} 1 & \lambda_0 \\ 1 & \lambda_1 \end{vmatrix}} \end{aligned} \tag{1.125}$$

Readers are cautioned that λ_0 and λ_1 in (1.125) are eigenvalues of matrix A, not A_1 or A_2. The solution for the two time segments are finally obtained as

$$x_1(t) = \begin{pmatrix} [\beta_0(t) \cdot I + \beta_1(t) \cdot A]X_1 - \\ A^{-1}\{I - [\beta_0(t) \cdot I + \beta_1(t) \cdot A]\}B\left(\frac{N_s}{N_p} V_{\text{in}} - V_D\right) \end{pmatrix}$$

$$[u(t) - u(t - D \cdot T_s)] \tag{1.126}$$

$$x_2(t) = \begin{pmatrix} [\beta_0(t - D \cdot T_s) \cdot I + \beta_1(t - D \cdot T_s) \cdot A]X_2 + \\ A^{-1}\{I - [\beta_0(t - D \cdot T_s) \cdot I + \beta_1(t - D \cdot T_s) \cdot A]\}B \cdot V_D \end{pmatrix}$$

$$[u(t - D \cdot T_s) - u(t - T_s)]$$

In (1.126), the inductor current is the first element of vector $x_1(t)$ and $x_2(t)$, while the output voltage is a superposition quantity consisting of both the first and the second elements of state vectors $x_1(t)$ and $x_2(t)$:

$$\begin{aligned} i(t) &= [x_1(t)]_0 + [x_2(t)]_0, \\ v_o(t) &= R_p \cdot i(t) + k\{[x_1(t)]_1 + [x_2(t)]_1\}, \quad k = \frac{R_L}{r_c + R_L} \end{aligned} \tag{1.127}$$

Appendix 2 gives the MathCAD listing and computation that confirms the same results using Laplace transformation. Readers proficient in MATLAB may prefer the tool. Using the alternative tool, one bypasses (1.120) through (1.126). However, (1.119) is still needed to compute X1 and X2. Then (1.118) follows. Appendix 3 gives the MATLAB program listing.

Chapter 2

Push–Pull Converter with Current-Mode Control and Slope Compensation

In Chapter 1, the basic forward converter was covered extensively. The topology treated is a single-transistor version of the forward family. As such, the power switch shoulders all the burden of providing load power and power consumed locally. The configuration therefore is limited to low-power applications, for instance, less than 100 W, due to concerns for transistor ratings and thermal management. In addition, the transformer used in the single transistor version utilizes only the first quadrant of the core *B–H* plane. It not only carries a DC current and underutilizes the magnetic core, but also requires external core reset. In light of these deficiencies, solutions must be found. This is where the push–pull configuration steps in.

The conventional push–pull topology employing a single-loop voltage-mode control comes in several forms. Figure 2.1 shows the two most popular implementations. Both configurations enjoy the benefits of dual transistors, four-quadrant core *B–H* utilization, and natural magnetic reset. However, they all suffer potential flux imbalance and, consequently, core saturation. Numerous techniques were conceived to prevent the threat. Most result in additional circuits but with only limited success. It turns out that peak-current current-mode control easily

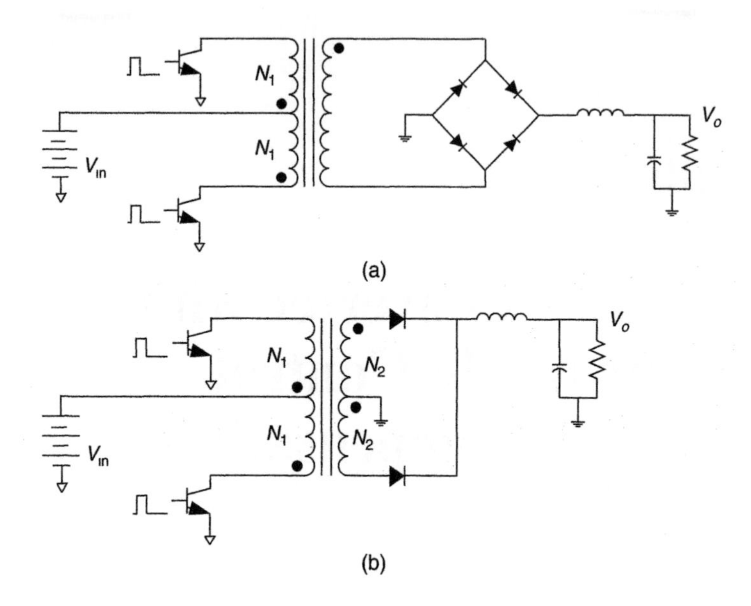

Figure 2.1: Push–pull converter power stages

alleviates the problem by, in effect, placing a current limit on the power switches. With this technique, albeit with minute volt–second imbalance, the power switch is never subjected to destructive overcurrent conditions. Given the advantage, this chapter gives the complete coverage of a center-tap push–pull converter with peak-current current-mode control, slope compensation, and the ability to operate in both the continuous and discontinuous conduction modes.

2.1 Power Stage of a Center-Tapped Push–Pull Converter

Figure 2.1 shows two push–pull configurations. Of the two, the center-tap is selected more often by designers. Therefore Figure 2.1(b) is the basis for this chapter. Also, to proceed with the analysis, other supporting circuits, including the current feedback, the external slope compensation, and the PWM comparator, are added in Figure 2.2. The reason

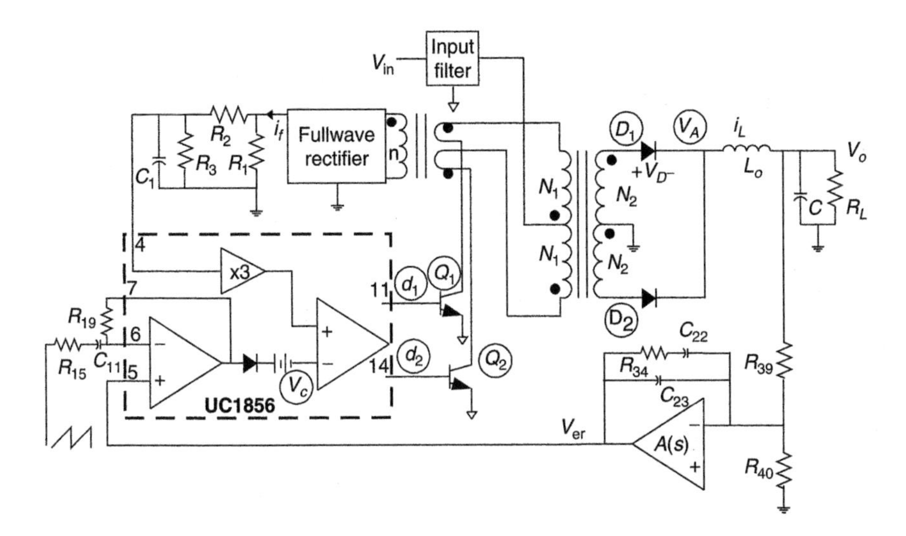

Figure 2.2: Push–pull converter with current-mode control

for adding the current feedback is self-evident. As for the external slope compensation, it is explained in a later section. Moreover, prior to the analysis, it is beneficial to summarize the steady-state waveforms surrounding the power stage. Here, two distinctive operation modes are presented. Figure 2.3 shows the center-tap configuration in the discontinuous conduction mode (DCM) while Figure 2.4 gives the more familiar continuous conduction-mode (CCM) operation. Readers should have no difficulty observing the difference between them.

2.2 Discontinuous Conduction-Mode Operation

Basically, the push–pull converters are forward-derived converters. Many concepts and processes presented in Chapter 1 are applicable in this instance. Among them, the selection of a minimum output filter inductor by Equation (1.17) needs more attention. As given in the equation, the minimum inductance required to maintain CCM operation is inversely proportional to the minimum load current. The inductance value, and consequently the inductor geometry, can easily become

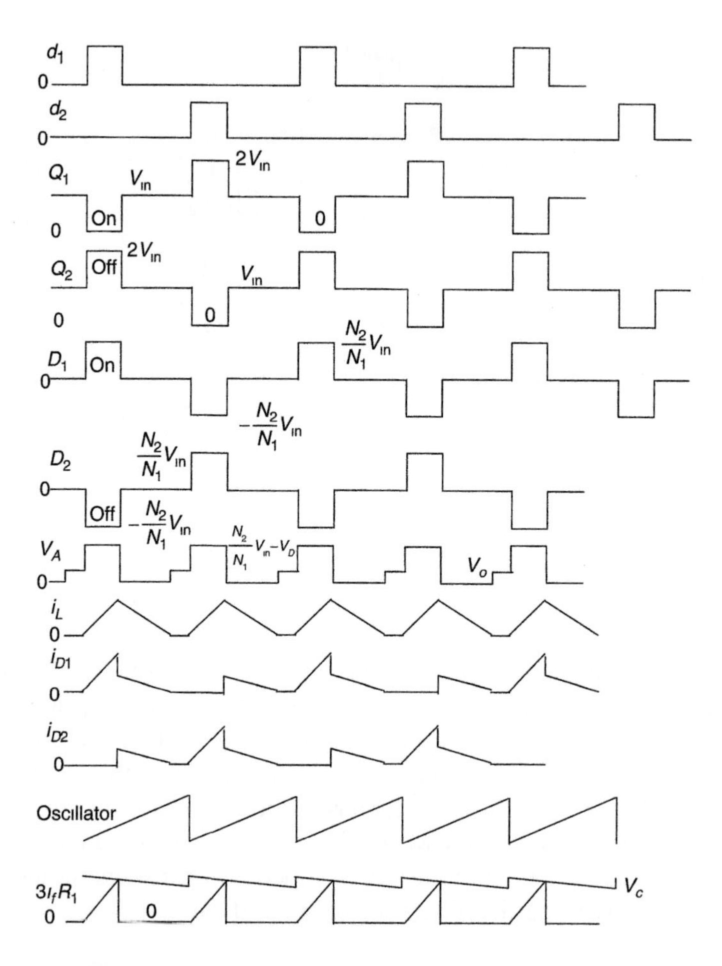

Figure 2.3: Push–pull converter waveforms in DCM

impractically large if an extremely low load must be accommodated. Therefore, a compromise must be struck between the selection of a realistic device size and the stringent requirement. In other words, it may become necessary to operate the converter in both the DCM and CCM, anticipating a wide dynamic load range. Fortunately, multiple-loop current-mode control techniques offer just that capability and versatility. With this understanding in mind, we first treat the converter in the DCM.

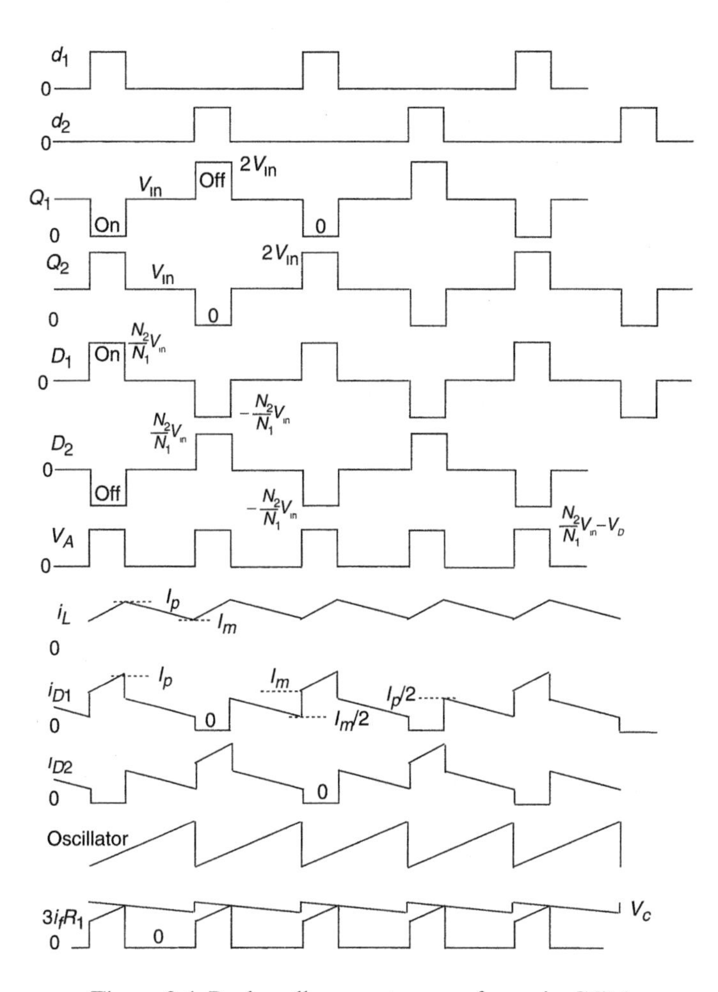

Figure 2.4: Push-pull converter waveforms in CCM

A. Pulse-Width Modulation and Gains

Under the DCM, the key voltage and current waveforms of the power stage are as shown in Figure 2.3. The switch current, i_p, can be written as

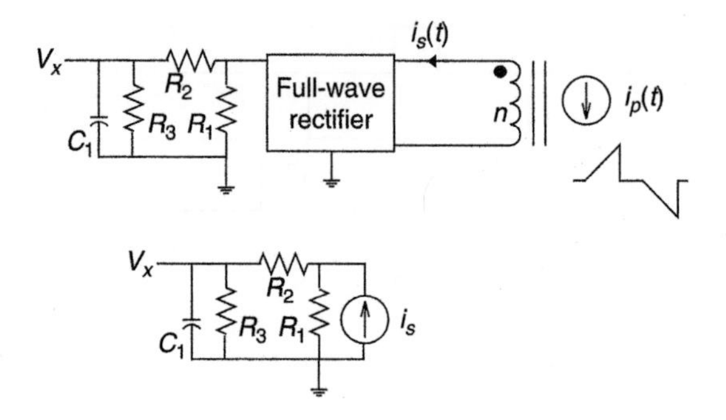

Figure 2.5: Current feedback processing in DCM

$$i_p(t) = \left(\frac{V_{\text{in}}}{L_p} + \frac{N_2}{N_1} \bullet \frac{\frac{N_2}{N_1} V_{\text{in}} - V_D - V_o}{L_o} \right) \bullet t = B \bullet t \tag{2.1}$$

This current is sensed by a current transformer with a turn ratio of $1/n$. It then feeds a simple RC filter, as shown in Figure 2.5.

The output voltage of the filter can be given in the following frequency-domain form:

$$V_x(s) = \frac{R_k}{R_T \bullet C_1 \bullet s + 1} \bullet I_s(s) \tag{2.2}$$

where

$$R_k = \frac{R_1 \bullet R_3}{R_1 + R_2 + R_3},\ R_T = \frac{(R_1 + R_2)R_3}{R_1 + R_2 + R_3},\ I_s(s) = \frac{B}{n \bullet s^2}$$

Given the transfer function, the time-domain output is easily proven to be

$$v_x(t) = \frac{B \bullet R_k}{n} \left[t + R_T \bullet C_1 \left(\mathrm{e}^{-\frac{t}{R_T \bullet C_1}} - 1 \right) \right] \tag{2.3}$$

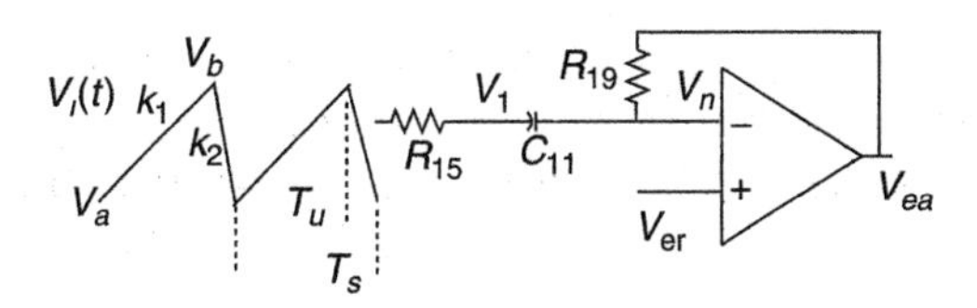

Figure 2.6: Slope-compensation processing

Next, the steady-state slope compensation output must be determined. For this, we begin by writing a differential equation for node V_1 of Figure 2.6:

$$\frac{dV_1}{dt} + \frac{1}{\tau_1} V_1 = \frac{1}{\tau_1} V_i, \ \tau_1 = R_{15} \cdot C_{11} \tag{2.4}$$

where V_i is the external oscillator input, given as

$$V_i(t) = \begin{cases} (V_a + k_1 \cdot t) & 0 \leq t \leq T_u \\ V_b - k_2 \cdot (t - T_u) & T_u \leq t \leq T_s \end{cases} \tag{2.5}$$

Note that the oscillation input is periodic and, by using approaches similar to those outlined in Chapter 1 and again invoking the concept of continuity of state, V_1 is obtained:

$$V_1(t) = \begin{cases} v_{1a} \cdot e^{-\frac{t}{\tau_1}} + V_a \left(1 - e^{-\frac{t}{\tau_1}}\right) + \\ k_1 \left[t + \tau_1 \left(e^{-\frac{t}{\tau_1}} - 1\right)\right] & 0 \leq t \leq T_u \\ v_{1b} \cdot e^{-\frac{t - T_u}{\tau_1}} + V_b \left(1 - e^{-\frac{t - T_u}{\tau_1}}\right) - & T_u \leq t \leq T_s \\ k_2 \left[t - T_u + \tau_1 \left(e^{-\frac{t - T_u}{\tau_1}} - 1\right)\right] \end{cases} \tag{2.6}$$

where v_{1a} and v_{1b} are the cyclic starting states of V_1 and the end result of applying the technique of continuity of states.

Once V_1 is obtained, another differential equation at the output node, V_{ea}, can be written

$$V_{ea}(t) = V_{\text{er}} - \tau_3 \frac{dV_1}{dt}, \ \tau_3 = R_{19} \cdot C_{11} \tag{2.7}$$

By further combining V_1 and the second differential equation, the potential at the inverting input of the PWM, TI UC1856, comparator is then expressed as

$$V_{\text{er}} - \tau_3 \left[\left(\frac{V_a - v_{1a}}{\tau_1} \right) \mathrm{e}^{-\frac{t}{\tau_1}} + k_1 \left(1 - \mathrm{e}^{-\frac{t}{\tau_1}} \right) \right] - 1.2 \quad 0 \leq t \leq T_u \tag{2.8}$$

The steady-state duty cycle is hence determined when the current feedback signal intercepts the inverting input. That is, the duty cycle is embedded in the following:

$$3 \cdot v_x(D \cdot T_s) = V_{\text{er}} - \tau_3 \left[\left(\frac{V_a - v_{1a}}{\tau_1} \right) \mathrm{e}^{-\frac{D \cdot T_s}{\tau_1}} + k_1 \left(1 - \mathrm{e}^{-t\frac{D \cdot T_s}{\tau_1}} \right) \right] - 1.2 \tag{2.9}$$

From this identity, a function can be defined:

$$f(V_{\text{er}}, V_{\text{in}}, V_o, D) = 3 \cdot v_x(D \cdot T_s) - V_{\text{er}} +$$

$$\tau_3 \left[\left(\frac{V_a - v_{1a}}{\tau_1} \right) \mathrm{e}^{-\frac{D \cdot T_s}{\tau_1}} + k_1 \left(1 - \mathrm{e}^{-\frac{D \cdot T_s}{\tau_1}} \right) \right] + 1.2 = 0 \tag{2.10}$$

The preceding function then yields the small-signal gains as

$$F_m = \frac{\partial D}{\partial V_{\text{er}}} = -\frac{\dfrac{\partial f}{\partial V_{\text{er}}}}{\dfrac{\partial f}{\partial D}}, \ F_v = \frac{\partial D}{\partial V_o} = -\frac{\dfrac{\partial f}{\partial V_o}}{\dfrac{\partial f}{\partial D}}, \ F_g = \frac{\partial D}{\partial V_{\text{in}}} = -\frac{\dfrac{\partial f}{\partial V_{\text{in}}}}{\dfrac{\partial f}{\partial D}} \tag{2.11}$$

These gain factors aree used in Section 2.1.C.

B. DC Closed Loop

As is shown in Figure 2.3, the output inductor current under DCM is a repetitive triangle with a peak magnitude of

$$I_{pk} = \frac{\left(\frac{N_2}{N_1} V_{\text{in}} - V_o\right) D}{L_o \bullet f_s} \tag{2.12}$$

In theory, the load current equals the average value of the output inductor current; that is,

$$\frac{\left(\frac{N_2}{N_1} V_{\text{in}} - V_o\right) D(D + D_2)}{2 \bullet L_o \bullet f_s} = \frac{V_o}{R_L} \tag{2.13}$$

Furthermore, volt–second balance also requires

$$\left(\frac{N_2}{N_1} V_{\text{in}} - V_o\right) D - (V_o + V_D) D_2 = 0 \tag{2.14}$$

Both equations can be combined and rearranged. The combination produces a quadratic equation:

$$a \bullet V_o^2 + b \bullet V_o + c = 0 \tag{2.15}$$

where

$$a = \frac{2L_o \bullet f_s}{R_L},\ b = \frac{2L_o \bullet f_s \bullet V_D}{R_L} + D^2\left(V_D + \frac{N_2}{N_1} V_{\text{in}}\right),$$
$$c = -\frac{N_2}{N_1} V_{\text{in}} \bullet D^2 \left(V_D + \frac{N_2}{N_1} V_{\text{in}}\right) \tag{2.16}$$

That is to say, the converter main output can be expressed in terms of the open-loop duty cycle, D, the input bus, V_{in}, the load, R_L, and other power stage components and parameters:

$$V_o = \frac{-b + \sqrt{b^2 - 4 \bullet a \bullet c}}{2 \bullet a} \tag{2.17}$$

Ideally, it would be even more desirable to express the main output in a closed form, similar to what was done in section 1.10. However, in this case, no solution in that form can be obtained, since (2.10) is a transcendental equation and would not yield the duty cycle in explicit form. This condition leaves us the remaining approach, the numerical approximation method and loop partition outlined in section 1.6. This latter technique is performed by first redefining function $f(V_{er}, V_{in}, V_o, D) = f[V_{er}(R_x, \ldots, V_o), V_{in}, V_o, D]$ because, in almost all cases, the feedback error voltage can be expressed as a function of feedback circuit components, R_x, and the main output. Then the other function, $g(X, \ldots, V_{in}, V_o, D)$, is also defined based on (2.17):

$$g(X, \ldots, V_{in}, V_o, D) = V_o - \frac{-b + \sqrt{b^2 - 4 \bullet a \bullet c}}{2 \bullet a} = 0 \tag{2.18}$$

where X stands for many component values and device parameters.

Once both functions are defined, the closed-loop numerical solution can be found by solving both simultaneously. The solution consists of the steady-state, closed-loop duty cycle and the main output, given a load condition R_L and input voltage V_{in}.

In addition to the steady-state solution, the output sensitivity can be derived by applying exactly the same process as outlined in section 1.6.

C. AC Closed Loop

Through a rather tedious procedure, the steady-state condition is established in the previous sections. The remaining task for DCM operation is again the AC studies. Without exception, the power stage transfer functions are the main focus. We once again invoke the state-space averaged model, Figure 2.7, that also includes isolation and input filter source impedance Z_s and transfer function $H(s)$. The input loop gives

$$\hat{i}_s = \frac{\frac{N_2}{N_1} H(s)}{\left(\frac{N_2}{N_1}\right)^2 Z_s} \hat{v}_{in} - \frac{1}{\left(\frac{N_2}{N_1}\right)^2 Z_s} \hat{v}_g \tag{2.19}$$

And the input and output nodes give

$$-\left(\frac{\left(\frac{N_2}{N_1}\right)^2 Z_s}{r_1}+1\right)\hat{v}_g+\left(\frac{N_2}{N_1}\right)^2 Z_s \bullet g_1 \bullet \hat{v}_o$$

$$=\frac{N_2}{N_1}H(s)\bullet \hat{v}_B+\left(\frac{N_2}{N_1}\right)^2 Z_s \bullet j_1 \bullet \hat{d},\ g_2 \bullet \hat{v}_g-\left(\frac{1}{r_2}+\frac{1}{R_L}+C\bullet s\right)\hat{v}_o$$
$$=-j_2 \bullet \hat{d} \tag{2.20}$$

By solving the two equations in determinant form and expanding and grouping terms, the output is found

$$\hat{v}_o=\frac{\left\{\frac{j_2}{g_2}\left[\frac{\left(\frac{N_2}{N_1}\right)^2 Z_s}{r_1}+1\right]-j_1\bullet\left(\frac{N_2}{N_1}\right)^2 Z_s\right\}\hat{d}+\frac{N_2}{N_1}H(s)\bullet \hat{v}_s}{\frac{1}{g_2}\left[\frac{\left(\frac{N_2}{N_1}\right)^2 Z_s}{r_1}+1\right]\left(\frac{1}{r_2}+\frac{1}{R_L}+C\bullet s\right)-\left(\frac{N_2}{N_1}\right)^2 Z_s \bullet g_1} \tag{2.21}$$

The power stage gain functions, therefore, are

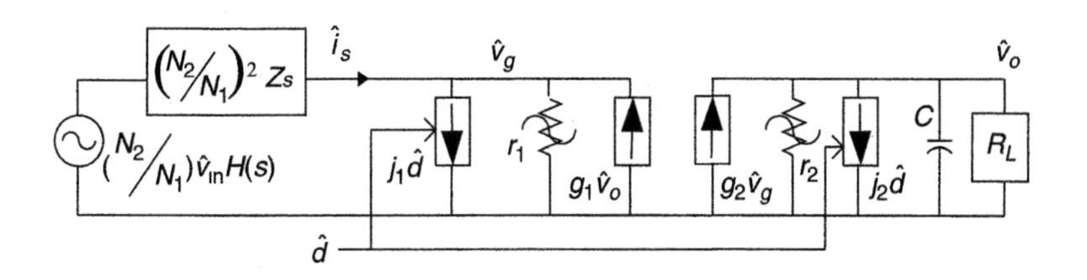

Figure 2.7: Canonical model of a push–pull converter power stage in DCM

$$G_{vd}(s) = \frac{\frac{j_2}{g_2}\left[\frac{\left(\frac{N_2}{N_1}\right)^2 Z_s}{r_1} + 1\right] - j_1 \bullet \left(\frac{N_2}{N_1}\right)^2 Z_s}{\frac{1}{g_2}\left[\frac{\left(\frac{N_2}{N_1}\right)^2 Z_s}{r_1} + 1\right]\left(\frac{1}{r_2} + \frac{1}{R_L} + C \bullet s\right) - \left(\frac{N_2}{N_1}\right)^2 Z_s \bullet g_1} \quad (2.22)$$

$$G_{vg}(s) = \frac{1}{\frac{1}{g_2}\left[\frac{\left(\frac{N_2}{N_1}\right)^2 Z_s}{r_1} + 1\right]\left(\frac{1}{r_2} + \frac{1}{R_L} + C \bullet s\right) - \left(\frac{N_2}{N_1}\right)^2 Z_s \bullet g_1} \quad (2.23)$$

These transfer functions and the PWM small-signal gains, F_m, F_b, and F_v (2.11), are then incorporated into the AC block diagram (Figure 2.8(a)). Based on this block diagram, and for developing the voltage loop loop-gain under constant input voltage, the inner current loop containing F_v is consolidated and the diagram is simplified to Figure 2.8(b). The loop-gain, $T(s)$, is obtained by inspection:

$$T(s) = -K_f \bullet E_A(s) \bullet A_e(s) \bullet F_m \bullet G_i(s) \quad (2.24)$$

where

$$G_i(s) = \frac{G_{vd}(s)}{1 - F_v \bullet G_{vd}(s)} \quad (2.25)$$

Figure 2.9 shows a very good match between the actual measurement (a) and the prediction (b) of loop-gain based on (2.24). Such a good match validates the analytical procedure just given.

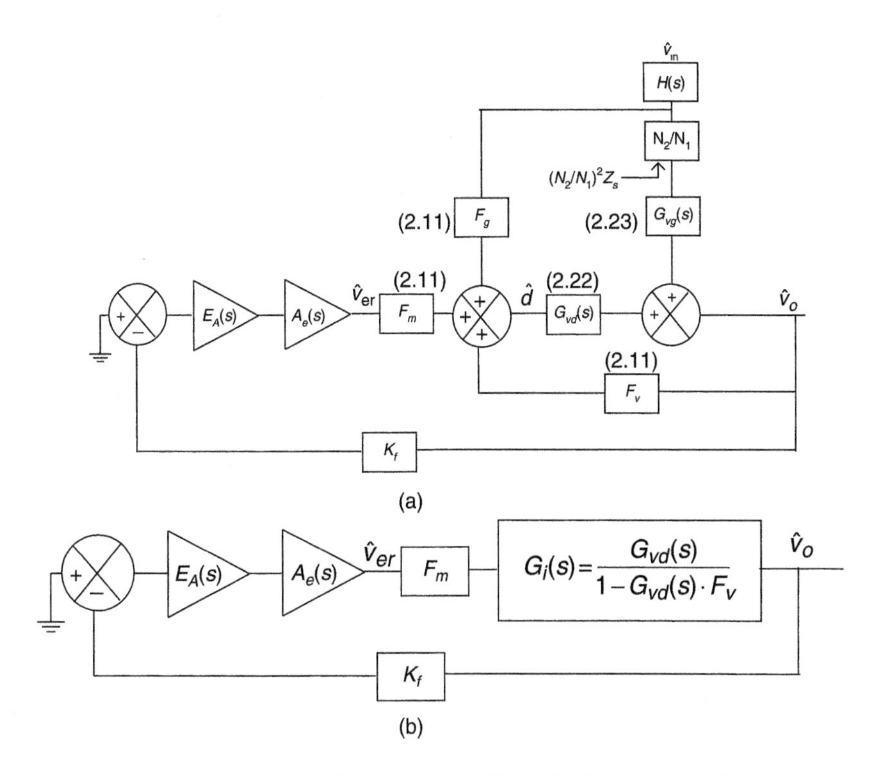

Figure 2.8: Push–pull converter AC block diagram in DCM

2.3 Continuous Conduction-Mode Operation

The push-pull converter possesses the capability of working in both the DCM and the CCM. A main cause making the converter operate in the two distinctive modes is the load current. Under a heavy load, the converter output inductor current is understood to take the profile shown in Figure 2.4. Obviously, significant changes take place and the analytical treatments are expected to change as well. We therefore follow almost the same flow as that of Chapter 1. However, some mathematical duplication is intentionally omitted, since identical key steps and equations have been presented.

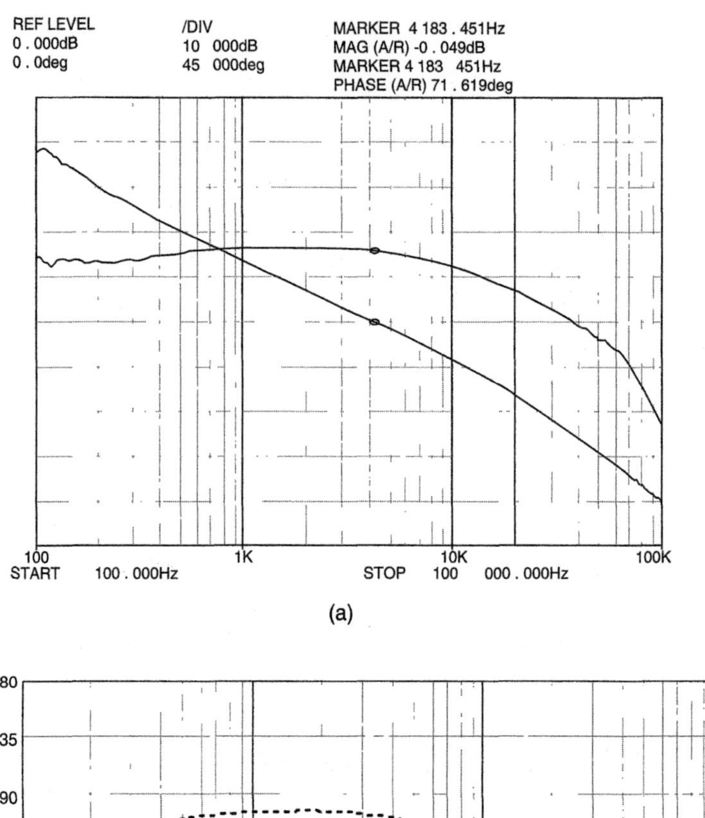

(a)

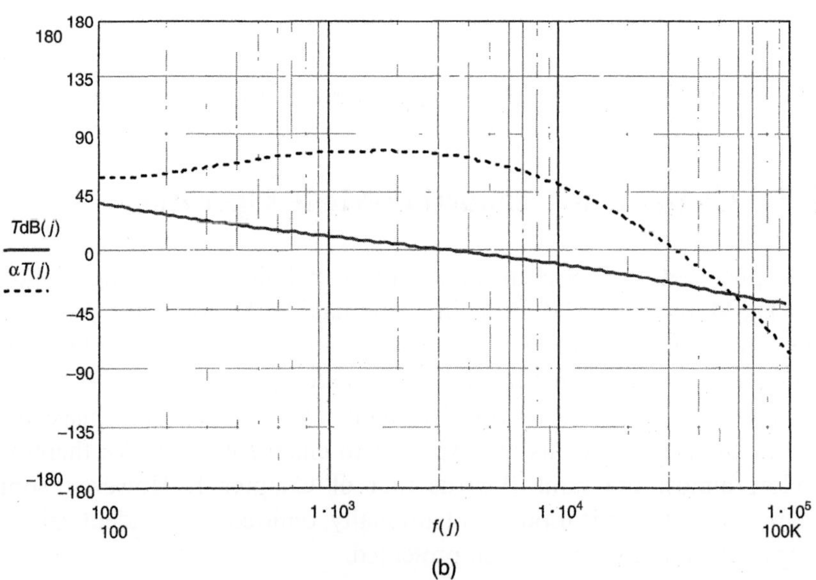

(b)

Figure 2.9: Loop gain of a push–pull converter in DCM

A. Pulse-Width Modulation and Gains

Under the CCM operation and in reference to Figure 2.4 in which important currents associated with the power stage are clearly identified and designated, the power switch current in a single on period can be described as

$$i_p(t) = A + B \cdot t \tag{2.26}$$

where

$$A = \frac{N_2}{N_1}\left[I_o - \frac{V_D + V_o}{2L_o}(1 - D)T_s\right],$$

$$B = \frac{N_2}{N_1}\frac{\left[\frac{N_2}{N_1}V_{\text{in}} - (V_D + V_o)\right]}{L_o} + \frac{V_{\text{in}}}{L_p} \tag{2.27}$$

The primary current is processed by the same sensing circuit and yields

$$v_x(t) = \frac{A \cdot R_k}{n}\left(1 - \mathrm{e}^{-\frac{t}{R_T \cdot C_1}}\right) + \frac{B \cdot R_k}{n} \times \left[t + R_T \cdot C_1\left(\mathrm{e}^{-\frac{t}{R_T \cdot C_1}} - 1\right)\right] \tag{2.28}$$

Then, by exactly the same operation as in (2.9) and (2.10), the CCM duty cycle is established by a different function:

$$g(v_{\text{er}}, V_{\text{in}}, V_o, D) = 3 \cdot v_x(D \cdot T_s) - v_{\text{er}} + \tau_3\left[\left(\frac{v_a - v_{1a}}{\tau_1}\right)\mathrm{e}^{-\frac{D \cdot T_s}{\tau_1}} + k_1\left(1 - \mathrm{e}^{-\frac{D \cdot T_s}{\tau_1}}\right)\right] + 1.2 = 0 \tag{2.29}$$

and so are the small-signal gains F_m, F_v, and F_g.

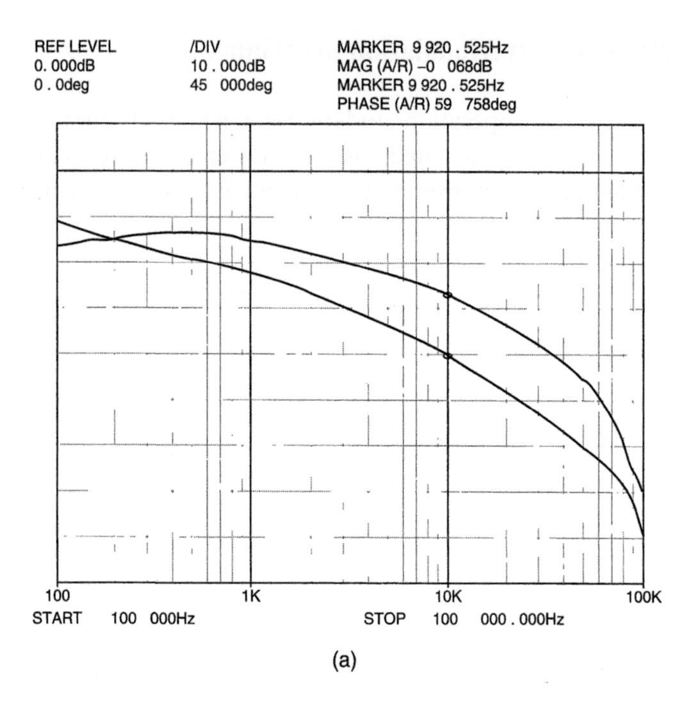

(a)

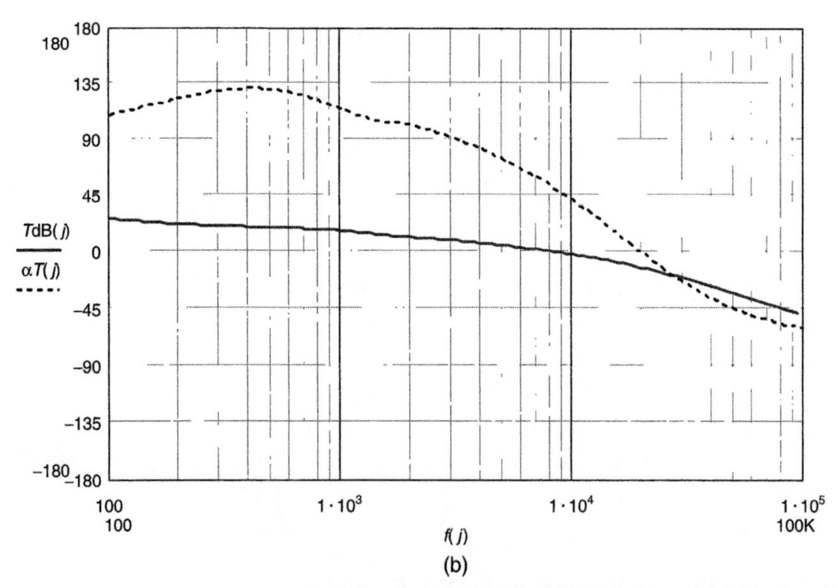

(b)

Figure 2.10: Loop gain of the push–pull converter in CCM

B. DC Closed Loop

Under the CCM operation, the output voltage as a function of the open-loop duty cycle is no different from that developed in Chapter 1, since the push–pull converter is basically a forward converter. The open-loop output therefore is given in the implicit function

$$h(X, \ldots, V_{\text{in}}, V_o, D) = V_o - \frac{N_2}{N_1}\left[V_{\text{in}} - (R_s \bullet D + R_w + R_{\text{on}})\frac{N_2}{N_1}\frac{V_o}{R_L}\right]D + V_D + R_l\frac{V_o}{R_L} = 0 \quad (2.30)$$

The closed-loop output, as in the DCM case, is hiding behind the two functions, $h(X, \ldots, V_{\text{in}}, V_o, D)$ and $g[v_{\text{er}}(R_x, \ldots, V_o), V_{\text{in}}, V_o, D]$. Again, readers are cautioned that g and h are two new, implicit functions.

C. AC Closed Loop

Under the CCM, not only is the steady-state analytical form altered in comparison with that of the DCM, but so is the AC model. Basically, the power stage model needs a new look. However, this has all been done in Chapter 1. We present just the end results, comparing the actual measurement (a) against the theoretical prediction (b) (Figure 2.10).

Chapter 3

Nonisolated Forward Converter with Average Current-Mode Control

The average current-mode control, in contrast to the peak current-mode, was claimed by Tang and Lee [3] to offer several advantages. However, in treating the topic, many existing studies did not handle the analytical procedure properly. This chapter gives an in-depth analysis of the subject. All discussions refer to Figure 3.1, which depicts a complete, nonisolated forward converter using average current-mode control.

3.1 Average Current Feedback

Under CCM operation, the output filter inductor, L, current is known to exhibit a waveform as shown in Figure 3.2.

The steady-state current feedback signal, including the sense resistor, R_i, can be described as

$$V_i(t) = \begin{cases} R_i \cdot \left(I_b + \frac{V_g - V_o}{L} \cdot t \right) = v_1(t) & 0 \leq t \leq D \cdot T_s \\ R_i \cdot \left[I_b + \delta i - \frac{V_o}{L} \cdot (t - D \cdot T_s) \right] = v_2(t) & D \cdot T_s \leq t \leq T_s \end{cases} \tag{3.1}$$

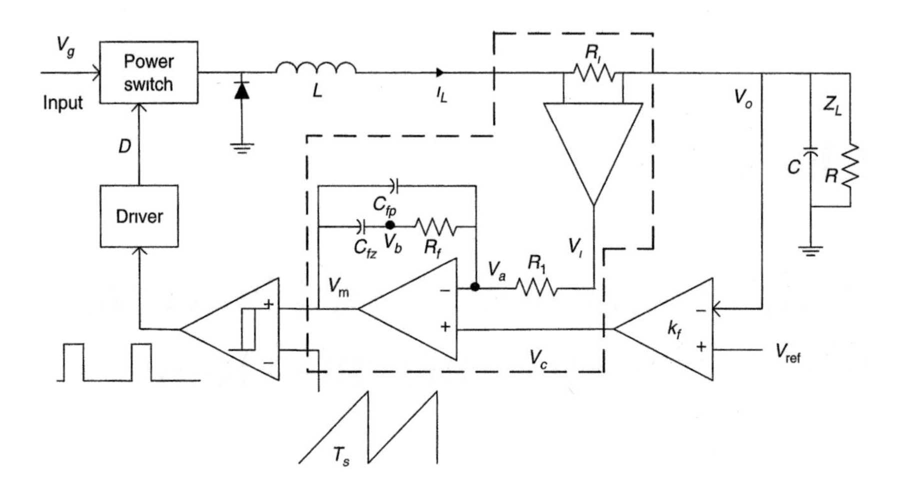

Figure 3.1: Nonisolated forward converter using average current-mode control

where

$$\delta i = \frac{V_g - V_o}{L} \cdot D \cdot T_s \qquad I_b = \frac{V_o}{R_L} - \frac{V_g - V_o}{2 \cdot L} \cdot D \cdot T_s$$

and V_g is the line input voltage; δi, the peak-to-peak ripple current; D, the steady-state duty cycle; T_s, the switching period; V_o the regulated output. Then, at the inverting input node, V_a, and the node V_b, two differential equations can be written:

$$\frac{dV_m}{dt} + \frac{1}{\tau_2} \cdot V_b = \frac{V_c}{\tau_1} - \frac{1}{\tau_3} \cdot V_i \quad \frac{dV_m}{dt} - \frac{dV_b}{dt} - \frac{1}{\tau_4} \cdot V_b = -\frac{V_c}{\tau_4} \tag{3.2}$$

where

$$\tau_1 = C_{fp} \cdot \frac{R_1 \cdot R_f}{R_1 + R_f}, \ \tau_2 = R_f \cdot C_{fp}, \ \tau_3 = R_1 \cdot C_{fp}, \ \tau_4 = R_f \cdot C_{fz}$$

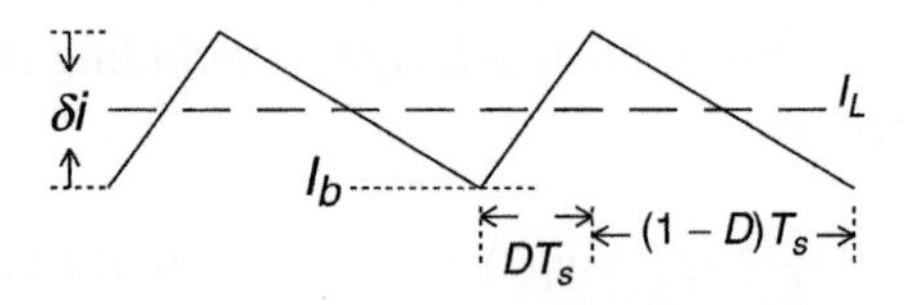

Figure 3.2: Inductor current profile

The equations in (3.2) are consolidated into a single one for node voltage V_b:

$$\frac{dV_b}{dt} + \omega_p \cdot V_b = \omega_x \cdot V_c - \frac{V_\iota}{\tau_3} \tag{3.3}$$

where

$$\omega_p = \frac{1}{\tau_4} + \frac{1}{\tau_2}, \qquad \omega_x = \frac{1}{\tau_4} + \frac{1}{\tau_1}$$

Equation (3.3), together with the cyclic input functions of (3.1), can be solved for the two time segments in the steady state with the assumption of two unknown, cyclic starting states, V_{bx} and V_{by}, at the beginning of each segment. The solutions for node voltage V_b are

$$V_{b1}(t) = \left\{ \begin{matrix} V_{bx} \cdot e^{-\omega_p \cdot t} + \dfrac{\omega_x}{\omega_p}(1 - e^{-\omega_p \cdot t}) V_c - \\ \dfrac{R_\iota \cdot I_b}{\tau_3}[B_1 \cdot t + A_1(1 - e^{-\omega_p \cdot t})] \end{matrix} \right\} \cdot u(t) \tag{3.4}$$

where

$$B_1 = \frac{V_g - V_o}{\omega_p \cdot L \cdot \left(\dfrac{V_o}{R_L} - \dfrac{V_g - V_o}{2 \cdot L} \cdot D \cdot T_s \right)}$$

$$A_1 = \frac{1}{\omega_p} \cdot \left[1 - \frac{V_g - V_o}{\omega_p \cdot L \cdot \left(\dfrac{V_o}{R_L} - \dfrac{V_g - V_o}{2 \cdot L} \cdot D \cdot T_s \right)} \right]$$

and

$$V_{b2}(t) = \left(\begin{matrix} V_{by} \cdot e^{-\omega_p \cdot (t - D \cdot T_s)} + \dfrac{\omega_x}{\omega_p}\left(1 - e^{-\omega_p \cdot (t - D \cdot T_s)}\right) V_c - \\ \dfrac{R_i \cdot (I_b + \delta i)}{\tau_3} \left\{ B_2 \cdot (t - D \cdot T_s) + A_2 \left[1 - e^{-\omega_p \cdot (t - D \cdot T_s)} \right] \right\} \end{matrix} \right) \cdot u(t - D \cdot T_s) \tag{3.5}$$

where

$$B_2 = \frac{-V_o}{\omega_p \bullet L \bullet \left(\dfrac{V_o}{R_L} + \dfrac{V_g - V_o}{2 \bullet L} \bullet D \bullet T_s\right)}$$

$$A_2 = \frac{1}{\omega_p} \bullet \left[1 + \frac{V_g - V_o}{\omega_p \bullet L \bullet \left(\dfrac{V_o}{R_L} + \dfrac{V_g - V_o}{2 \bullet L} \bullet D \bullet T_s\right)}\right]$$

However, under the steady state, continuity of state requires that, at $t = D \bullet T_s$,

$$V_{b1}(D \bullet T_s) = V_{by} \tag{3.6}$$

and, at $t = T_s$,

$$V_{b1}(D \bullet T_s) = V_{by} \tag{3.7}$$

These two constraints at the two time boundaries further give

$$\begin{aligned} a_{11} \bullet V_{bx} + a_{12} \bullet V_{by} &= b_1 \\ a_{21} \bullet V_{bx} + a_{22} \bullet V_{by} &= b_2 \end{aligned} \tag{3.8}$$

where

$$a_{11} = \mathrm{e}^{-\omega_p \bullet D \bullet T_s}, \qquad a_{12} = -1,$$

$$b_1 = -\frac{\omega_x}{\omega_p}\left(1 - \mathrm{e}^{-\omega_p \bullet D \bullet T_s}\right)V_c + \frac{R_i \bullet I_b}{\tau_3}\left[B_1 \bullet D \bullet T_s + A_1\left(1 - \mathrm{e}^{-\omega_p \bullet D \bullet T_s}\right)\right],$$

$$a_{21} = -1, \qquad a_{22} = \mathrm{e}^{-\omega_p \bullet (1-D) \bullet T_s},$$

$$b_2 = -\frac{\omega_x}{\omega_p}\left(1 - \mathrm{e}^{-\omega_p \bullet (1-D) \bullet T_s}\right)V_c$$

$$+ \frac{R_i \bullet (I_b + \delta i)}{\tau_3}\left[B_2 \bullet (1 - D) \bullet T_s + A_2\left(1 - \mathrm{e}^{-\omega_p \bullet 1 - D) \bullet T_s}\right)\right]$$

Eventually, the unknown starting states, V_{bx} and V_{by}, are given as

$$\begin{bmatrix} V_{bx} \\ V_{by} \end{bmatrix} = \begin{bmatrix} a_{11} & a_{12} \\ a_{21} & a_{22} \end{bmatrix}^{-1} \begin{bmatrix} b_1 \\ b_2 \end{bmatrix} = \frac{\begin{bmatrix} e^{-\omega_p \cdot (1-D) \cdot T_s} \bullet b_1 + b_2 \\ b_1 + e^{-\omega_p \cdot D \cdot T_s} \bullet b_2 \end{bmatrix}}{e^{-\omega_p \cdot T_s} - 1} \tag{3.9}$$

By plugging in the starting states in (3.4) and (3.5), the node voltage at V_b in a steady state over one cycle is completely determined:

$$V_b(t) = V_{b1}(t) + V_{b2}(t) \tag{3.10}$$

And, consequently, the output of the summing amplifier is given as

$$V_m(t) = V_b(t) - \frac{1}{C_{fz}} \int_0^t \frac{V_c - V_b(\tau)}{R_f} d\tau \tag{3.11}$$

3.2 Duty Cycle Determination

Next, the open-loop duty cycle is determined. However, $V_{b1}(t)$ is rearranged to make it easier for processing, using equation (3.11). The step leads to

$$V_{b1}(t) = (k_1 \bullet e^{-\omega_p \cdot t} - k_2 \bullet t + k_3)[u(t) - u(t - D \bullet T_s)] \tag{3.12}$$

where

$$k_1 = -\frac{\omega_x}{\omega_y} \bullet V_c + V_{bx} + \frac{R_i}{\tau_3}\left[\frac{V_o}{R_L} - \frac{V_g - V_o}{2 \bullet L} \bullet D \bullet T_s\right] A_1,$$

$$k_2 = \frac{R_i}{\tau_3}\left[\frac{V_o}{R_L} - \frac{V_g - V_o}{2 \bullet L} \bullet D \bullet T_s\right] B_1,$$

$$k_3 = \frac{\omega_x}{\omega_y} \bullet V_c - \frac{R_i}{\tau_3}\left[\frac{V_o}{R_L} - \frac{V_g - V_o}{2 \bullet L} \bullet D \bullet T_s\right] A_1$$

And the corresponding summing amplifier output is given

$$V_{m1}(t) = k_1 \bullet e^{-\omega_p t} - k_2 \bullet t + k_3 - \frac{V_c}{R_f \bullet C_{fz}} t + \frac{1}{R_f \bullet C_{fz}} \left[-\frac{k_1}{\omega_p}(e^{-\omega_p t} - 1) - \frac{k_2}{2} t^2 + k_3 \bullet t \right] \tag{3.13}$$

In general, an external sawtooth clock is also provided for PWM operation. It is easily described as $(V_{os} + S_e t)$, in which V_{os} is the triangle wave offset and S_e is the ramp slope. The open-loop, pulsewidth-modulated duty cycle is then determined by the intercept of the ramp and the summing amplifier output, $V_{m1}(t)$:

$$V_{m1}(V_o, V_g, V_c, D, R_L, \ldots) = V_{os} + S_e \bullet D \bullet T_s \tag{3.14}$$

3.3 Steady-State Closed Loop

Referring to Figure 3.1, as in previous chapters, we can establish the following two equations representing the negative feedback with an external precision reference voltage, V_{ref}, and high gain error amplifier, k_{fb}, and the power stage:

$$V_c = k_{fb}(V_{ref} - V_o) \tag{3.15}$$

$$V_o = D \bullet V_g \tag{3.16}$$

It is understood that (3.14)–(3.16), when combined, constitute the closed-loop description for the converter. However, neither the output nor the duty cycle is given in explicit function form. Instead, the output, the control voltage, and the duty cycle are given in implicit function form. All three variables can be solved numerically, but they do not offer symbolic, closed-form solutions so desired by some. It shall also be understood that A_1, B_1, A_2, B_2, b_1, b_2, V_{bx}, V_{by}, k_1, k_2, and k_3 are all functions of output V_o, source V_g, error V_c, duty cycle D, and load R_L.

3.4 Closed-Loop Regulation and Output Sensitivity

Power supply output, in general, is given a specified regulation range against load and line (input) variation. It is therefore essential to know the output sensitivity against load resistance and input change. These performance figures are mathematically equivalent to the partial derivatives $\partial V_o/\partial R_L$ and $\partial V_o/\partial V_g$. However, as indicated in the previous section, the output voltage cannot be easily expressed in explicit, compact closed form, and the desire to have a simple derivative is hindered. We again employ the Jacobian determinant. First, three implicit functions are defined:

$$\begin{aligned}
&f(V_o, V_g, V_c, D, R_L) = V_{m1}(V_o, V_g, V_c, D, vR_L) - (V_{os} + S_e \bullet D \bullet T_s) = 0 \\
&g(V_o, V_g, V_c, D, R_L) = V_c - k_{fb}(V_{ref} - V_o) = 0 \\
&h(V_o, V_g, V_c, D, R_L) = V_o - D \bullet V_g = 0
\end{aligned} \tag{3.17}$$

The output sensitivity against input with the load held constant is given as

$$\frac{\partial V_o}{\partial V_g} = -\frac{\begin{vmatrix} \partial f/\partial V_g & \partial f/\partial V_c & \partial f/\partial D \\ \partial g/\partial V_g & \partial g/\partial V_c & \partial g/\partial D \\ \partial h/\partial V_g & \partial h/\partial V_c & \partial h/\partial D \end{vmatrix}}{\begin{vmatrix} \partial f/\partial V_o & \partial f/\partial V_c & \partial f/\partial D \\ \partial g/\partial V_o & \partial g/\partial V_c & \partial g/\partial D \\ \partial h/\partial V_o & \partial h/\partial V_c & \partial h/\partial D \end{vmatrix}} \tag{3.18}$$

By the same token, the output sensitivity against the load with input held constant is given as

$$\frac{\partial V_o}{\partial R_L} = -\frac{\begin{vmatrix} \partial f/\partial R_L & \partial f/\partial V_c & \partial f/\partial D \\ \partial g/\partial R_L & \partial g/\partial V_c & \partial g/\partial D \\ \partial h/\partial R_L & \partial h/\partial V_c & \partial h/\partial D \end{vmatrix}}{\begin{vmatrix} \partial f/\partial V_o & \partial f/\partial V_c & \partial f/\partial D \\ \partial g/\partial V_o & \partial g/\partial V_c & \partial g/\partial D \\ \partial h/\partial V_o & \partial h/\partial V_c & \partial h/\partial D \end{vmatrix}} \tag{3.19}$$

Other sensitivity figures against components, of course, can be obtained through a similar procedure, providing the associated implicit function $f(V_o, \ldots)$ is modified accordingly to become $f(V_o, \ldots, R_x, \ldots)$. Then, the total regulation is expressed as

$$\sqrt{\left(\frac{\partial V_o}{\partial V_g} \bullet dV_g\right)^2 + \left(\frac{\partial V_o}{\partial R_L} \bullet dR_L\right)^2 + \left(\frac{\partial V_o}{\partial R_x} \bullet dR_x\right)^2 + \ldots} \tag{3.20}$$

3.5 Small-Signal Loop Gain and Stability

Given the steady state obtained so far, the examination for the converter's small-signal behavior can then proceed. The small-signal block diagram is given in Figure 3.3.

Three important pulsewidth-modulation gains, $F_m = \partial D/\partial V_c$, $F_{vo} = \partial D/\partial V_o$, and $F_{vg} = \partial D/\partial V_g$, must be derived first. But, as can be seen from (3.14), the duty cycle is again in implicit function form. The single-function Jacobian determinant is therefore enlisted once more to obtain those gain factors. They are

$$F_m = -\frac{\partial f/V_c}{\partial f/D}, \quad F_{vo} = -\frac{\partial f/V_o}{\partial f/D}, \quad F_{vg} = -\frac{\partial f/V_g}{\partial f/D} \tag{3.21}$$

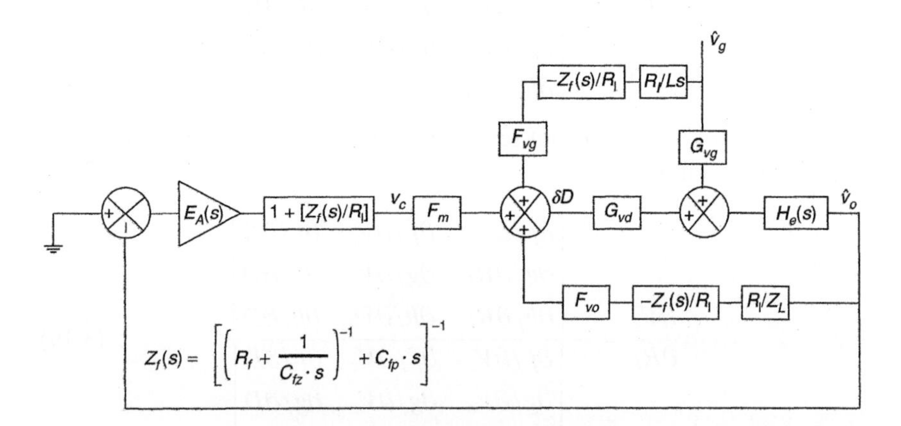

Figure 3.3: Small-signal block diagram

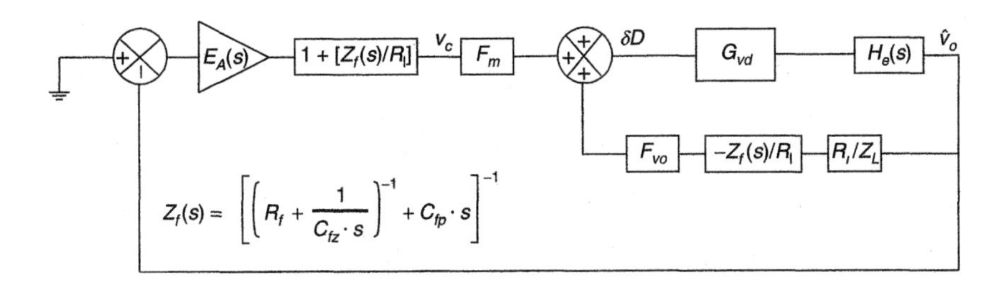

Figure 3.4: Simplified block diagram

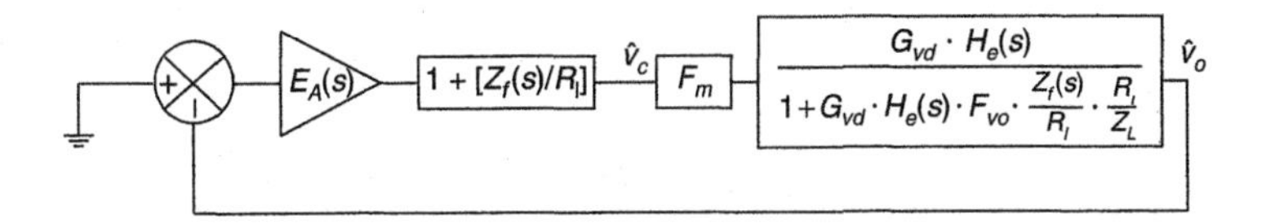

Figure 3.5: Block diagram with current loop absorbed

where f is the first implicit function in (3.17). The three gain factors stem from the realization that the total differential, dD, of the duty cycle is expressible as

$$dD = \frac{\partial D}{\partial V_c} \bullet dV_c + \frac{\partial D}{\partial V_o} \bullet dV_o + \frac{\partial D}{\partial V_g} \bullet dV_g \tag{3.22}$$

Given constant source, the block diagram in Figure 3.3 can be simplified to Figure 3.4. By absorbing the inner loop, Figure 3.4 is reduced to a single loop (Figure 3.5). From the single-loop figure, the loop gain is given as

$$T(s) = E_A(s) \bullet \left[1 + \frac{Z_f(s)}{R_1}\right] \bullet F_m \bullet \frac{G_{vd} \bullet H_e(s)}{1 + G_{vd} \bullet H_e(s) \bullet F_{vo} \bullet \frac{Z_f(s)}{R_1} \bullet \frac{R_l}{Z_L}} \tag{3.23}$$

3.6 Example

Tang and Lee [3] give an example with $V_g = 14$, $V_o = 5$, $R = 1$, $L = 37.5\,\mu\text{H}$, $C = 380\,\mu\text{F}$, $f_s = 50\,\text{KHz}$, $R_i = 0.1$, $R_1 = 2.2\,\text{K}$,

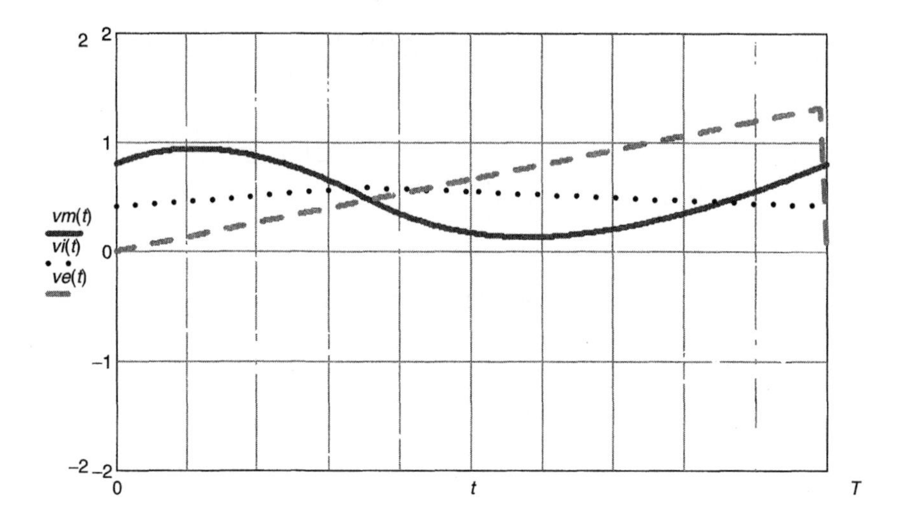

Figure 3.6: Average current-mode duty cycle determination

$R_f = 30.5\,\mathrm{K}$, $C_{fz} = 5.8\,\mathrm{nF}$, and $C_{fp} = 220\,\mathrm{pF}$. Based on the example, (3.14) yields Figure 3.6.

The closed loop numerical solution for (3.14), (3.15), and (3.16) gives $D = 0.357$, $V_o = 5$, and $V_c = 0.596$, if a 50-KHz clock (oscillator ramp) swinging between 1 V and 3.5 V is used.

3.7 State Transition Technique

In Chapter 1's last section, state transition techniques were applied to obtain the steady-state output for a buck converter, bypassing the Laplace transform. The procedure enjoys an advantage in that the input drive to the output filter, Figure 1.42, alternates between two constant values that can be moved out of an (mathematical) integration process. However, for the current treatment, the input V_i to the average amplifier of Figure 3.1 is no longer a constant. This makes the mathematical process more cumbersome. We shall see how a time-varying input raises the complexity level of the analytical treatment. Furthermore, to reduce the symbolic complexity to a manageable level, we use the component values given in section 3.6.

Now, based on (3.3) and (3.2), the two state equations are rewritten in the standard matrix form:

$$\frac{d}{dt}x(t) = \frac{d}{dt}\begin{bmatrix} V_b(t) \\ V_m(t) \end{bmatrix} = A \cdot x(t) + B \cdot E(t) \tag{3.24}$$

where

$$A = \begin{bmatrix} -\omega_p & 0 \\ -\dfrac{1}{\tau_2} & 0 \end{bmatrix}, \quad B = \begin{bmatrix} -\omega_x & \dfrac{1}{\tau_3} \\ -\dfrac{1}{\tau_1} & \dfrac{1}{\tau_3} \end{bmatrix}, \quad E(t) = \begin{bmatrix} V_c \\ V_i(t) \end{bmatrix} \tag{3.25}$$

in which all elements are defined in section 3.1. Matrix A has two eigenvalues, $\lambda_0 = 0$ and $\lambda_1 = -1.547 \cdot 10^5$. Following the same procedure outlined in the last section of Chapter 1, two matrix exponentials are expressed in polynomial form:

$$\begin{aligned} e^{A \cdot t} = I + \beta_1(t) \cdot A &= \begin{bmatrix} 1 & 0 \\ 0 & 1 \end{bmatrix} + \beta_1(t) \cdot \begin{bmatrix} -\omega_p & 0 \\ -\dfrac{1}{\tau_2} & 0 \end{bmatrix} \\ &= \begin{bmatrix} 1 - \omega_p \cdot \beta_1(t) & 0 \\ -\dfrac{1}{\tau_2}\beta_1(t) & 1 \end{bmatrix} \end{aligned} \tag{3.26}$$

$$e^{A(t-\tau)} = \begin{bmatrix} 1 - \omega_p \cdot \beta_1(t-\tau) & 0 \\ -\dfrac{1}{\tau_2}\beta_1(t-\tau) & 1 \end{bmatrix} \tag{3.27}$$

where

$$\beta_1(t) = \frac{1}{\lambda_1}(e^{\lambda_1 \cdot t} - 1) \tag{3.28}$$

Again, the general solution for (3.24) is given as

$$x(t) = e^{A(t-t_0)}x(t_0) + \int_{t_0}^{t} e^{A(t-\tau)} \cdot B \cdot E(\tau)d\tau \tag{3.29}$$

The integrant in (3.29) must be expanded as

$$e^{A(t-\tau)} \bullet B \bullet E(\tau) = \left\{ \begin{array}{c} \dfrac{[1 - \omega_p \bullet \beta_1(t-\tau)][\omega_x V_c \tau_3 - V_i(\tau)]}{\tau_3} \\ -\dfrac{[V_c \tau_3 \beta_1(t-\tau)\omega_x \tau_1 - V_c \tau_3 \tau_2 - V_i(\tau)\tau_1 \beta_1(t-\tau) + V_i(\tau)\tau_1 \tau_2]}{\tau_1 \tau_3 \tau_2} \end{array} \right\} \tag{3.30}$$

During the turn-on time and assuming $t_o = 0$, the inductor current is ramping up and the input signal feeding the average amplifier is represented by $v_1(t)$ of (3.1). Equation (3.29) gives

$$\begin{aligned} x_1(t) &= e^{A \bullet t} X_1 + \int_0^t e^{A(t-\tau)} \bullet B \bullet E(\tau) d\tau \\ &= M_1(t) X_1 + M_2(t) \end{aligned} \tag{3.31}$$

where

$$M_1(t) = \begin{bmatrix} 1 - \omega_p \bullet \beta_1(t) & 0 \\ -\dfrac{1}{\tau_2} \beta_1(t) & 1 \end{bmatrix} \tag{3.32}$$

$$M_2(t) = \left\{ \begin{array}{c} \displaystyle\int_0^t \dfrac{[1 - \omega_p \bullet \beta_1(t-\tau)]\,[\omega_x V_c \tau_3 - v_1(\tau)]}{\tau_3} d\tau \\ \displaystyle -\int_0^t \dfrac{[V_c \tau_3 \beta_1(t-\tau)\omega_x \tau_1 - V_c \tau_3 \tau_2 - v_1(\tau)\tau_1 \beta_1(t-\tau) + v_1(\tau)\tau_1 \tau_2]}{\tau_1 \tau_3 \tau_2} d\tau \end{array} \right\} \tag{3.33}$$

During the turn-off time, the inductor current is ramping down and the input signal feeding the average amplifier is represented by $v_2(t)$ of (3.1). Equation (3.29) gives

$$\begin{aligned} x_2(t) &= e^{A(t-DT_s)} X_2 + \int_{DT_s}^t e^{A(t-\tau)} \bullet B \bullet E(\tau) d\tau \\ &= M_3(t) X_2 + M_4(t) \end{aligned} \tag{3.34}$$

where

$$M_3(t) = \begin{bmatrix} 1 - \omega_p \bullet \beta_1(t - DT_s) & 0 \\ -\frac{1}{\tau_2}\beta_1(t - DT_s) & 1 \end{bmatrix} \tag{3.35}$$

$$M_4(t) = \left\{ \begin{array}{c} \displaystyle\int_{DT_s}^{t} \frac{[1 - \omega_p \bullet \beta_1(t - \tau)]\,[\omega_x V_c \tau_3 - v_2(\tau)]}{\tau_3} d\tau \\ \displaystyle -\int_{DT_s}^{t} \frac{[V_c \tau_3 \beta_1(t - \tau)\omega_x \tau_1 - V_c \tau_3 \tau_2 - v_2(\tau)\tau_1 \beta_1(t - \tau) + v_2(\tau)\tau_1 \tau_2]}{\tau_1 \tau_3 \tau_2} d\tau \end{array} \right\} \tag{3.36}$$

At the steady state, the following holds:

$$\begin{aligned} M_1(DT_s)X_1 + M_2(DT_s) &= X_2 \\ M_3(T_s)X_2 + M_4(T_s) &= X_1 \end{aligned} \tag{3.37}$$

In other words, both cyclic starting states X_1 and X_2 can be solved, using (3.38). Once both are solved, the average amplifier's steady-state output, $V_m(t)$, can be extracted from $x_1(t)$ and $x_2(t)$ column vectors. The remaining procedure, similar to (3.14), (3.15), and (3.16), follows:

$$\begin{aligned} X_1 &= [I - M_3(T_s)M_1(DT_s)]^{-1}[M_3(T_s)M_2(DT_s) + M_4(T_s)] \\ X_2 &= M_1(DT_s)X_1 + M_2(DT_s) \end{aligned} \tag{3.38}$$

Unfortunately, in computing the matrix inverse for X_1, a singular condition occurs. Equation (3.38) fails to solve both starting state vectors. The condition can be remedied by adding a very large resistor, $R_a = 100$ Mega-ohm, across C_{fp} of Figure 3.1. Not only is the trick effective in avoiding singularity, it is also physically permissible, since in general, the capacitor has leakage associated with it. Once this is done, the circuit matrix of (3.25) changes slightly:

$$A = \begin{bmatrix} -\omega_p & -\frac{1}{\tau_5} \\ -\frac{1}{\tau_2} & -\frac{1}{\tau_5} \end{bmatrix}, \quad B = \begin{bmatrix} \omega_q & \frac{1}{\tau_3} \\ \omega_r & \frac{1}{\tau_3} \end{bmatrix}, \quad E(t) = \begin{bmatrix} V_c \\ V_i(t) \end{bmatrix},$$

$$\omega_q = \frac{1}{\tau_2} + \frac{1}{\tau_3} + \frac{1}{\tau_5} + \frac{1}{\tau_4}, \quad \omega_r = \frac{1}{\tau_2} + \frac{1}{\tau_3} + \frac{1}{\tau_5}, \quad \tau_5 = R_a C_{fp} \tag{3.39}$$

The modified matrix A now has two nonzero eigenvalues, λ_0 and λ_1. Two time functions are then found:

$$\beta_0(t) = \frac{\begin{vmatrix} e^{\lambda_0 t} & \lambda_0 \\ e^{\lambda_1 t} & \lambda_1 \end{vmatrix}}{\begin{vmatrix} 1 & \lambda_0 \\ 1 & \lambda_1 \end{vmatrix}}, \qquad \beta_0(t) = \frac{\begin{vmatrix} 1 & e^{\lambda_0 t} \\ 1 & e^{\lambda_1 t} \end{vmatrix}}{\begin{vmatrix} 1 & \lambda_0 \\ 1 & \lambda_1 \end{vmatrix}} \tag{3.40}$$

The matrix exponentials also change:

$$e^{A \cdot t} = \beta_0(t) \cdot I + \beta_1(t) \cdot A = \begin{bmatrix} \beta_0(t) - \omega_p \cdot \beta_1(t) & -\frac{\beta_1(t)}{\tau_5} \\ -\frac{1}{\tau_2}\beta_1(t) & \beta_0(t) - \frac{\beta_1(t)}{\tau_5} \end{bmatrix} \tag{3.41}$$

$$e^{A \cdot (t-\tau)} = \begin{bmatrix} \beta_0(t-\tau) - \omega_p \cdot \beta_1(t-\tau) & -\frac{\beta_1(t-\tau)}{\tau_5} \\ -\frac{1}{\tau_2}\beta_1(t-\tau) & \beta_0(t-\tau) - \frac{\beta_1(t-\tau)}{\tau_5} \end{bmatrix}$$

$$e^{A(t-\tau)} \cdot B \cdot E(\tau) = \begin{bmatrix} \beta_0(t-\tau) - \omega_p \cdot \beta_1(t-\tau) & -\frac{\beta_1(t-\tau)}{\tau_5} \\ -\frac{1}{\tau_2}\beta_1(t-\tau) & \beta_0(t-\tau) - \frac{\beta_1(t-\tau)}{\tau_5} \end{bmatrix} \cdot \begin{bmatrix} \omega_q & \frac{1}{\tau_3} \\ \omega_r & \frac{1}{\tau_3} \end{bmatrix} \begin{bmatrix} V_c \\ V_i(t) \end{bmatrix} \tag{3.42}$$

Given the new matrix exponentials, equations (3.30) through (3.38) also are modified. However, the expansion of the integrand, (3.42), is too much to type in this small page. We omit it and assure readers that eventually it does lead to the final solution and confirm Figure 3.6. (See Appendix D.).

Anyway, by comparing this with the procedure given in section 3.1, it is not clear which can be considered a more efficient algorithm. We therefore leave it to the individual reader to pick and choose. For higher-order systems, the approach using matrix exponentials may have an edge over the approach using Laplace transformation.

Chapter 4

Phase-Shifted Full-Bridge Converter

Basic forward converters are covered in the previous chapters. All those converters share a common trait: The duty cycle of the main switches varies significantly depending on the input voltage level and the load demand. Because of that, the Fourier harmonic contents of switching waveforms also change wildly and produce an undesirable electromagnetic interference environment and generate more local heat dissipation. The phase-shifted full-bridge converter shown in Figure 4.1 solves a major part of the problem by limiting the duty cycle to nearly a constant 50% for the left leg switches, Q_A and Q_B, while providing PWM by phase shifting the right leg switches, Q_C and Q_D, also in a near 50% duty cycle. In other words, all switches are not pulsewidth-modulated. Instead, the transformer winding and core volt–second are pulsewidth-modulated. Readers are referred to Texas Instrument's design considerations [4] for complete circuits.

Since the converter is still a buck converter in almost all aspects, we do not duplicate previous efforts in analyzing the circuit. Instead, we focus on the unique feature given for the first time in the circuit. As mentioned in the figure caption, a current-doubling filter is employed. By sharing the load current between two identical inductors and operating the isolation transformer core truly in four quadrants (*B–H*), magnetic material is efficiently utilized and the thermal environment is better managed.

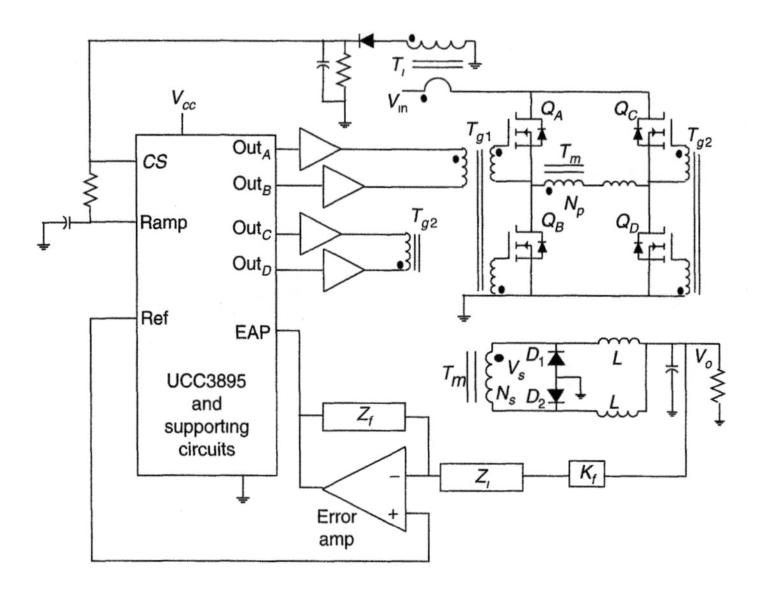

Figure 4.1: Phase-shifted full-bridge converter with current-doubling filter

4.1 Power-Stage Operation

Prior to the mathematical treatment for the current doubler, we must understand how the main circuit works. The four switches allow four permissible states: states *AD* (Q_A and Q_D on), *CB*, *AC*, and *BD*. State *AD* places V_{in} across the transformer, T_m, primary with the dot-end positive. State *CB* reverses the primary voltage with the dot-end grounded. States *AC* and *BD* place a short across the primary. Therefore, as far as the transformer's secondary side is concerned, there are three states across N_s: positive (dot-end), zero, and negative. In a continuous operation, the secondary state sequences through positive, zero, negative, and zero, and the cycle repeats in rapid succession.

4.2 Current Doubler

To give a detailed study of the subject matter, the current-doubling output circuit is redrawn as Figure 4.2 with loss elements included.

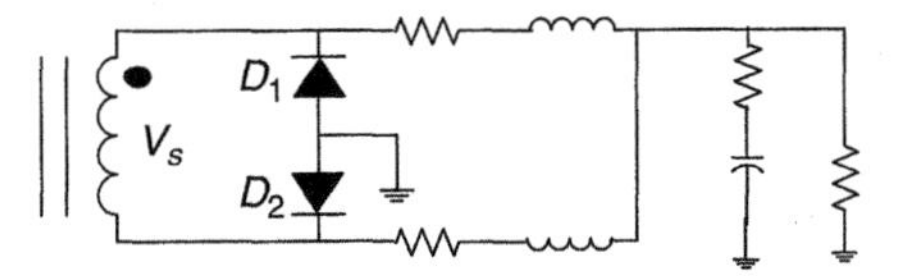

Figure 4.2: Current doubling filter

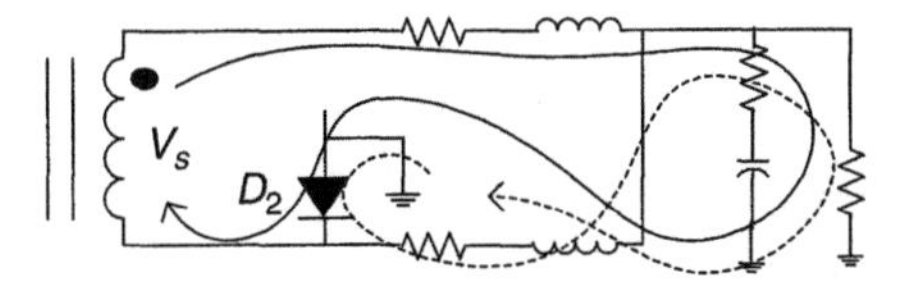

Figure 4.3: Current-doubling filter in a positive state

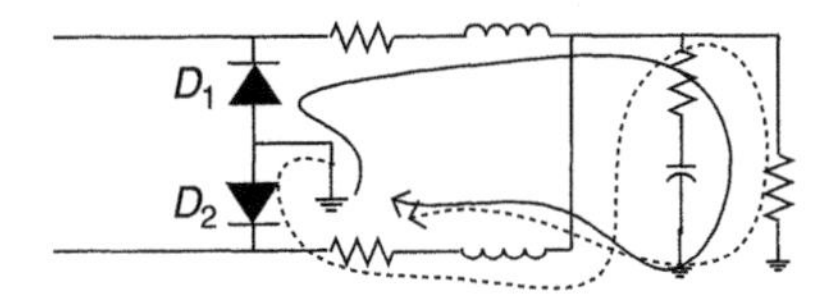

Figure 4.4: Current-doubling filter in a zero state

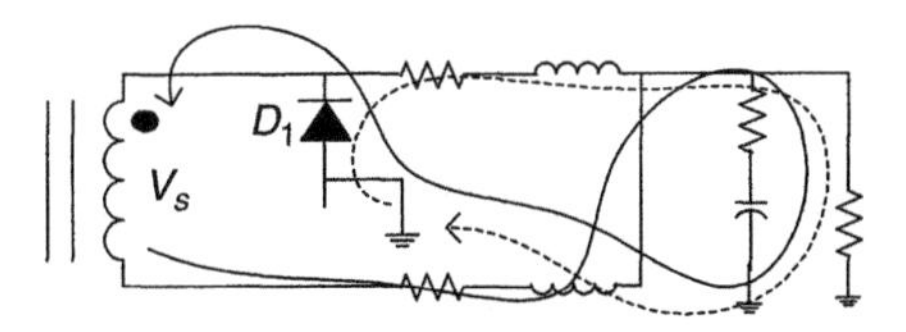

Figure 4.5: Current-doubling filter in a negative state

During the positive state, Figure 4.2 reduces to Figure 4.3, in which diode D_2 is conducting and carries two current components, the forward transferring current (solid line) and the free-wheeling current (dashed line).

During the zero state, Figure 4.4 commences. In this state, both diodes conduct in a freewheeling mode and share half the load.

During the negative state, Figure 4.2 reduces to Figure 4.5, in which diode D_1 is conducting and carries two current components, the forward transferring current (solid line) and the freewheeling current (dashed line).

4.3 Steady-State Duty Cycle

For an in-depth mathematical analysis, Figure 4.3 is redrawn as Figure 4.6 with both the forward transferring current and the freewheeling current given the symbolic names, i_1 and i_2.

As explained in section 4.1, the transformer secondary voltage, V_s, has the form shown in Figure 4.7. Based on the concept of volt–second balance and using the definition of oscillator time, T_{osc}, the following is established for each inductor. It is noted that the freewheeling action lasts $(2 - D)T_{osc}$, and V_D stands for diode drop.

$$D \bullet T_{osc} \bullet \left[V_{in} \bullet \frac{N_s}{N_p} - (V_D + V_o) \right] - (V_D + V_o)[(1 - D) \bullet T_{osc} + T_{osc}] = 0 \quad (4.1)$$

This equation gives the duty cycle referred to oscillator time as

$$D = \frac{2(V_D + V_o)}{V_{in} \bullet \frac{N_s}{N_p}} \quad (4.2)$$

If resistive losses are considered, (4.2) becomes

$$D = \frac{2\left(V_D + V_o + \frac{1}{2} \frac{V_o}{R_L} r_1 \right)}{V_{in} \bullet \frac{N_s}{N_p}} \quad (4.3)$$

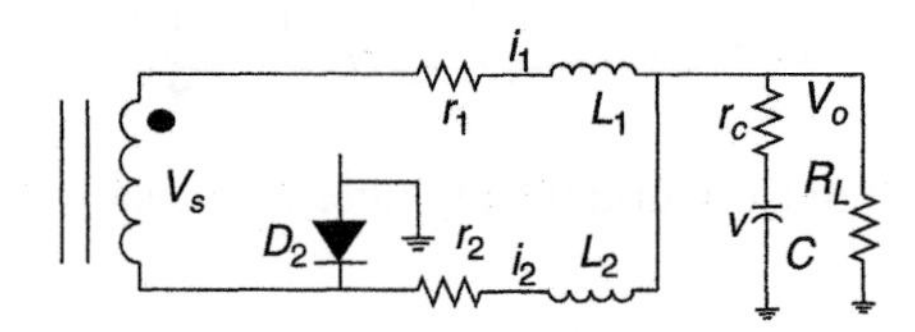

Figure 4.6: Current-doubling filter in a positive state, with mathematical symbols

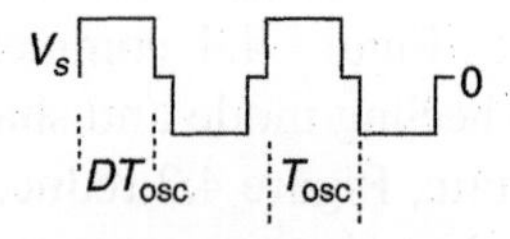

Figure 4.7: Source voltage feeding the current-doubling filter

4.4 Steady-State Output Waveforms

Now, referring to Figure 4.6, we can write two (inductor) voltage-loop equations and one (capacitor) current-node equation:

$$\begin{aligned}
&\frac{di_1}{dt}+\frac{r_1+R_p}{L_1}i_1+\frac{R_p}{L_1}i_2+\frac{k}{L_1}v=\frac{V_s-V_D}{L_1}\\
&\frac{R_p}{L_2}i_1+\frac{di_2}{dt}+\frac{r_2+R_p}{L_2}i_2+\frac{k}{L_2}v=\frac{-V_D}{L_2}\\
&-\frac{R_p}{r_c \bullet C}i_1-\frac{R_p}{r_c \bullet C}i_2+\frac{dv}{dt}+\frac{1-k}{r_c \bullet C}v=0
\end{aligned} \tag{4.4}$$

where

$$k=\frac{R_L}{r_c+R_L},\qquad R_p=\frac{r_c \bullet R_L}{r_c+R_L},\qquad V_s=\frac{N_s}{N_p}V_{\text{in}}$$

During the positive state, and by taking Laplace transformation, (4.4) is transformed to

$$\begin{aligned}
&\left(s+\frac{r_1+R_p}{L_1}\right)I_{1a}(s)+\frac{R_p}{L_1}I_{2a}(s)+\frac{k}{L_1}V_a(s)=I_{10a}+\frac{V_s-V_D}{L_1 \bullet s}\\
&\frac{R_p}{L_2}I_{1a(s)}+\left(s+\frac{r_2+R_p}{L_2}\right)I_{2a}(s)+\frac{k}{L_2}V_a(s)=I_{20a}+\frac{-V_D}{L_2 \bullet s}\\
&-\frac{R_p}{r_c \bullet C}I_{1a}(s)-\frac{R_p}{r_c \bullet C}I_{2a}(s)+\left(s+\frac{1-k}{r_c \bullet C}\right)V_a(s)=V_{0a}
\end{aligned} \tag{4.5}$$

During the zero state, (4.4) is transformed to

$$\begin{aligned}
&\left(s+\frac{r_1+R_p}{L_1}\right)I_{1b}(s)+\frac{R_p}{L_1}I_{2b}(s)+\frac{k}{L_1}V_b(s)=\left(I_{10b}+\frac{-V_D}{L_1 \bullet s}\right)\mathrm{e}^{-D \bullet T_{\text{osc}} \bullet s}\\
&\frac{R_p}{L_2}I_{1b(s)}+\left(s+\frac{r_2+R_p}{L_2}\right)I_{2b}(s)+\frac{k}{L_2}V_b(s)=\left(I_{20b}+\frac{-V_D}{L_2 \bullet s}\right)\mathrm{e}^{-D \bullet T_{\text{osc}} \bullet s}\\
&-\frac{R_p}{r_c \bullet C}I_{1b}(s)-\frac{R_p}{r_c \bullet C}I_{2b}(s)+\left(s+\frac{1-k}{r_c \bullet C}\right)V_b(s)=V_{0b} \bullet \mathrm{e}^{-D \bullet T_{\text{osc}} \bullet s}
\end{aligned} \tag{4.6}$$

where I_{10a}, I_{20a}, V_{0a}, I_{10b}, I_{20b}, and V_{0b} are unknown starting conditions for time interval $D \bullet T_{\text{osc}}$ and $(1-D)T_{\text{osc}}$.

Equation set (4.5), for instance, yields the first inductor loop current in the transformed domain:

$$I_{1a}(s) = \frac{\begin{vmatrix} I_{10a} + \dfrac{V_s - V_D}{L_1 \bullet s} & \dfrac{R_p}{L_1} & \dfrac{k}{L_1} \\ I_{20a} + \dfrac{-V_D}{L_2 \bullet s} & \left(s + \dfrac{r_2 + R_p}{L_2}\right) & \dfrac{k}{L_2} \\ V_{0a} & -\dfrac{R_p}{r_c \bullet C} & \left(s + \dfrac{1-k}{r_c \bullet C}\right) \end{vmatrix}}{D(s)} \tag{4.7}$$

$$I_{1a}(s) = \frac{\begin{vmatrix} \left(s + \dfrac{r_2 + R_p}{L_2}\right) & \dfrac{k}{L_2} \\ -\dfrac{R_p}{r_c \bullet C} & \left(s + \dfrac{1-k}{r_c \bullet C}\right) \end{vmatrix}}{D(s)} I_{10a}$$

$$+ \frac{-\begin{vmatrix} \dfrac{R_p}{L_1} & \dfrac{k}{L_1} \\ -\dfrac{R_p}{r_c \bullet C} & \left(s + \dfrac{1-k}{r_c \bullet C}\right) \end{vmatrix}}{D(s)} I_{20a} + \frac{\begin{vmatrix} \dfrac{R_p}{L_1} & \dfrac{k}{L_1} \\ \left(s + \dfrac{r_2 + R_p}{L_2}\right) & \dfrac{k}{L_2} \end{vmatrix}}{D(s)} V_{0a}$$

$$+ \frac{\begin{vmatrix} \left(s + \dfrac{r_2 + R_p}{L_2}\right) & \dfrac{k}{L_2} \\ -\dfrac{R_p}{r_c \bullet C} & \left(s + \dfrac{1-k}{r_c \bullet C}\right) \end{vmatrix}}{L_1 \bullet s \bullet D(s)} (V_s - V_D) + \frac{\begin{vmatrix} \dfrac{R_p}{L_1} & \dfrac{k}{L_1} \\ -\dfrac{R_p}{r_c \bullet C} & \left(s + \dfrac{1-k}{r_c \bullet C}\right) \end{vmatrix}}{L_2 \bullet s \bullet D(s)} V_D,$$

$$D(s) = \begin{vmatrix} \left(s + \dfrac{r_1 + R_p}{L_1}\right) & \dfrac{R_p}{L_1} & \dfrac{k}{L_1} \\ \dfrac{R_p}{L_2} & \left(s + \dfrac{r_2 + R_p}{L_2}\right) & \dfrac{k}{L_2} \\ -\dfrac{R_p}{r_c \bullet C} & -\dfrac{R_p}{r_c \bullet C} & \left(s + \dfrac{1-k}{r_c \bullet C}\right) \end{vmatrix} \tag{4.8}$$

For further processing, (4.8) is rewritten with the understanding that individual transfer functions $F_1(s)$ through $F_5(s)$ are identified in (4.8):

$$I_{1a}(s) = F_1(s) \bullet I_{10a} + F_2(s) \bullet I_{20a} + F_3(s) \bullet V_{0a} + F_4(s,\ V_s) + F_5(s) \quad (4.9)$$

With a little patience, we can carry out the same procedures for $I_{2a}(s)$ and $V_a(s)$ to obtain

$$\begin{aligned} I_{2a}(s) &= G_1(s) \bullet I_{10a} + G_2(s) \bullet I_{20a} + G_3(s) \bullet V_{0a} + G_4(s,\ V_s) + G_5(s), \\ V_a(s) &= H_1(s) \bullet I_{10a} + H_2(s) \bullet I_{20a} + H_3(s) \bullet V_{0a} + H_4(s,\ V_s) + H_5(s) \end{aligned} \quad (4.10)$$

Next we proceed with equation set (4.6), with the understanding $V_s = 0$, to obtain

$$\begin{aligned} I_{1b}(s) &= [F_1(s) \bullet I_{10b} + F_2(s) \bullet I_{20b} + F_3(s) \bullet V_{0b} + F_6(s, 0) + F_5(s)]e^{-D \bullet T_{osc} \bullet s}, \\ I_{2b}(s) &= [G_1(s) \bullet I_{10b} + G_2(s) \bullet I_{20b} + G_3(s) \bullet V_{0b} + G_6(s, 0) + G_5(s)]e^{-D \bullet T_{osc} \bullet s}, \\ V_b(s) &= [H_1(s) \bullet I_{10b} + H_2(s) \bullet I_{20b} + H_3(s) \bullet V_{0b} + H_6(s, 0) + H_5(s)]e^{-D \bullet T_{osc} \bullet s} \end{aligned} \quad (4.11)$$

By taking the inverse Laplace transformation of (4.9), (4.10), and (4.11), the following is established in matrix form:

$$\begin{bmatrix} i_{1a}(t) \\ i_{2a}(t) \\ v_a(t) \end{bmatrix} = \left\{ \begin{bmatrix} f_1(t) & f_2(t) & f_3(t) \\ g_1(t) & g_2(t) & g_3(t) \\ h_1(t) & h_2(t) & h_3(t) \end{bmatrix} \begin{bmatrix} I_{10a} \\ I_{20a} \\ V_{0a} \end{bmatrix} + \begin{bmatrix} f_4(t) + f_5(t) \\ g_4(t) + g_5(t) \\ h_4(t) + h_5(t) \end{bmatrix} \right\} [u(t) - u(t - D \bullet T_{osc})] \quad (4.12)$$

$$\begin{bmatrix} i_{1b}(t) \\ i_{2b}(t) \\ v_b(t) \end{bmatrix} = \left\{ \begin{array}{l} \begin{bmatrix} f_1(t-D \bullet T_{osc}) & f_2(t-D \bullet T_{osc}) & f_3(t-D \bullet T_{osc}) \\ g_1(t-D \bullet T_{osc}) & g_2(t-D \bullet T_{osc}) & g_3(t-D \bullet T_{osc}) \\ h_1(t-D \bullet T_{osc}) & h_2(t-D \bullet T_{osc}) & h_3(t-D \bullet T_{osc}) \end{bmatrix} \begin{bmatrix} I_{10b} \\ I_{20b} \\ V_{0b} \end{bmatrix} + \\ \begin{bmatrix} f_6(t-D \bullet T_{osc}) + f_5(t-D \bullet T_{osc}) \\ g_6(t-D \bullet T_{osc}) + g_5(t-D \bullet T_{osc}) \\ h_6(t-D \bullet T_{osc}) + h_5(t-D \bullet T_{osc}) \end{bmatrix} \end{array} \right\} [u(t-D \bullet T_{osc}) - u(t - T_{osc})] \quad (4.13)$$

From (4.12) at $t = D \cdot T_{osc}$, the following is true:

$$\begin{bmatrix} f_1(D \cdot T_{osc}) & f_2(D \cdot T_{osc}) & f_3(D \cdot T_{osc}) \\ g_1(D \cdot T_{osc}) & g_2(D \cdot T_{osc}) & g_3(D \cdot T_{osc}) \\ h_1(D \cdot T_{osc}) & h_2(D \cdot T_{osc}) & h_3(D \cdot T_{osc}) \end{bmatrix} \begin{bmatrix} I_{10a} \\ I_{20a} \\ V_{0a} \end{bmatrix} + \begin{bmatrix} f_4(D \cdot T_{osc}) + f_5(D \cdot T_{osc}) \\ g_4(D \cdot T_{osc}) + g_5(D \cdot T_{osc}) \\ h_4(D \cdot T_{osc}) + h_5(D \cdot T_{osc}) \end{bmatrix} = \begin{bmatrix} I_{10b} \\ I_{20b} \\ V_{0b} \end{bmatrix} \tag{4.14}$$

That is,

$$A_1 \cdot \begin{bmatrix} I_{10a} \\ I_{20a} \\ V_{0a} \end{bmatrix} + B_1 = \begin{bmatrix} I_{10b} \\ I_{20b} \\ V_{0b} \end{bmatrix} \tag{4.15}$$

At $t = T_{osc}$, and from (4.13), the other boundary condition is also established:

$$\begin{bmatrix} f_1(T_{osc} - D \cdot T_{osc}) & f_2(T_{osc} - D \cdot T_{osc}) & f_3(T_{osc} - D \cdot T_{osc}) \\ g_1(T_{osc} - D \cdot T_{osc}) & g_2(T_{osc} - D \cdot T_{osc}) & g_3(T_{osc} - D \cdot T_{osc}) \\ h_1(T_{osc} - D \cdot T_{osc}) & h_2(T_{osc} - D \cdot T_{osc}) & h_3(T_{osc} - D \cdot T_{osc}) \end{bmatrix} \begin{bmatrix} I_{10b} \\ I_{20b} \\ V_{0b} \end{bmatrix} + \begin{bmatrix} f_6(T_{osc} - D \cdot T_{osc}) + f_5(T_{osc} - D \cdot T_{osc}) \\ g_6(T_{osc} - D \cdot T_{osc}) + g_5(T_{osc} - D \cdot T_{osc}) \\ h_6(T_{osc} - D \cdot T_{osc}) + h_5(T_{osc} - D \cdot T_{osc}) \end{bmatrix} = \begin{bmatrix} I_{20a} \\ I_{10a} \\ V_{0a} \end{bmatrix} \tag{4.16}$$

That is,

$$A_2 \begin{bmatrix} I_{10b} \\ I_{20b} \\ V_{0b} \end{bmatrix} + B_2 = \begin{bmatrix} I_{20a} \\ I_{10a} \\ V_{0a} \end{bmatrix} \tag{4.17}$$

We substitute (4.15):

$$A_2 \left\{ A_1 \cdot \begin{bmatrix} I_{10a} \\ I_{20a} \\ V_{0a} \end{bmatrix} + B_1 \right\} + B_2 = \begin{bmatrix} I_{20a} \\ I_{10a} \\ V_{0a} \end{bmatrix} \tag{4.18}$$

The implication of (4.18) is that, under the steady state, all the cyclic starting conditions I_{10a}, I_{20a}, V_{0a}, I_{10b}, I_{20b}, and V_{0b} can be determined. Once all are known, the individual inductor loop current and capacitor node voltage can be given a steady-state, one-cycle expression:

$$i_{1\_1T}(t) = \left\{ \begin{bmatrix} I_{10a} \\ I_{20a} \\ V_{0a} \end{bmatrix}^T \begin{bmatrix} f_1(t) \\ f_2(t) \\ f_3(t) \end{bmatrix} + f_4(t) + f_5(t) \right\} [u(t) - u(t - D \cdot T_{osc})]$$
$$+ \left\{ \begin{bmatrix} I_{10b} \\ I_{20b} \\ V_{0b} \end{bmatrix}^T \begin{bmatrix} f_1(t - D \cdot T_{osc}) \\ f_2(t - D \cdot T_{osc}) \\ f_3(t - D \cdot T_{osc}) \end{bmatrix} + f_6(t - D \cdot T_{osc}) + f_5(t - D \cdot T_{osc}) \right\}$$
$$[u(t - D \cdot T_{osc}) - u(t - T_{osc})] \tag{4.19}$$

$$i_{2\_1T}(t) = \left\{ \begin{bmatrix} I_{10a} \\ I_{20a} \\ V_{0a} \end{bmatrix}^T \begin{bmatrix} g_1(t) \\ g_2(t) \\ g_3(t) \end{bmatrix} + g_4(t) + g_5(t) \right\} (u(t) - u(t - D \cdot T_{osc}))$$
$$+ \left\{ \begin{bmatrix} I_{10b} \\ I_{20b} \\ V_{0b} \end{bmatrix}^T \begin{bmatrix} g_1(t - D \cdot T_{osc}) \\ g_2(t - D \cdot T_{osc}) \\ g_3(t - D \cdot T_{osc}) \end{bmatrix} + g_6(t - D \cdot T_{osc}) + g_5(t - D \cdot T_{osc}) \right\}$$
$$[u(t - D \cdot T_{osc}) - u(t - T_{osc})] \tag{4.20}$$

By the same token, the (ideal) capacitor voltage is expressed as

$$v(t) = \left\{ \begin{bmatrix} I_{10a} \\ I_{20a} \\ V_{0a} \end{bmatrix}^T \begin{bmatrix} h_1(t) \\ h_2(t) \\ h_3(t) \end{bmatrix} + h_4(t) + h_5(t) \right\} [u(t) - u(t - D \cdot T_{osc})]$$
$$+ \left\{ \begin{bmatrix} I_{10b} \\ I_{20b} \\ V_{0b} \end{bmatrix}^T \begin{bmatrix} h_1(t - D \cdot T_{osc}) \\ h_2(t - D \cdot T_{osc}) \\ h_3(t - D \cdot T_{osc}) \end{bmatrix} + h_6(t - D \cdot T_{osc}) + h_5(t - D \cdot T_{osc}) \right\}$$
$$[u(t - D \cdot T_{osc}) - u(t - T_{osc})] \tag{4.21}$$

However, (4.19) and (4.20) do not completely depict the cyclic nature of inductor loop current yet, since the current periodicity encompasses two oscillator cycles. Therefore, both inductor currents are represented by

$$i_1(t) = i_{1\_1T}(t) + i_{2\_1T}(t + T), \quad i_2(t) = i_1(t - T) \tag{4.22}$$

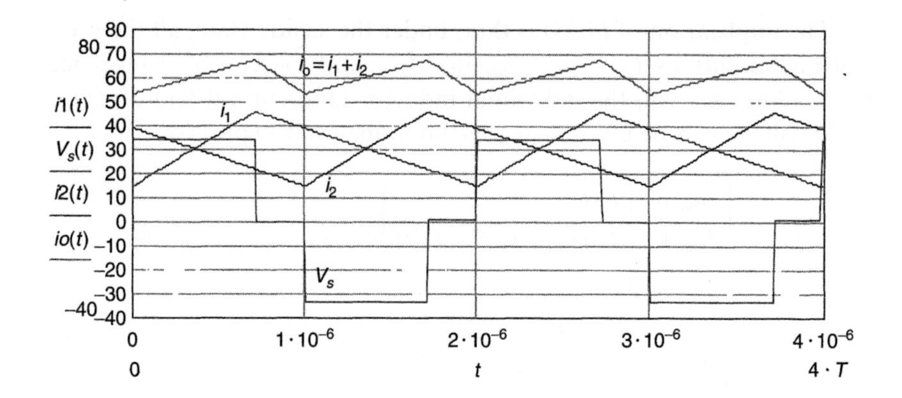

Figure 4.8: Example waveforms of a current doubler

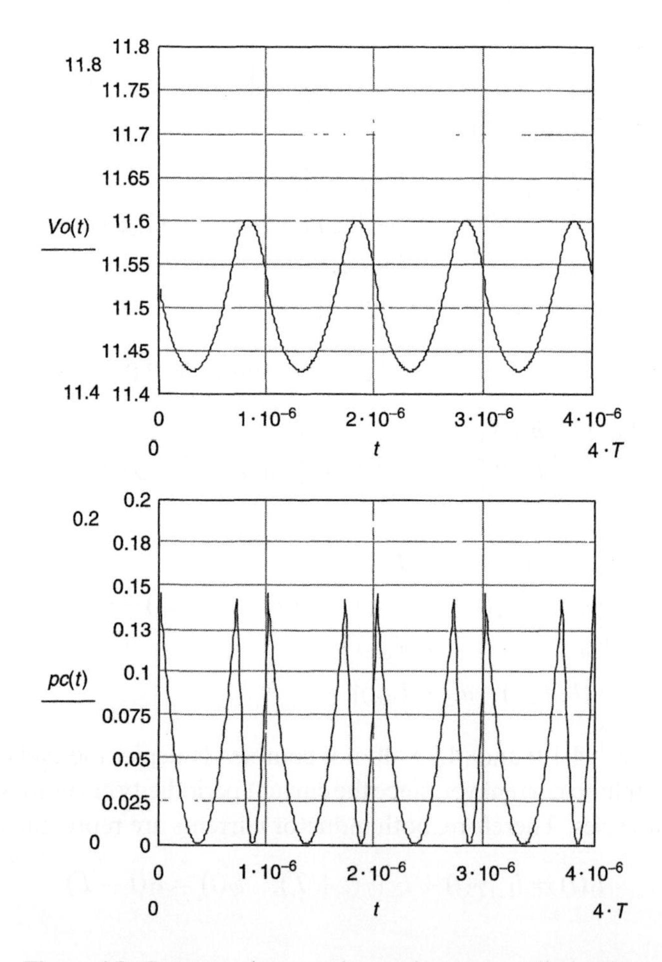

Figure 4.9: Output voltage and capacitor power dissipation

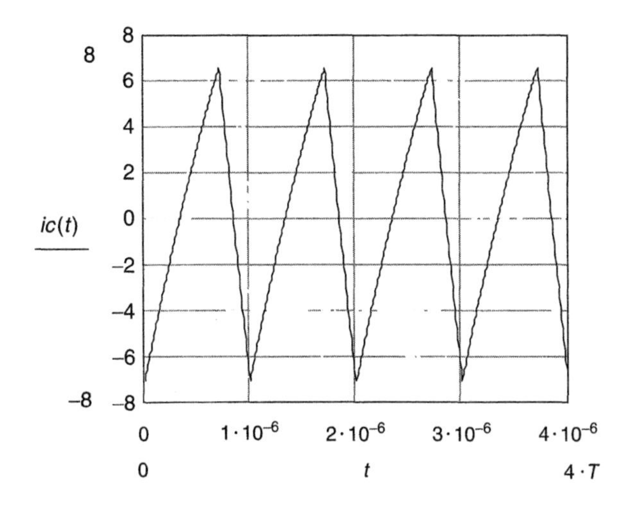

Figure 4.10: Capacitor current

Again, (4.22) has periodicity of $2 \bullet T_{\text{osc}}$. And the output voltage is

$$V_o(t) = R_p[i_1(t) + i_2(t)] + k \bullet v(t) \tag{4.23}$$

The AC ripple current through the capacitor is

$$i_c(t) = i_1(t) + i_2(t) - \frac{V_o(t)}{R_L} \tag{4.24}$$

As a result, the capacitor power dissipation is

$$p_C(t) = r_c \bullet \left[i_1(t) + i_2(t) - \frac{V_o(t)}{R_L} \right]^2 \tag{4.25}$$

4.5 Steady-State Output Waveforms Example

Given a design with $V_{\text{in}} = 270\,\text{V}$, $V_o = 11.5\,\text{V}$, $I_o = 60\,\text{amp}$, $N_p = 16$, $N_s = 2$, $r_c = 0.003$, $r_1 = r_2 = 0.001$, $L_1 = L_2 = \mu\text{H}$, $C = 10\,\mu\text{F}$, oscillator frequency = 1 MHz, rectifier drop = 0.5 V. The duty cycle based on (4.3) is 0.713. Figure 4.8 shows the steady-state waveforms. Figure 4.9 gives the output voltage and capacitor power dissipation profile. Figure 4.10 gives the capacitor current.

Figure 4.10: Capacitor current

Again, (4.22) has periodicity of $2 \cdot T_{osc}$. And the output voltage is

$$V_o(t) = K_3[i_1(t) + i_2(t)] + K_4 v_C(t) \tag{4.23}$$

The AC ripple current through the capacitor is

$$i_C(t) = i_1(t) + i_2(t) - \frac{V_o(t)}{R_L} \tag{4.24}$$

As a result, the capacitor power dissipation is

$$p_C(t) = r_c \cdot \left[i_1(t) + i_2(t) - \frac{V_o(t)}{R_L} \right]^2 \tag{4.25}$$

4.5 Steady-State Output Waveforms Example

Given a design with $V_{in} = 270$ V, $V_o = 11.5$ V, $I_o = 60$ amp, $N_p = 16$, $N_s = 2$, $r_s = 0.003$, $r_1 = r_2 = 0.001$, $L_1 = L_2 =$ [illegible] μH, $C = 10$ μF, oscillator frequency = 1 MHz, rectifier drop = 0.5 V. The duty cycle based on (4.3) is 0.71. Figure 4.8 shows the steady-state waveforms. Figure 4.9 gives the output voltage and capacitor power dissipation profile. Figure 4.10 gives the capacitor current.

Chapter 5

Current-Fed Push–Pull Converters

All buck-derived converters presented so far share one more common trait, the input line current pulsates if no additional input filter is incorporated. Coupled with the action of PWM, the pulsating current, equivalent to a high *di/dt*, can create real problems, for instance, conducted emission generated by AC current rushing back and forth along the input wire/connection. The problem can be mitigated with the addition of an input filter including a choke (inductor). Well, if that is all it takes, why don't we just move the output filter inductor to the input side? This is the origin of current-fed (current-driven) converters in contrast to the voltage-fed (voltage-driven) converters given in Chapters 1–4.

In Figure 5.1, we show four current-fed converters. But we focus our attention on only 5.1(a).

Again, as in the case of voltage-fed push–pull converters, the clock frequency runs at twice the frequency both switches are subjected to. Then, because of the selection of the serial input inductor and permissible duty cycle range, all four topologies have four operating modes. Based on the switch frequency (one-half the clock frequency), switch duty cycle, D, defined as the ratio of switch on-time to the switch period $T_s(=2T_{osc})$ can be larger or smaller than 50%. Operations with a larger than 50% steady-state duty cycle are considered in an overlapping mode, whereas those in a smaller than 50% cycle are in a nonoverlapping mode. In addition, the

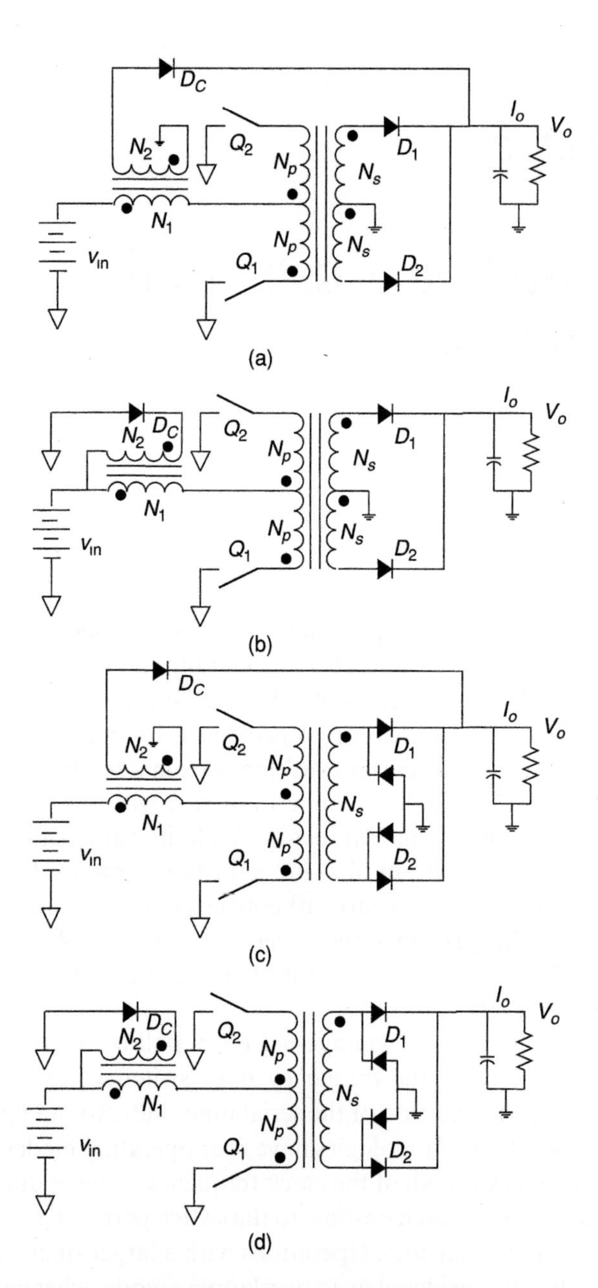

Figure 5.1: Current-fed converters

instantaneous inductor current may exhibit two cases: continuous conduction and discontinuous conduction. Continuous conduction is characterized by an inductor current that never runs dry, while discontinuous conduction dwells in a zero inductor current state for a definite time duration. As a consequence, the four modes of operations are overlapping continuous conduction, overlapping discontinuous conduction, nonoverlapping continuous conduction, and nonoverlapping discontinuous conduction. Due to limitations of size, we give extensive coverage to only overlapping continuous conduction and nonoverlapping continuous conduction. Readers can treat the other operating modes and configurations following the same procedure outlined in the following.

5.1 Overlapping Continuous-Conduction Mode

This operating mode is understood to alternate between two structures, as shown in Figure 5.2.

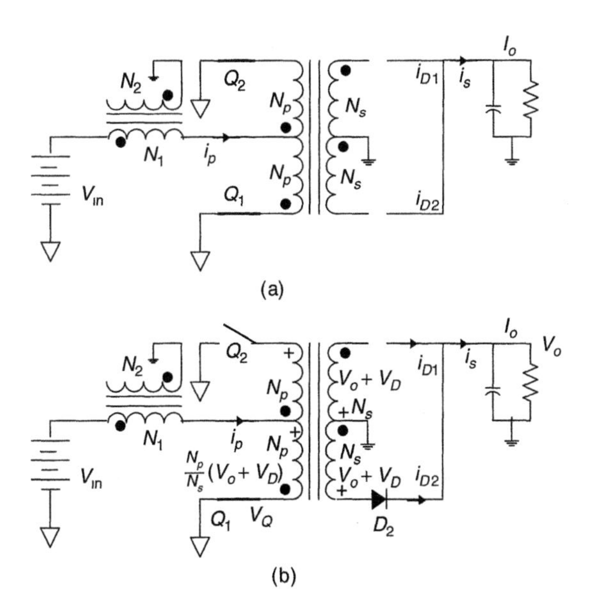

Figure 5.2: Current-fed converter with overlapping continuous conduction

The dwell time in Figure 5.2(a), in which both switches are on, is $(D - 0.5)T_s$ if the oscillator period is used as the time base. Then, the other dwell time is $(1 - D)T_s$ in which only one switch is on. While in $(D - 0.5)T_s$, a short is placed across the transformer primary. Input V_{in} is impressed across L_{N1}. During $(1 - D)T_s$, the output V_o is reflected across one half primary. The volt–second balance requirement across L_{N1} gives

$$(V_{\text{in}} - V_Q)(D - 0.5)T_s + \left[V_{\text{in}} - \frac{N_p}{N_s}(V_o + V_D) - V_Q\right](1 - D)T_s = 0 \quad (5.1)$$

This leads to the open-loop duty cycle:

$$D = 1 - \frac{N_s \bullet (V_{\text{in}} - V_Q)}{2 \bullet N_p \bullet (V_o + V_D)} \quad (5.2)$$

The input inductor, identified as i_p or i_L, and the output side rectifiers are also understood to carry continuous pulsating currents as shown in Figure 5.3.

From the output-current, i_s, waveform, the following is established:

$$I_M = \frac{I_o}{2(1 - D)} \quad (5.3)$$

From the primary-side-current, i_p, waveform, the ramp-down decrement δi_p, during $(1 - D)T_s$ is given by the volt–second changes divided by inductance:

$$\delta i_p = \frac{\left[\frac{N_p}{N_s}(V_o + V_D) - V_{\text{in}} + V_Q\right](1 - D)T_s}{L_{N1}} \quad (5.4)$$

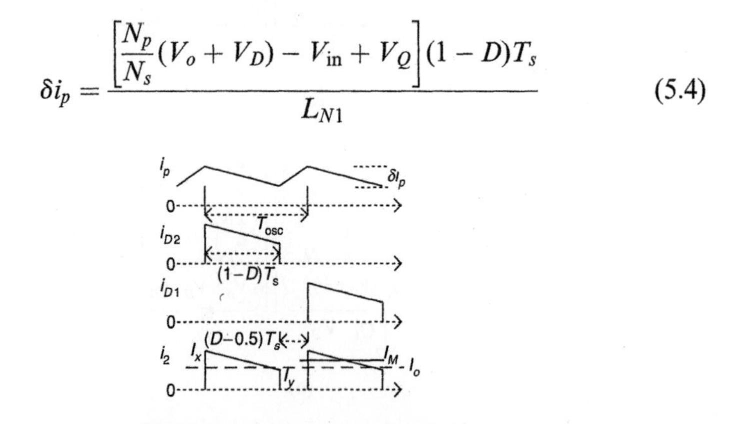

Figure 5.3: Current waveforms

Ampere's law dictates that

$$\delta i_s = \frac{N_p}{N_s}\delta i_p = \frac{N_p}{N_s} \bullet \frac{\left[\frac{N_p}{N_s}(V_o + V_D) - V_{\text{in}} + V_Q\right](1 - D)T_s}{L_{N1}} \tag{5.5}$$

The secondary side's current extreme values are

$$\begin{aligned} I_x &= I_M + \frac{\delta i_s}{2} \\ &= \frac{I_o}{2(1-D)} + \frac{N_p}{2N_s} \bullet \frac{\left[\frac{N_p}{N_s}(V_o + V_D) - V_{\text{in}} + V_Q\right](1 - D)T_s}{L_{N1}} \end{aligned} \tag{5.6}$$

$$I_y = I_M - \frac{\delta i_s}{2} = \frac{I_o}{2(1-D)} - \frac{N_p}{2N_s} \bullet \frac{\left[\frac{N_p}{N_s}(V_o + V_D) - V_{\text{in}} + V_Q\right](1 - D)T_s}{L_{N1}}$$

The primary-side current peak and trough are given:

$$\begin{aligned} I_{N1\_\max} &= \frac{N_s}{N_p} I_x \\ &= \frac{N_s}{N_p}\frac{I_o}{2(1-D)} + \frac{\left[\frac{N_p}{N_s}(V_o + V_D) - V_{\text{in}} + V_Q\right](1 - D)T_s}{2 \bullet L_{N1}} \end{aligned} \tag{5.7}$$

$$I_{N1\_\min} = \frac{N_s}{N_p} I_y = \frac{N_s}{N_p}\frac{I_o}{2(1-D)} - \frac{\left[\frac{N_p}{N_s}(V_o + V_D) - V_{\text{in}} + V_Q\right](1 - D)T_s}{2 \bullet L_{N1}}$$

Equations (5.7) are rewritten; for instance, the first is rewritten as

$$I_{N1\_\max} = \frac{P_o(V_o + V_D)}{V_o(V_{\text{in}} - V_Q)} + \frac{(V_{\text{in}} - V_Q)T_s}{4 \bullet L_{N1}}\left[1 - \frac{N_s(V_{\text{in}} - V_Q)}{N_p(V_o + V_D)}\right] \tag{5.8}$$

From Figure 5.2(b), the output rectifier sees a peak inverse voltage, PIV:

$$\text{PIV}_{D1} = 2V_o + V_D \tag{5.9}$$

while the input choke clamping diode sees

$$\mathrm{PIV_{DC}} = \frac{N_2}{N_1}(V_{\mathrm{in}} - V_Q) + V_o \tag{5.10}$$

From (5.2),

$$V_o = \frac{N_s \bullet (V_{\mathrm{in}} - V_Q)}{2 \bullet N_p \bullet (1 - D)} - V_D \tag{5.11}$$

As a result, we obtain the power stage gain:

$$\frac{\partial V_o}{\partial D} = \frac{2N_p \bullet (V_o + V_D)^2}{N_s \bullet (V_{\mathrm{in}} - V_Q)} \tag{5.12}$$

One more step establishes the critical inductance required to maintain continuous conduction mode: that is,

$$\begin{aligned} I_{N1\_\mathrm{min}} &= \frac{P_o(V_o + V_D)}{V_o(V_{\mathrm{in}} - V_Q)} - \frac{(V_{\mathrm{in}} - V_Q)T_s}{4 \bullet L_{N1\_\mathrm{cri}}}\left[1 - \frac{N_s(V_{\mathrm{in}} - V_Q)}{N_p(V_o + V_D)}\right] = 0 \\ L_{N1\_\mathrm{cri}} &= \frac{V_o(V_{\mathrm{in}} - V_Q)^2 T_s}{4P_o(V_o + V_D)}\left[1 - \frac{N_s(V_{\mathrm{in}} - V_Q)}{N_p(V_o + V_D)}\right] \end{aligned} \tag{5.13}$$

Also, in order to ensure overlapping mode of operation, the duty cycle must be larger than 50%. Therefore,

$$D = 1 - \frac{N_s \bullet (V_{\mathrm{in}} - V_Q)}{2 \bullet N_p \bullet (V_o + V_D)} \geq 0.5 \tag{5.14}$$

And the constraint leads to the transformer's turn ratio:

$$\frac{N_p}{N_s} \geq \frac{V_{\mathrm{in}} - V_Q}{V_o + V_D} \tag{5.15}$$

Figure 5.2(b) also clearly shows the minimum breakdown voltage rating for both switches.

5.2 Overlapping Continuous Conduction, Steady State

For steady-state time-domain studies, Figure 5.2 can be translated into the equivalent circuits in Figure 5.4, which correspond to the two alternating dwell times. When both switches are on, Figure 5.4(a), two equations, one voltage loop, and one current node can be established, describing the circuit behavior:

$$\frac{di_1}{dt} + \frac{r_L}{L} i_1 = \frac{V_{\text{in}} - V_Q}{L}, \quad \frac{dv_1}{dt} = -\frac{1}{(r_C + R_L)C} v_1 \tag{5.16}$$

When only one switch is on, as in Figure 5.4(b), and by reflecting the primary circuit to the secondary side, a second set of differential equations is given:

$$\begin{aligned} &\frac{di_2}{dt} + \frac{r_L + n^2 \bullet R_p}{L} i_2 + \frac{n \bullet k_r}{L} v_2 = \frac{V_{\text{in}} - V_Q}{L} \\ &-\frac{n \bullet R_p}{r_C \bullet C} i_2 + \frac{dv_2}{dt} + \frac{1 - k_r}{r_C \bullet C} v_2 = 0 \end{aligned} \tag{5.17}$$

where

$$k_r = \frac{R_L}{r_C + R_L}, \quad R_p = \frac{r_C \bullet R_L}{r_C + R_L}$$

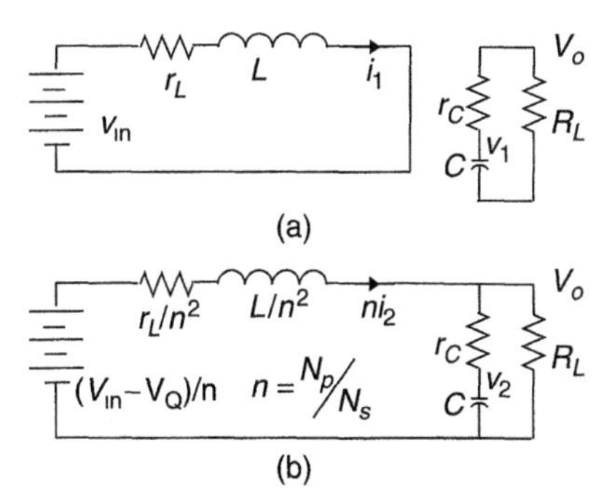

Figure 5.4: Current-fed converter with overlapping continuous conduction

Taking a Laplace transformation of (5.16) and solving both transformed equations yields

$$I_1(s) = \frac{I_{10}}{\left(s + \frac{r_L}{L}\right)} + \frac{V_{in} - V_Q}{L \cdot s \cdot \left(s + \frac{r_L}{L}\right)},$$

$$V_1(s) = \frac{V_{10}}{s + \frac{1}{(r_C + R_L)C}} \tag{5.18}$$

where I_{10} and V_{10} are unknown starting states yet to be found.

The inverse Laplace transform of (5.18), employing a matrix form, leads us to

$$\begin{bmatrix} i_1(t) \\ v_1(t) \end{bmatrix} = \left\{ \begin{bmatrix} f_1(t) & 0 \\ 0 & g_2(t) \end{bmatrix} \begin{bmatrix} I_{10} \\ V_{10} \end{bmatrix} + \begin{bmatrix} f_3(t) \\ 0 \end{bmatrix} \right\} \{u(t) - u[t - (D - 0.5)T_s]\} \tag{5.19}$$

The same procedure is applied to (5.17):

$$\left(s + \frac{r_L + n^2 \cdot R_p}{L}\right) I_2(s) + \frac{n \cdot k_r}{L} V_2(s)$$
$$= \left(I_{20} + \frac{V_{in} - V_Q}{L \cdot s}\right) e^{-(D-0.5)T_s \cdot s},$$
$$-\frac{n \cdot R_p}{r_C \cdot C} I_2(s) + \left(s + \frac{1 - k_r}{r_C \cdot C}\right) V_2(s) = V_{20} e^{-(D-0.5)T_s \cdot s}, \tag{5.20}$$

$$I_2(s) = \left[\frac{\left(s + \frac{1 - k_r}{r_C \cdot C}\right)}{D_e(s)} I_{20} + \frac{-\frac{n \cdot k_r}{L}}{D_e(s)} V_{20} + \frac{(V_{in} - V_Q)\left(s + \frac{1 - k_r}{r_C \cdot C}\right)}{L \cdot s \cdot D_e(s)} \right] e^{-(D-0.5)T_s \cdot s},$$

$$V_2(s) = \left[\frac{\frac{n \cdot R_p}{r_C \cdot C}}{D_e(s)} I_{20} + \frac{s + \frac{r_L + n^2 \cdot R_p}{L}}{D_e(s)} V_{20} + \frac{(V_{in} - V_Q)\frac{n \cdot R_p}{r_C \cdot C}}{L \cdot s \cdot D_e(s)} \right] e^{-(D-0.5)T_s \cdot s}$$

$$\tag{5.21}$$

where the denominator function is

$$D_e(s) = \begin{vmatrix} \left(s + \frac{r_L + n^2 \bullet R_p}{L}\right) & \frac{n \bullet k_r}{L} \\ -\frac{n \bullet R_p}{r_C \bullet C} & \left(s + \frac{1 - k_r}{r_C \bullet C}\right) \end{vmatrix}$$

Taking an inverse Laplace transform of (5.21) and placing it in matrix form, we get

$$\begin{bmatrix} i_2(t) \\ v_2(t) \end{bmatrix} = \left\{ \begin{aligned} &\begin{bmatrix} h_1[t-(D-0.5)T_s] & h_2[t-(D-0.5)T_s \\ p_1[t-(D-0.5)T_s] & p_2[t-(D-0.5)T_s] \end{bmatrix} \begin{bmatrix} I_{20} \\ V_{20} \end{bmatrix} \\ &+ \begin{bmatrix} h_3[t-(D-0.5)T_s] \\ p_3[t-(D-0.5)T_s] \end{bmatrix} \end{aligned} \right\}$$
$$\{u[t-(D-0.5)T_s] - u[t-T_s]\} \tag{5.22}$$

Equation (5.19) is evaluated at time boundary $t = (D - 0.5)T_s$:

$$\begin{bmatrix} i_1[(D-0.5)T_s] \\ v_1[(D-0.5)T_s] \end{bmatrix} = \begin{bmatrix} f_1[(D-0.5)T_s] & 0 \\ 0 & g_2[(D-0.5)T_s] \end{bmatrix} \begin{bmatrix} I_{10} \\ V_{10} \end{bmatrix} + \begin{bmatrix} f_3[(D-0.5)T_s] \\ 0 \end{bmatrix}$$
$$X_2 = A_1 \bullet X_1 + B_1 \tag{5.23}$$

Equation (5.22) is evaluated at time boundary $t = T_s/2$, and we consider the continuity of states:

$$\begin{bmatrix} i_2\left(\frac{T_s}{2}\right) \\ v_2\left(\frac{T_s}{2}\right) \end{bmatrix} = \begin{bmatrix} h_1[(1-D)T_s] & h_2[(1-D)T_s] \\ p_1[(1-D)T_s] & p_2[(1-D)T_s] \end{bmatrix} \begin{bmatrix} I_{20} \\ V_{20} \end{bmatrix} + \begin{bmatrix} h_3[(1-D)T_s] \\ p_3[(1-D)T_s] \end{bmatrix}$$
$$A_2 \bullet X_2 + B_2 = X_1 \tag{5.24}$$

The unknown starting states are solved

$$X_1 = (I - A_2 \bullet A_1)^{-1}(A_2 \bullet B_1 + B_2)$$
$$X_2 = A_2 \bullet X_1 + B_1 \tag{5.25}$$

Equations (5.19) and (5.22) then give the complete description of the circuit under a steady state.

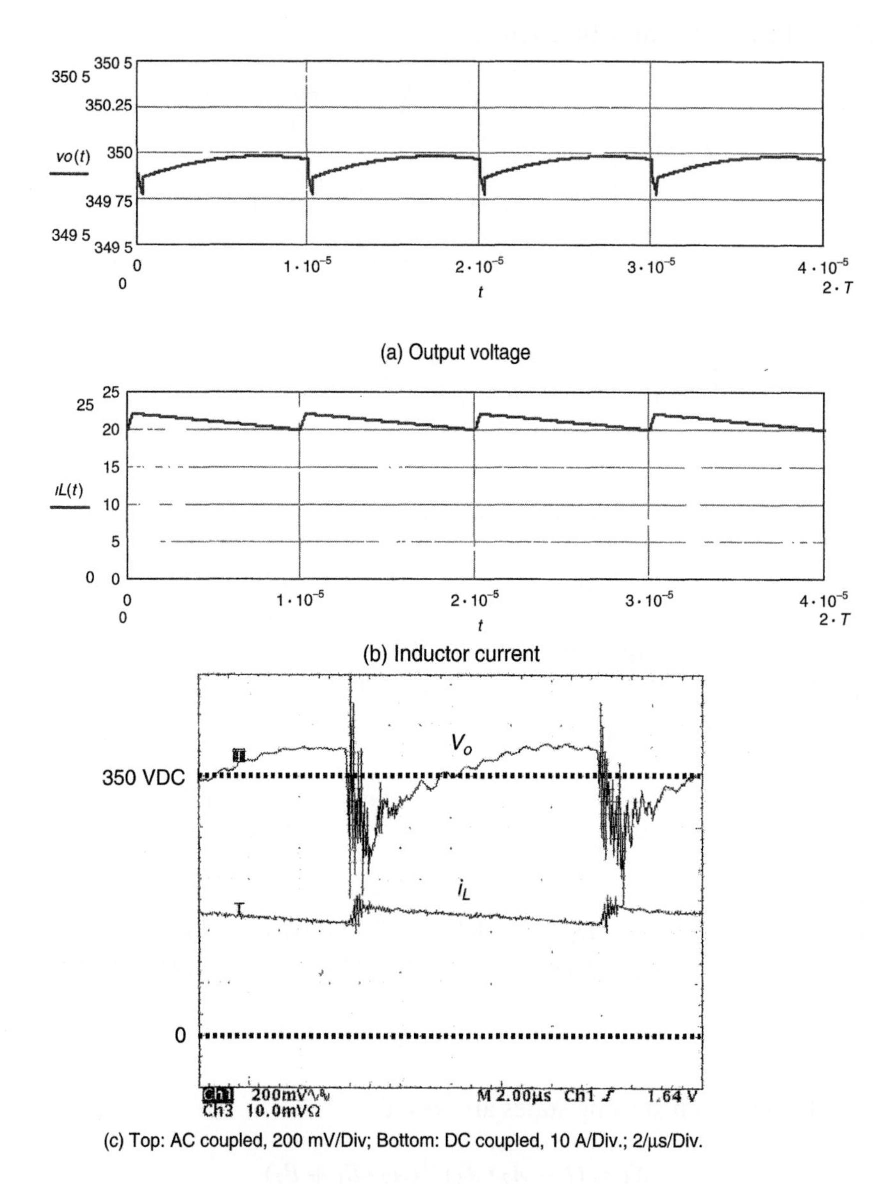

Figure 5.5: Current-fed converter with overlapping continuous conduction, voltage and inductor current: (a) and (b) theoretical; (c) actual

5.3 Overlapping Continuous Conduction, Example

Given a design using the configuration in Figure 5.1(c)—$N_p = 9$, $N_s = 44$, $V_{in} = 70$, $V_o = 350$, L = 8.5 μH, $r_L = 0.005$, $C = 10\,\mu F$, $r_C = 0.02$, $P_{out} = 1.46\,KW$, $V_Q = 0.2$, V_D(Schottkydiodes) = 0.5, and $f_s = 50\,KHz$ ($f_{osc} = 100\,KHz$)—the output voltage v_o and the inductor current i_L are given by Figure 5.5. The switch and transformer secondary currents are given by Figure 5.6.

5.4 Nonoverlapping Continuous-Conduction Mode

In the case of nonoverlapping continuous-conduction operation, two alternating topologies exist: one switch on and two switches off. The dwell time for one switch on is DT_s, and while for two switches off it is $(0.5 - D)T_s$. Clearly, the switch duty cycle on the basis of $2T_{osc}$ must be less than 50%; otherwise, the factor $(0.5 - D)$ is negative. Figure 5.7 shows the two alternating configurations of the converter cycles through repeated use in normal operation. During the dwell time DT_s, one switch and one rectifier are conducting. During the dwell time $(0.5 - D)T_s$, diode D_c and both rectifiers are conducting. All currents, inductor, energy recovery, and rectifiers are understood to look like those in Figure 5.8.

Again, the application of conservation of flux linkage gives

$$\left[V_{in} - \frac{N_p}{N_s}(V_o + V_D) - V_Q\right] D \cdot T_s = \frac{N_1}{N_2}(V_o + V_D)(0.5 - D)T_s \quad (5.26)$$

which leads to the open-loop duty cycle:

$$D = \frac{\frac{N_1}{N_2}(V_o + V_D)}{2\left[V_{in} + \left(\frac{N_1}{N_2} - \frac{N_p}{N_s}\right)(V_o + V_D) - V_Q\right]} \quad (5.27)$$

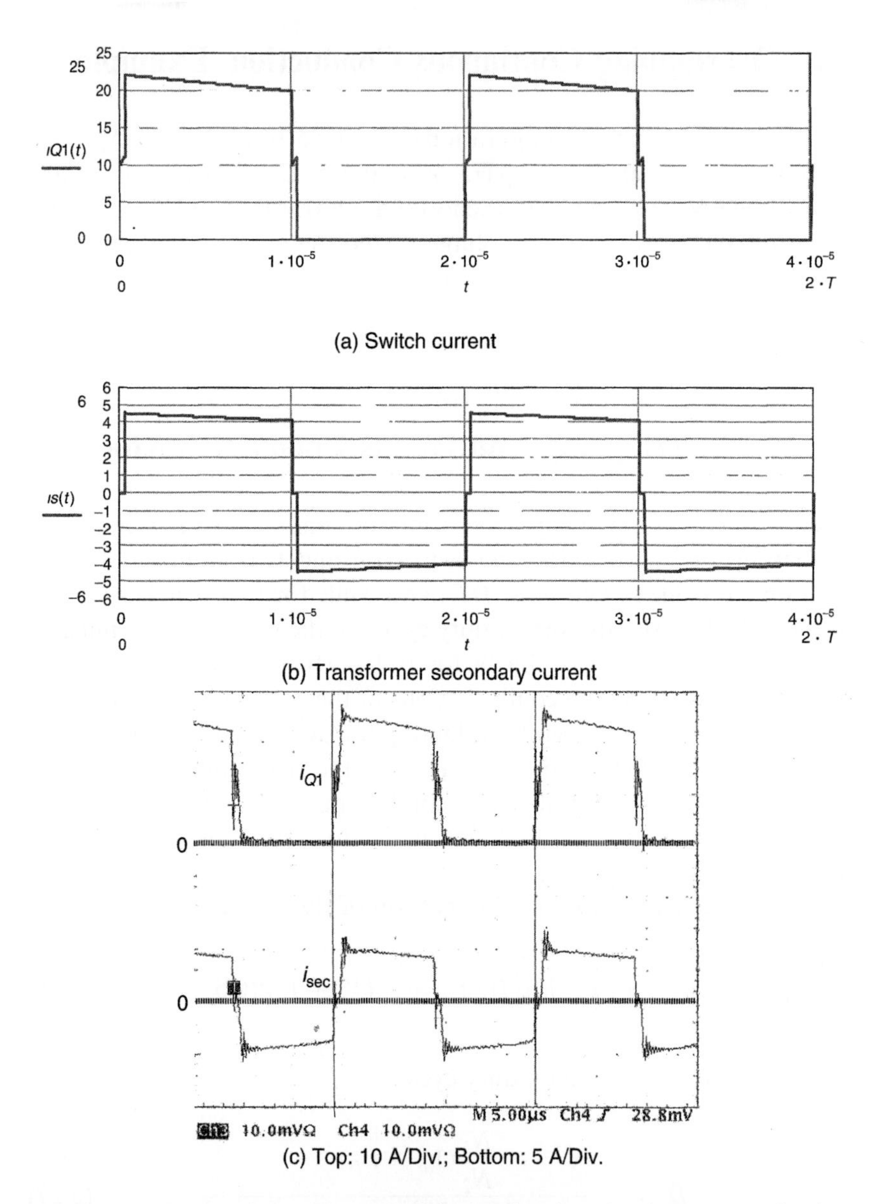

(a) Switch current

(b) Transformer secondary current

(c) Top: 10 A/Div.; Bottom: 5 A/Div.

Figure 5.6: Current-fed converter with overlapping continuous conduction, switch and transformer secondary currents: (a) and (b) theoretical; (c) actual

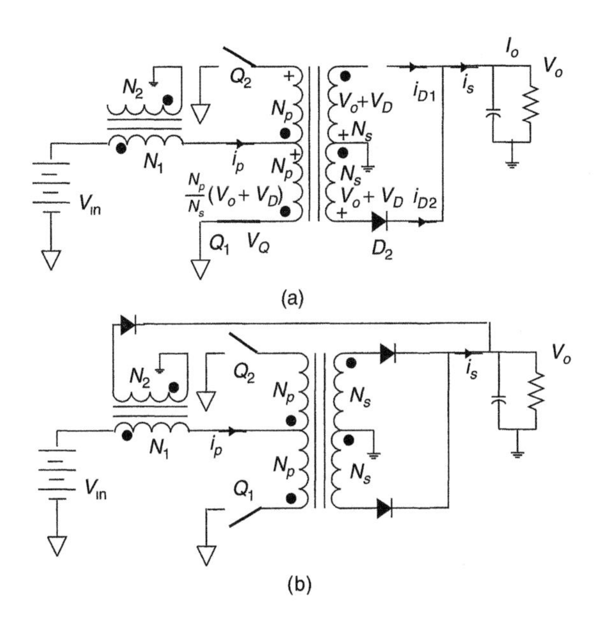

Figure 5.7: Current-fed converter with nonoverlapping continuous conduction

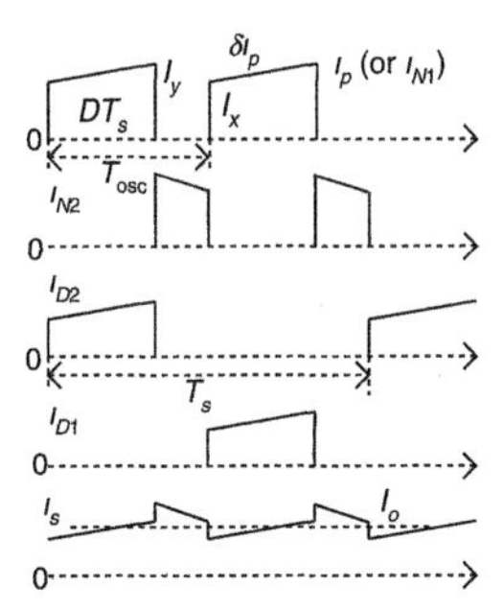

Figure 5.8: Current waveforms for nonoverlapping continuous conduction

From the current waveforms, the DC load current I_o can be expressed as the sum of the averaged i_{N2} and the averaged rectifier currents i_{D1} and i_{D2}:

$$< i_{N2} > = \frac{N_1}{N_2}(I_x + I_y)(0.5 - D)$$

$$< i_{D1} + i_{D2} > = \frac{N_p}{N_s}(I_x + I_y)D$$

$$\left[\frac{N_1}{N_2}(0.5 - D) + \frac{N_p}{N_s}D\right](I_x + I_y) = I_o$$

$$I_x + I_y = \frac{I_o}{\frac{N_1}{N_2}(0.5 - D) + \frac{N_p}{N_s}D} \tag{5.28}$$

The ripple current, δi_p, is given by

$$I_y - I_x = \frac{\left[V_{\text{in}} - \frac{N_p}{N_s}(V_o + V_D) - V_Q\right]D \cdot T_s}{L_{N1}} \tag{5.29}$$

We can follow the procedure of (5.6) to (5.14) and obtain the PIV for switches, PIV for rectifiers, open-loop output, output gain, and critical inductance. However, there is no point repeating those steps, and we go to the next important topic.

5.5 Load Current Sharing and Parallel Operation

The current-fed converters possess one very desirable quality all converters presented so far lack. By having an input inductor but no output inductor, not only is the input current made nonpulsating, the output behaves effectively as a current-source-feeding load impedance shared by other identical current-fed converters. In other words, we can place multiple current-fed converters in parallel, feeding one common load such that the load current is almost equally shared among converters. We use Figures 5.9 and 5.10 to discuss this important subject: current sharing. We begin with Figure 5.9.

Figure 5.9 intentionally omits showing the output load impedance, since that is to be shared by other units. In this figure, the output is a current source. But it is to be understood that the voltage feedback is

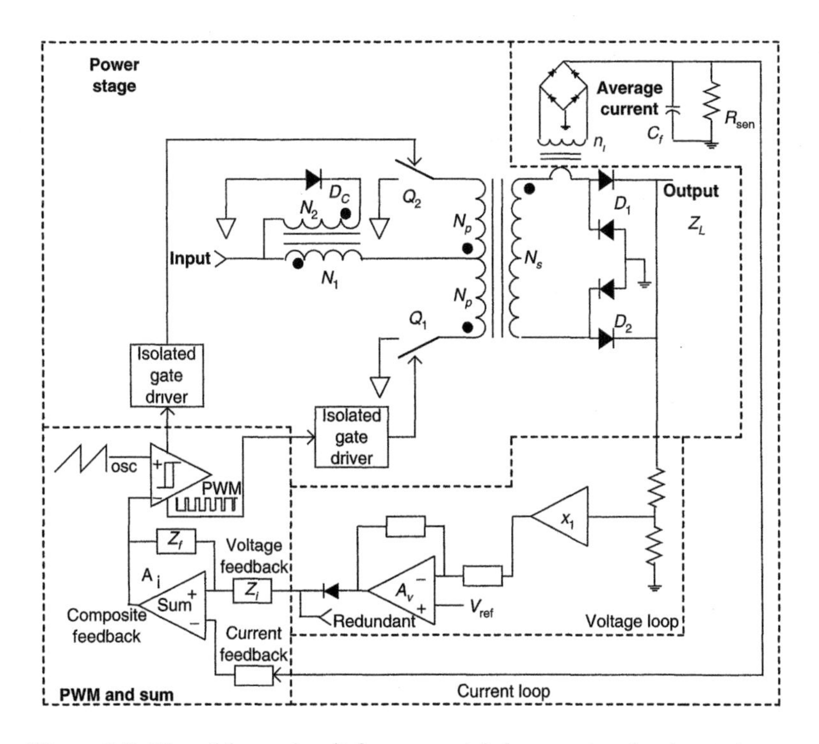

Figure 5.9: Closed-loop circuit for current-fed converter, load not shown

taken from the output voltage derived from the output current feeding the load impedance. The output voltage feeds a voltage divider and a unity gain buffer. The high-gain-error voltage amplifier creates the voltage feedback. The secondary side current is sensed and the full-wave is rectified. With a simple RC network, the average current feedback is generated. With the voltage feedback acting as a current command, a composite control signal that reflects the status of output current and voltage is also generated. With an oscillator of fixed frequency, the PWM block produces two alternating drives for both switches. By so doing, the loop is closed. In the following, we treat the current sensing in detail.

From Section 5.1, equations (5.5) and (5.6) in particular, the current feeding the current sensing filter, R_{sen} and C_f, in Figure 5.9, can be given as

$$i_{sen}(t) = \begin{cases} \dfrac{I_x}{n_i} - \dfrac{S_m}{n_i}t & 0 < t < (1 - \mathrm{D})T_s, \text{ time } a \\ 0 & (1 - \mathrm{D})T_s < t < \frac{T_s}{2}, \text{ time } b \end{cases} \tag{5.30}$$

where I_x is given by (5.6), n_t is the current transformer, and the slope is

$$S_m = \frac{N_p}{N_s} \bullet \frac{\left[\frac{N_p}{N_s}(V_o + V_D) - V_{\text{in}} + V_Q\right]}{L_{N1}} \tag{5.31}$$

At the output node of the current-sensing, full-wave rectifier, the following is established for the voltage across the $R_{\text{sen}} C_f$ filter:

$$\frac{dv}{dt} + \frac{v}{R_{\text{sen}} C_f} = \frac{dv}{dt} + \frac{v}{\tau} = \frac{i_{\text{sen}}(t)}{C_f} \tag{5.32}$$

Taking a Laplace transform and considering the two cyclic time intervals given in (5.30), the following is further established:

$$\begin{bmatrix} V_a(s) \\ V_b(s) \end{bmatrix} = \begin{Bmatrix} \dfrac{1}{s+\frac{1}{\tau}} V_{0a} + \dfrac{1}{C_f}\left[\dfrac{I_x}{n_t} \bullet \dfrac{1}{s\left(s+\frac{1}{\tau}\right)} - \dfrac{S_m}{n_t} \bullet \dfrac{1}{s^2\left(s+\frac{1}{\tau}\right)}\right], & \text{time } a \\ \dfrac{1}{s+\frac{1}{\tau}} e^{-1(1-D)T \bullet s_s} V_{0b}, & \text{time } b \end{Bmatrix} \tag{5.33}$$

Inverse Laplace transformation gives

$$\begin{bmatrix} v_a(t) \\ v_b(t) \end{bmatrix} = \begin{Bmatrix} e^{-t/\tau} V_{0a} + \dfrac{1}{C_f}[f_1(t) - f_2(t)], & \text{time } a \\ e^{-[t-(1-D)T_s]/\tau} V_{0b}, & \text{time } b \end{Bmatrix} \tag{5.34}$$

Boundary conditions and the requirement of continuity of states gives

$$\begin{aligned} e^{-[(1-D)T_s]/\tau} V_{0a} - V_{0b} &= -\frac{1}{C_f}\{f_1[(1-D)T_s] - f_2[(1-D)T_s]\} \\ - V_{0a} + e^{-[(1-D)T_s]/\tau} V_{0b} &= 0 \end{aligned} \tag{5.35}$$

Equation (5.35) yields the cyclic starting states, V_{0a} and V_{0b}, and the current sensor output voltage is completely determined. For example, given $n_i = 100$, the example given in section 5.3 also produces a sensing current, $i_{\text{sen}}(t)$, and a current feedback voltage, $v_f(t)$, as Figure 5.10 shows.

We can also conduct a quick numerical check for the current sensing filter output voltage, given $R_{\text{sen}} = 50$ and $C_f = 0.033\,\mu\text{F}$:

$$\frac{(I_x + I_y)(1 - D)T_s}{2 \bullet \dfrac{T_s}{2}} \bullet \frac{1}{n_i} R_{\text{sen}} = 2.086 \tag{5.36}$$

Clearly, (5.36) gives what engineers consider a quite acceptable approximation for practical design purposes. Equation (5.6) gives I_x and I_y.

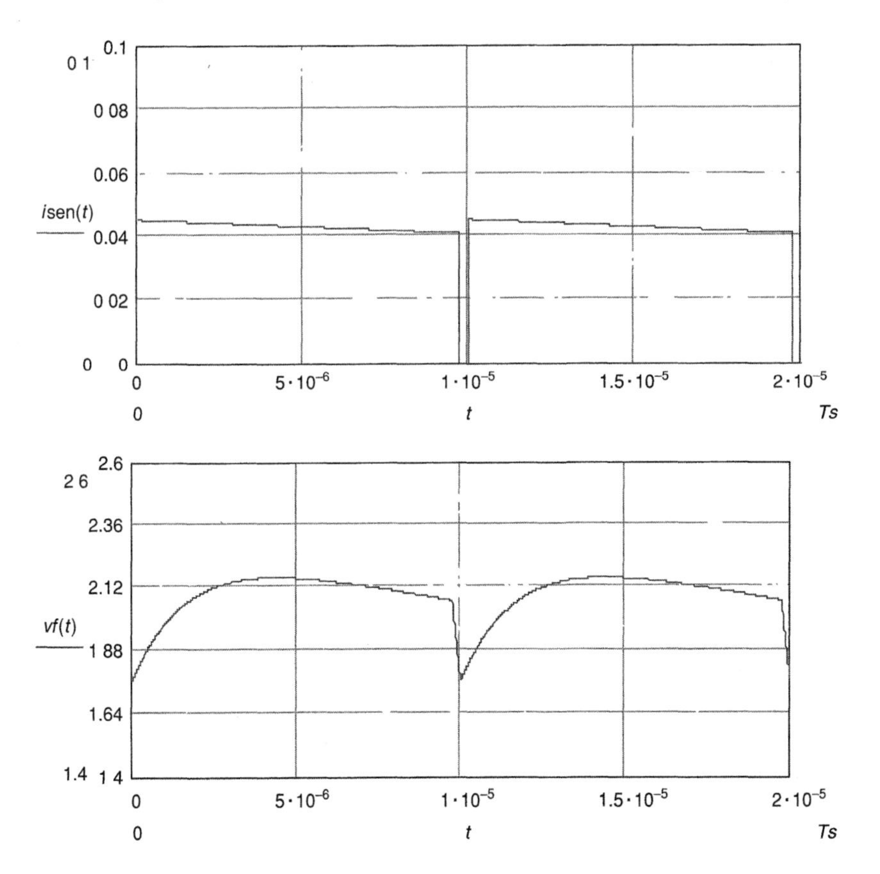

Figure 5.10: Sensing current and current feedback; $R_{\text{sen}} = 50$, $C_f = 0.033\,\mu\text{F}$

Next we show in Figure 5.11 how multiple converters are placed in parallel. Note that two voltage loops are provided for redundancy. Their individual voltage feedbacks are diode-ORed to give added reliability. They also serve as the common current control command, so that all parallel power stages share current properly. It may be necessary to also use a single master clock for all parallel units.

In theory, individual current contributions, or load current share I_n, can be estimated by a set of equations consisting of multiple two-equation subsets. We describe each two-equation subset as follows:

$$D_n = \left(\frac{\left\{ A_{v\_n}\left[V_{\mathrm{ref}\_n} - R_L(\sum I_n)K_{f\_n}\right] - V_{D\_n} - \frac{I_n}{n_i}R_{s\_n} \right\} A_{i\_n} - V_{\mathrm{offset}}}{m_{\mathrm{clk}\_n}} + \frac{T_s}{2} \right) \frac{1}{T_s} \tag{5.37}$$

$$R_L\left(\sum I_n\right) = \frac{N_s(V_{\mathrm{in}} - V_Q)}{2(1 - D_n)N_p} - V_{Dn} \tag{5.38}$$

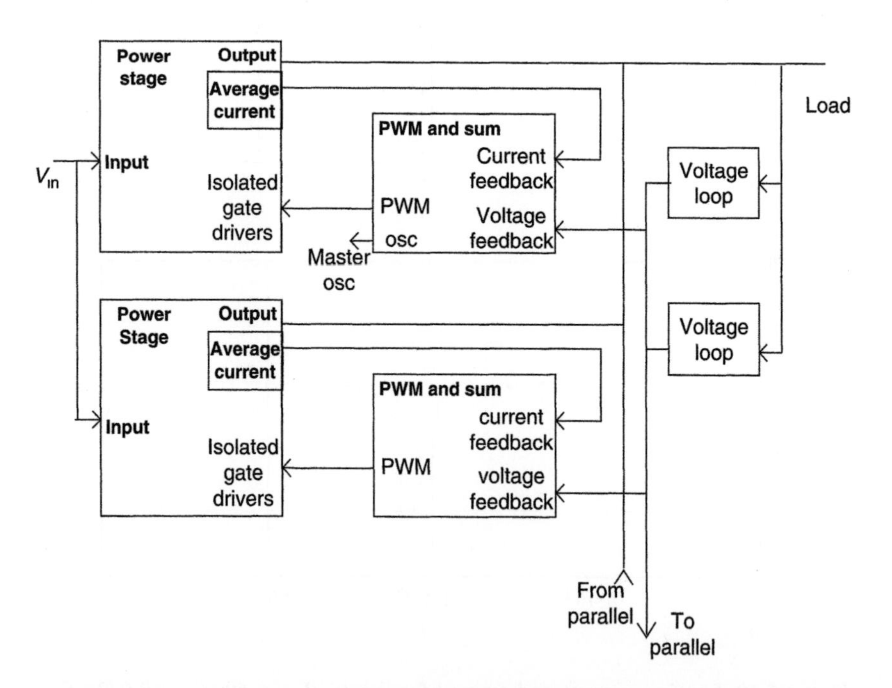

Figure 5.11: Parallel converters with overlapping continuous conduction

where the subscript n stands for parallel module count, A_v is voltage error amplifier gain, V_{ref} is precision reference, R_L is load, K_f is voltage feedback ratio, V_D is diode drop, R_s is current sensor, A_ι is current error amplifier gain, V_{offset} is oscillator trough voltage, m_{clk} is oscillator ramp-up slope, and D is duty cycle. For a nonoverlapping operating mode, (5.37) must be modified.

5.6 AC Small-Signal Studies Using State-Space Averaging

The state-space average technique was conceived in the 1970s by Robert Middlebrook and Slobodan Cuk [1]. In this section, we use the technique for Figure 5.1(a) with a minor modification and in a nonoverlapping continuous-conduction mode. The modification is that diode D_c is fed back to the input. In the meantime, two subtopologies for a nonoverlapping operation correspond to one switch on and both switches off. Figure 5.12 shows the equivalent circuits. Many parasitic components are also included: inductor series resistance r_L, primary winding resistance r_p, secondary winding resistance r_s, capacitor series resistance r_e, and L_{mp} for primary magnetizing inductance. From Figure 5.12(a), equations (5.39) are established with the secondary circuit reflected to the primary side:

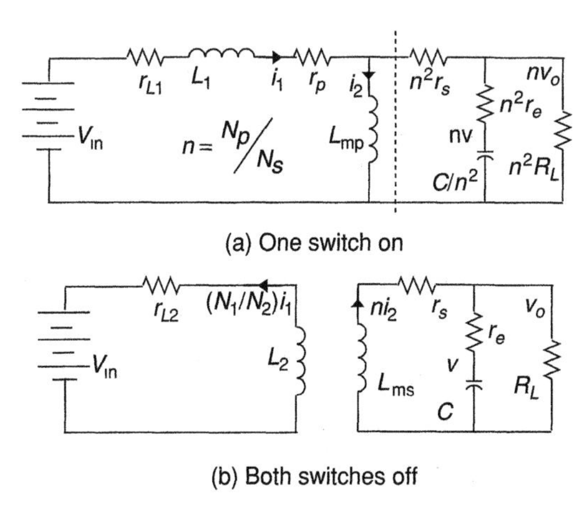

Figure 5.12: Equivalent circuits, nonoverlapping continuous conduction

$$\frac{d}{dt}\begin{bmatrix} i_1 \\ i_2 \\ v \end{bmatrix} = \frac{d}{dt}X_1 = \begin{bmatrix} \frac{r_{L1}+r_p+n^2(r_s+R_p)}{L_1} & \frac{n^2(r_s+R_p)}{L_1} & \frac{-k \cdot n}{L_1} \\ \frac{n^2(r_s+R_p)}{L_{\text{mp}}} & \frac{n^2(r_s+R_p)}{L_{\text{mp}}} & \frac{k \cdot n}{L_{\text{mp}}} \\ \frac{n \cdot R_p}{r_e \cdot C} & \frac{-n \cdot R_p}{r_e \cdot C} & \frac{-(1-k)}{r_e \cdot C} \end{bmatrix} \begin{bmatrix} i_1 \\ i_2 \\ v \end{bmatrix} + \begin{bmatrix} \frac{1}{L_1} \\ 0 \\ 0 \end{bmatrix} V_{\text{in}}$$

$$= A_1 \cdot X_1 + B_1 \cdot V_{\text{in}}$$

$$V_o = \begin{bmatrix} n \cdot R_p - n \cdot R_p k \end{bmatrix} \begin{bmatrix} i_1 \\ i_2 \\ v \end{bmatrix} = M_1 \cdot X_1 \tag{5.39}$$

For Figure 5.12(b), the circuit equation changes to

$$\frac{d}{dt}\begin{bmatrix} i_1 \\ i_2 \\ v \end{bmatrix} = \frac{d}{dt}X_2 = \begin{bmatrix} \frac{-r_{L2}}{L_2} & 0 & 0 \\ 0 & \frac{-(r_s+R_p)}{L_{\text{ms}}} & \frac{-k}{n \cdot L_{\text{ms}}} \\ 0 & \frac{n \cdot R_p}{r_e \cdot C} & \frac{-(1-k)}{r_e \cdot C} \end{bmatrix} \begin{bmatrix} i_1 \\ i_2 \\ v \end{bmatrix} + \begin{bmatrix} \frac{1}{L_2} \\ 0 \\ 0 \end{bmatrix}$$

$$V_{\text{in}} = A_2 \cdot X_2 + B_2 \cdot V_{\text{in}} \tag{5.40}$$

$$V_o = \begin{bmatrix} 0 n \cdot R_p k \end{bmatrix} \begin{bmatrix} i_1 \\ i_2 \\ v \end{bmatrix} = M_2 \cdot X_2$$

where

$$R_p = \frac{r_e \cdot R_L}{r_e + R_L}, \qquad k = \frac{R_L}{r_e + R_L}$$

The steady-state duty cycle, D, based on $T_s/2$, is then solved by

$$-[D \cdot M_1 + (1-D)M_2][D \cdot A_1 + (1-D)A_2]^{-1}[D \cdot B_1 + (1-D)B_2]V_{\text{in}} = V_o \tag{5.41}$$

Once the duty cycle is known, the state-space averaged model is defined as

$$\frac{d}{dt}X = A \bullet X + B \bullet V_{\text{in}},$$
$$X = D \bullet X_1 + (1 - D)X_2, \quad A = D \bullet A_1 + (1 - D)A_2,$$
$$B = D \bullet B_1 + (1 - D)B_2, \quad M = D \bullet M_1 + (1 - D)M_2 \quad (5.42)$$

The state vector at steady state is then given by

$$X = -A^{-1} \bullet B \bullet V_{\text{in}} \quad (5.43)$$

The output vector is $Y = M \bullet X$, and the power stage duty cycle to output transfer function is

$$G_{vd}(s) = M \bullet (s \bullet I - A)^{-1}[(A_1 - A_2) \bullet X + (B_1 - B_2) \bullet V_{\text{in}}] + (M_1 - M_2) \bullet X \quad (5.44)$$

That is to say, given a duty cycle perturbation δd, the output varies, $\delta v_o = G_{vd}\delta d$, accordingly. In turn, the disturbed output produces a sensing current change given by

$$\delta i_{\text{sen}} = \frac{\delta v_o}{Z_L \bullet n_i} \quad (5.45)$$

The sensing current is processed by the following circuit (Figure 5.13).

It is easy to show that the average current sensing transfer function is

$$\frac{\delta v_{oi}}{\delta v_o} = \frac{R_{\text{sen}}}{Z_L \bullet n_i} \bullet \frac{1}{R_{\text{sen}} \bullet C_f \bullet s + 1} \quad (5.46)$$

Next, using the process that leads to (1.28) and (1.46), we get the PWM gain:

$$F_m = \frac{\partial D}{\partial V_c} = \frac{0.98}{V_H - V_L} \quad (5.47)$$

With the most important block gains identified and using Figure 5.9 as a basis, the AC block diagram shown in Figure 5.14 is obtained for the condition of constant input.

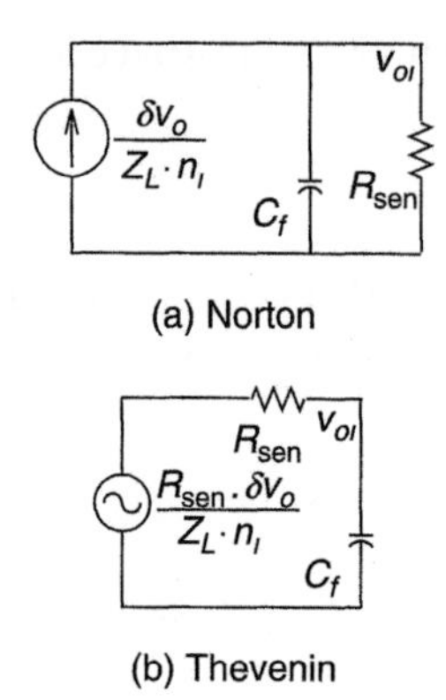

Figure 5.13: Average current sensing circuits

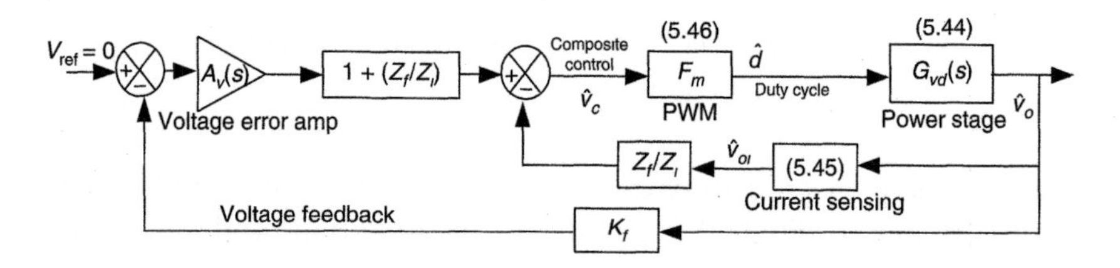

Figure 5.14: AC block diagram for nonoverlapping continuous conduction

5.7 State-Transition Technique

Based on Figure 5.4, symbols in section 5.2, and a procedure similar to section 1.14, the following is established:

$$A_1 = \begin{bmatrix} \dfrac{-r_L}{L} & 0 \\ 0 & \dfrac{-1}{(r_C + R_L)} \end{bmatrix}, \quad B_1 = \begin{bmatrix} \dfrac{1}{L} \\ 0 \end{bmatrix}, \quad E = V_{\text{in}} - V_Q,$$

$$A_2 = \begin{bmatrix} \dfrac{-(r_L + n^2 R_p)}{L} & \dfrac{-n \bullet k_r}{L} \\ \dfrac{n \bullet R_p}{r_C \bullet C} & \dfrac{k_r - 1}{r_C \bullet C} \end{bmatrix}, \quad B_2 = B_1, \tag{5.48}$$

$$\beta_0(t) = \frac{\begin{vmatrix} e^{\lambda 1_0 t} & \lambda 1_0 \\ e^{\lambda 1_1 t} & \lambda 1_1 \end{vmatrix}}{\begin{vmatrix} 1 & \lambda 1_0 \\ 1 & \lambda 1_1 \end{vmatrix}}, \quad \beta_1(t) = \frac{\begin{vmatrix} 1 & e^{\lambda 1_0 t} \\ 1 & e^{\lambda 1_1 t} \end{vmatrix}}{\begin{vmatrix} 1 & \lambda 1_0 \\ 1 & \lambda 1_1 \end{vmatrix}},$$

$$\alpha_0(t) = \frac{\begin{vmatrix} e^{\lambda 2_0 t} & \lambda 2_0 \\ e^{\lambda 2_1 t} & \lambda 2_1 \end{vmatrix}}{\begin{vmatrix} 1 & \lambda 2_0 \\ 1 & \lambda 2_1 \end{vmatrix}}, \quad \alpha_1(t) = \frac{\begin{vmatrix} 1 & e^{\lambda 2_0 t} \\ 1 & e^{\lambda 2_1 t} \end{vmatrix}}{\begin{vmatrix} 1 & \lambda 2_0 \\ 1 & \lambda 2_1 \end{vmatrix}},$$

$$M_1(t) = \beta_0(t) \bullet I + \beta_1(t) \bullet A_1, \; M_2(t) = \alpha_0(t) \bullet I + \alpha_1(t) \bullet A_2$$

where $\lambda 1_0$ and $\lambda 1_1$ are eigenvalues of A_1 and $\lambda 2_0$ and $\lambda 2_1$ of A_2.

The general solution was given in the form of (1.117). Again, if two unknown cyclical starting states, X_1 and X_2, are assumed, the following are also established at the transition boundaries, $t = (D - 0.5)T$ and $t = \mathrm{T}/2$:

$$M_1[(D - 0.5)T]X_1 - A_1^{-1}\{I - M_1[(D - 0.5)T]\}B_1 \bullet E = X_2$$
$$M_2[0.5T - (D - 0.5)T]X_2 - A_2^{-1}\{I - M_2[0.5T - (D - 0.5)T]\}B_2 \bullet E = X_1$$

In other words, X_1 and X_2 can be solved, which leads to the solutions for the two time segments:

$$x_1(t) = \{M_1(t)X_1 - A_1^{-1}[I - M_1(t)]B_1 \bullet E\}\{u(t) - u[t - (D - 0.5)T]\}$$
$$x_2(t) = (M_2[t - (D - 0.5)T]X_2 - A_2^{-1}\{I - M_2[t - (D - 0.5)T]\}B_2 \bullet E)$$
$$\bullet \{u[t - (D - 0.5)T] - u(t - 0.5T)\}$$

The inductor current and the output voltage are therefore given by

$$i(t) = [x_1(t)]_0 + [x_2(t)]_0$$
$$v(t) = k_r[x_1(t)]_1 + k_r[x_2(t)]_1 + n \bullet [x_2(t)]_0 \bullet R_p$$

where the subscripts of the square brackets stand for the element number of the column matrix (vector).

Chapter 6

Isolated Flyback Converters

For power levels less than a few hundred watts, a flyback converter is a less-expensive choice due to its simplicity in hardware. Figure 6.1 illustrates a typical converter with the most representative features. Again, like the forward converter and depending on the loading and design considerations, the converter's main energy storage inductor may operate in two modes: DCM and CCM. As for the control technique, two

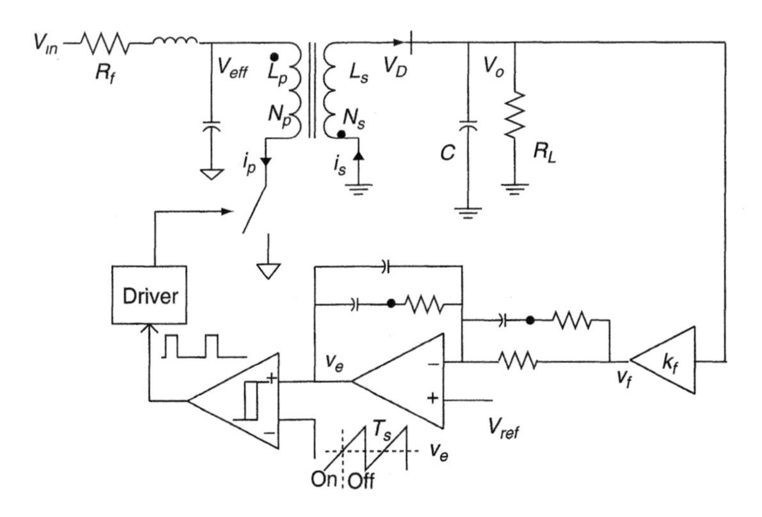

Figure 6.1: Flyback converter with voltage-mode control

approaches, voltage-mode control and current-mode control, are applicable. Wu [2] treated the subject in DCM using the voltage-mode and the peak current-mode controls. We do not duplicate those efforts but provide a refinement. We focus our coverage in the determination of critical inductance, the AC small-signal studies for DCM with peak-current mode, the DC studies for CCM with both voltage-mode and current-mode, the steady-state time-domain output, and the like.

6.1 DCM Duty-Cycle Determination, Another Approach

In Wu [2], the DCM duty cycle was determined based on the concept of total energy transfer. The book gives no information regarding the duration of the conducting and nonconducting flyback phases, identified as D_2 and D_3 in Figure 6.2, in which the relevant currents of a flyback converter operating in DCM are given. We fill the gap here.

Based on simple geometry, the DC load current, I_o, can be derived by

$$I_o = \frac{V_o}{R_L} = \frac{1}{2} \bullet I_{sp} \bullet D_2 = \frac{1}{2} \bullet \frac{N_p}{N_s} I_{pp} \bullet D_2 = \frac{1}{2} \bullet \frac{N_p}{N_s} \bullet \frac{V_{\text{eff}} \bullet D_{\text{DCM}}}{L_p \bullet f_s} \bullet D_2 \quad (6.1)$$

where L_p is the primary inductance and f_s is the switching frequency. However, (6.1) contains two unknowns, D_{DCM} and D_2. We need one more equation. And it is provided by the volt–second balance across the isolated inductor:

$$V_{\text{eff}} \bullet D_{\text{DCM}} \bullet T_s = \frac{N_p}{N_s}(V_o + V_D) D_2 \bullet T_s \quad (6.2)$$

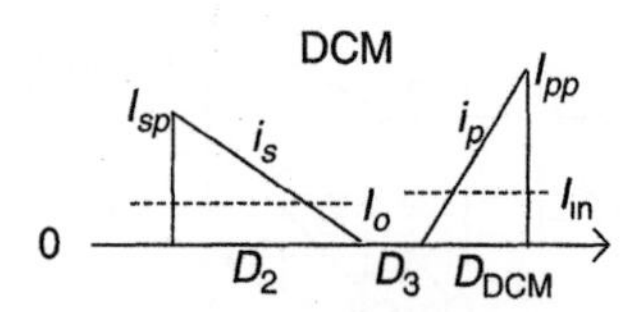

Figure 6.2: Currents for a flyback converter in DCM

Both equations can be consolidated and solved for both unknowns. The procedure leads to

$$D_{\mathrm{DCM}}^2 = \frac{2 \cdot L_p \cdot f_s \cdot V_o \cdot (V_o + V_D)}{V_{\mathrm{eff}}^2 \cdot R_L} \tag{6.3}$$

Because of the input line resistance losses, the effective input is less than the input:

$$V_{\mathrm{eff}} = V_{\mathrm{in}} - I_{\mathrm{in}} R_f = V_{\mathrm{in}} - \frac{1}{2} I_{pp} \cdot D_{\mathrm{DCM}} = V_{\mathrm{in}} - \frac{1}{2} \cdot \frac{V_{\mathrm{eff}} \cdot D_{\mathrm{DCM}}}{L_p \cdot f_s} \cdot D_{\mathrm{DCM}}$$

$$V_{\mathrm{eff}} = \frac{2 \cdot L_p \cdot f_s \cdot V_{\mathrm{in}}}{2 \cdot L_p \cdot f_s + D_{\mathrm{DCM}}^2} \tag{6.4}$$

In other words, (6.3) does not give the open-loop duty cycle yet. Equation (6.4) needs to be plugged into (6.3) to yield the end result:

$$D_{\mathrm{DCM}}^2 = \frac{2 \cdot L_p \cdot f_s \cdot V_o \cdot (V_o + V_D)}{\left(\frac{2 \cdot L_p \cdot f_s \cdot V_{\mathrm{in}}}{2 \cdot L_p \cdot f_s + D_{\mathrm{DCM}}^2} \right)^2 \cdot R_L} \tag{6.5}$$

Evidently, it is not a trivial matter to find the DCM duty cycle in symbolic closed form. We just leave it as it is with the understanding that the open-loop output and the open-loop duty cycle for the converter are now intimately related. In addition, it is interesting to interpret (6.5) in another way. We note that the factor $V_o(V_o + V_D)/R_L$ actually represents the output power including rectifier losses. Therefore, (6.5) also ties together the open-loop duty cycle and the output power.

6.2 CCM Duty-Cycle Determination

When CCM operation commences, the converter current waveforms change shape, as shown in Figure 6.3, in which the following is established:

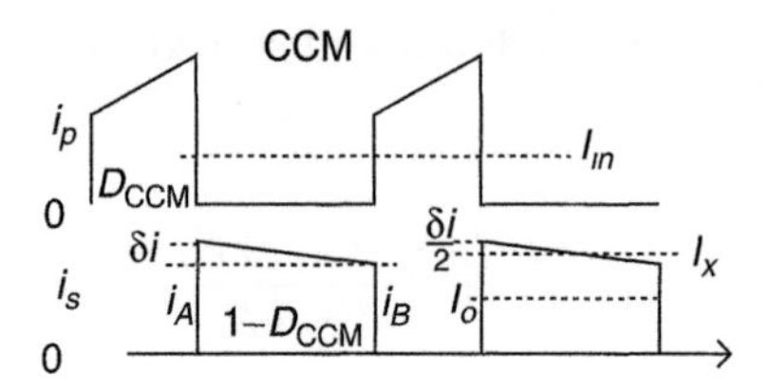

Figure 6.3: Currents for a flyback converter in CCM

$$
\begin{aligned}
i_A &= \frac{I_o}{1 - D_{\mathrm{CCM}}} + \frac{(V_o + V_D)(1 - D_{\mathrm{CCM}})}{2 \cdot L_s \cdot f_s} \\
i_B &= \frac{I_o}{1 - D_{\mathrm{CCM}}} - \frac{(V_o + V_D)(1 - D_{\mathrm{CCM}})}{2 \cdot L_s \cdot f_s} \\
I_{\mathrm{in}} &= \frac{N_s}{N_p} \cdot \frac{D_{\mathrm{CCM}}}{2}(i_A + i_B) = \frac{N_s}{N_p} \cdot \frac{D_{\mathrm{CCM}}}{1 - D_{\mathrm{CCM}}} \cdot I_o
\end{aligned}
\tag{6.6}
$$

Considering the input line resistance losses, the effective voltage feeding the converter is reduced to

$$
V_{\mathrm{eff}} = V_{\mathrm{in}} - I_{\mathrm{in}} \cdot R_f = V_{\mathrm{in}} - \frac{N_s}{N_p} \cdot \frac{D_{\mathrm{CCM}}}{1 - D_{\mathrm{CCM}}} \cdot \frac{V_o}{R_L} \cdot R_f \tag{6.7}
$$

Volt–second balance demands that

$$
V_{\mathrm{eff}} \cdot D_{\mathrm{CCM}} = \frac{N_p}{N_s}(V_o + V_D)(1 - D_{\mathrm{CCM}}) \tag{6.8}
$$

Equation (6.7) is plugged into (6.8) to find the open-loop duty cycle:

$$
\left(V_{\mathrm{in}} - \frac{N_s}{N_p} \cdot \frac{D_{\mathrm{CCM}}}{1 - D_{\mathrm{CCM}}} \cdot \frac{V_o}{R_L} \cdot R_f \right) D_{\mathrm{CCM}} = \frac{N_p}{N_s}(V_o + V_D)(1 - D_{\mathrm{CCM}}) \tag{6.9}
$$

Again, we do not attempt to solve the duty cycle in a symbolic, explicit form. It is quite baffling if one compares (6.5) with (6.9). In the case of DCM, more elements are involved, primary inductance and switching frequency in particular. But there is no sign of involvement for the two components in (6.9).

6.3 Critical Inductance

The critical inductance, which marks the transition boundary between CCM and DCM, is obtained by setting $i_B = 0$:

$$\begin{aligned} i_B &= \frac{I_o}{1 - D_{\text{CCM}}} - \frac{(V_o + V_D)(1 - D_{\text{CCM}})}{2 \bullet L_s \bullet f_s} = 0 \\ L_{s\_\text{cri}} &= \frac{(V_o + V_D)(1 - D_{\text{CCM}})^2}{2 \bullet I_o \bullet f_s} = \frac{(V_o + V_D)(1 - D_{\text{CCM}})^2}{2 \bullet \frac{V_o}{R_L} \bullet f_s} \end{aligned} \tag{6.10}$$

It is understood that D_{CCM} in L_{s_cri} is constrained by (6.9). Also, be cautioned that (6.10) points to the critical inductance from the viewpoint of the secondary side. The primary-side inductance is related to the secondary-side inductance by turn ratio squared. Taking either side, we are confronted by the utter complexity of such an outrageous equation form. One may wonder if it can be reduced to a simpler form. Moreover, (6.9) involves the turn ratio. We are still short one equation. The additional constraint is given by the requirement that the switch voltage rating must be larger than $[(V_o + V_D)N_p/N_s] + V_{\text{in}}$.

6.4 Voltage-Mode DCM Closed Loop

In Wu [2], this subject was briefly covered. Here, we give it better coverage. Under closed-loop operation, the complete converter block diagram can be as shown in Figure 6.4. Refer to Chapter 1 for the derivation of the voltage loop error signal and PWM.

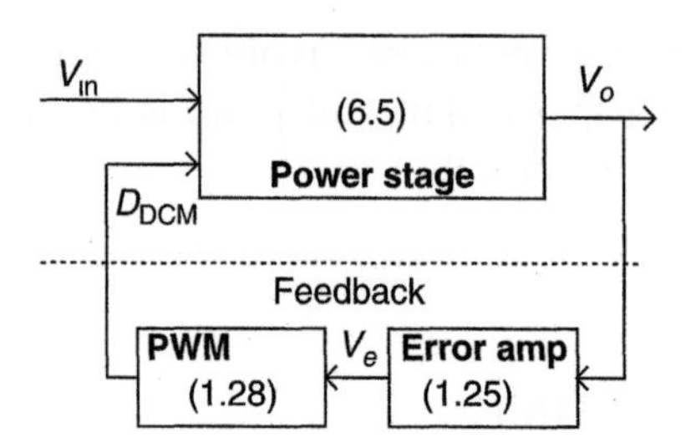

Figure 6.4: Block diagram of a flyback converter DCM in closed-loop operation

6.5 Voltage-Mode DCM Small-Signal Stability

In Chapter 2, a DCM canonical model, including a line input filter, was developed. Figure 2.7 and equations (2.19) through (2.23) are universal for all topologies. With a minor modification corresponding to Figure 6.1, we arrive at the AC block diagram in Figure 6.5 for the voltage-mode in DCM. The block diagram simplifies to Figure 6.6 for the condition of constant input. In the figures, $E_A(s) = (1.45)$, $F_m = (1.46)$. However, we should be careful to use the correct canonical parameters, j_1, g_1, r_1, and so forth in $G_{vd}(s)$ for flyback converters.

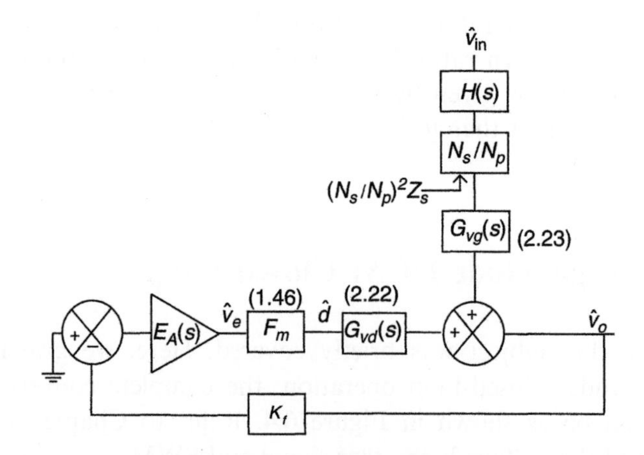

Figure 6.5: AC block diagram for the voltage-mode in DCM; transfer functions numbered have only the form, not the content, identified

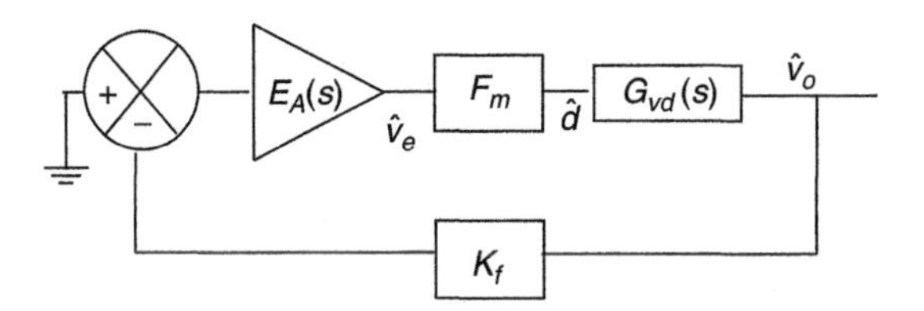

Figure 6.6: AC block diagram for the voltage-mode in DCM, constant input

The loop gain is then simply

$$T(s) = -K_f \cdot E_A(s) \cdot F_m \cdot G_{vd}(s) \tag{6.11}$$

and the conducted susceptibility is

$$C_s(s) = H(s) \cdot \frac{N_s}{N_p} \cdot G_{vg}(s) \frac{1}{1 + K_f \cdot E_A(s) \cdot F_m \cdot G_{vd}(s)} \tag{6.12}$$

6.6 Voltage-Mode CCM Closed Loop

For CCM operation, the major change taking place is in the power stage as outlined by (6.6) to (6.9). Other than that, there is no change in the way the error amplifier and the PWM operate. We simply take Figure 6.4 and change the content for the power stage block. We obtain the steady-state block diagram, Figure 6.7, for the subject.

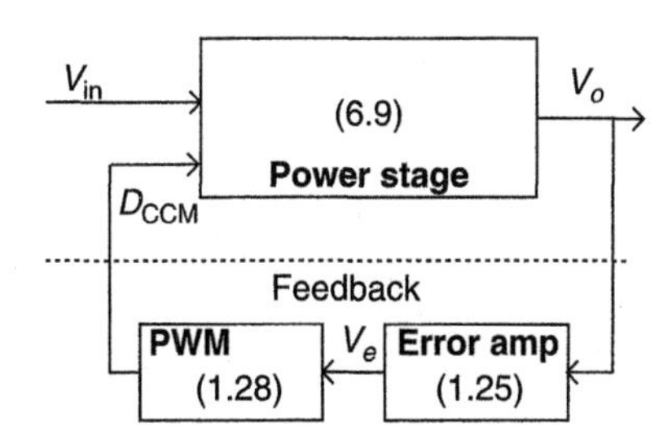

Figure 6.7: Block diagram of a flyback converter CCM in closed-loop operation

6.7 Voltage-Mode CCM Small-Signal Stability

In Chapter 1, a CCM canonical model including line input filter was developed. Figure 1.14 and equations (1.34) through (1.40) are universal for all topologies. With a minor modification corresponding to Figure 6.1, we arrive at the AC block diagram, Figure 6.8, for voltage-mode in CCM. Again, it reduces to Figure 6.9 for the condition with constant input. It is a simple matter to write down the loop gain and the conducted susceptibility given the two block diagrams. We just skip it.

6.8 Peak Current-Mode DCM Closed Loop

Current feedback is implemented by replacing the external oscillator in Figure 6.1 with an embedded switch current that is properly isolated and processed, as in Figure 6.10.

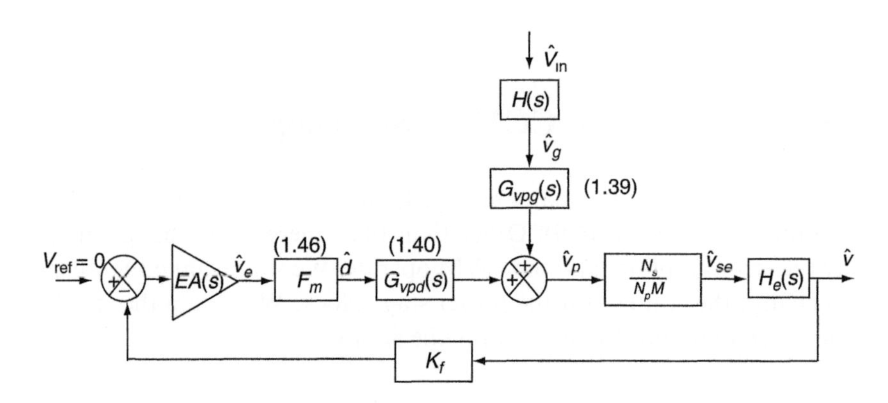

Figure 6.8: AC block diagram for the voltage-mode in CCM; transfer functions numbered have only the form, not the content, identified

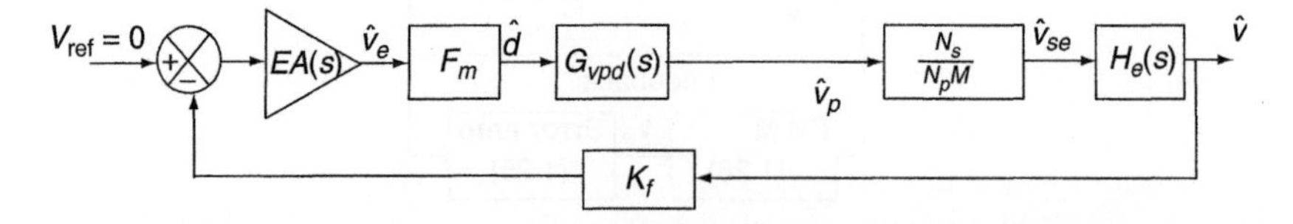

Figure 6.9: AC block diagram for the voltage-mode in CCM, constant input

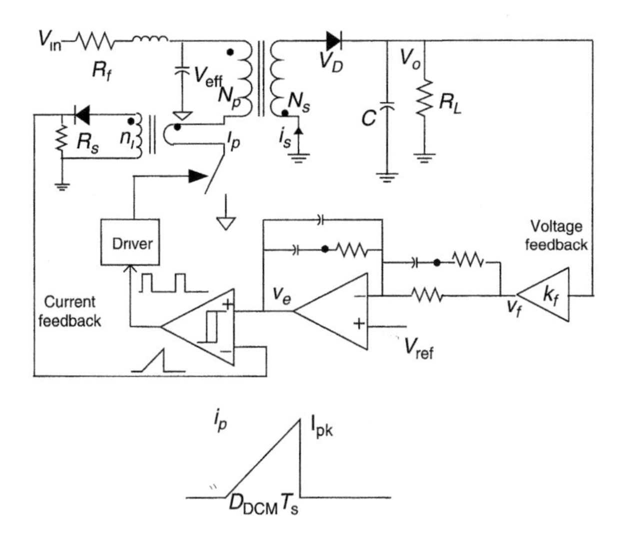

Figure 6.10: Peak current-mode control for a flyback converter in DCM

With the implementation, a minor modification takes place in the PWM action. This is now reflected by the fact that the switch current has a profile:

$$i_p = \frac{V_{\text{in}}}{L_p} t \tag{6.13}$$

The steady-state duty cycle is determined when the current feedback intercepts the error voltage generated by the voltage feedback; that is,

$$\begin{gathered}\frac{R_s}{n_i} \frac{V_{\text{in}}}{L_p} D_{\text{DCM}} \bullet T_s = \frac{R_s \bullet V_{\text{in}} \bullet D_{\text{DCM}}}{n_i \bullet L_p \bullet f_s} = V_e, \\ D_{\text{DCM}} = \frac{n_i \bullet L_p \bullet f_s \bullet V_e}{R_s \bullet V_{\text{in}}}\end{gathered} \tag{6.14}$$

With the new PWM mechanism, Figure 6.4 becomes Figure 6.11.

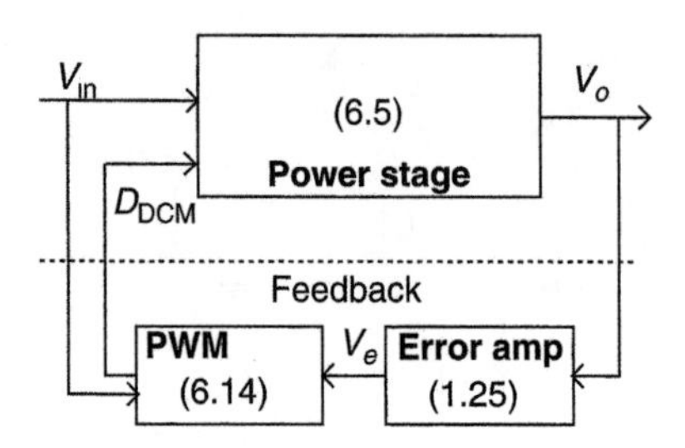

Figure 6.11: Block diagram of peak current-mode in DCM

6.9 Peak Current-Mode DCM Small-Signal Stability

Because of the changes in the PWM mechanism for peak current-mode control, the gain factors also change. From (6.14), we get

$$\begin{aligned} dD_{\text{DCM}} &= \frac{\partial D_{\text{DCM}}}{\partial V_e} dV_e + \frac{\partial D_{\text{DCM}}}{\partial V_{\text{in}}} dV_{\text{in}} \\ &= \frac{n_i \bullet L_p \bullet f_s}{R_s \bullet V_{\text{in}}} dV_e - \frac{n_i \bullet L_p \bullet f_s \bullet V_e}{R_s \bullet V_{\text{in}}^2} dV_{\text{in}} = F_m \bullet dV_e + F_g \bullet dV_{\text{in}} \end{aligned} \tag{6.15}$$

Clearly, the input voltage now plays a direct role in setting the pulse width. We name the effect feedforward. It offers improvements in line regulation. With it, Figure 6.5 is modified and the modification results in Figure 6.12.

The open-loop gain is again easily identified as

$$T(s) = -K_f \bullet E_A(s) \bullet F_m \bullet G_{vd}(s) \tag{6.16}$$

The conducted susceptibility is derived by a slightly complicated procedure:

$$\begin{aligned} \hat{v}_o &= G_{vd}(s) \bullet \hat{d} + G_{vg}(s) \bullet \frac{N_s}{N_p} \bullet H(s) \bullet \hat{v}_{\text{in}} \\ &= G_{vd}(s) \bullet \left[-K_f \bullet E_A(s) \bullet F_m \bullet \hat{v}_o + F_g \bullet H(s) \bullet \hat{v}_{\text{in}} \right] + G_{vg}(s) \bullet \frac{N_s}{N_p} \bullet H(s) \bullet \hat{v}_{\text{in}} \end{aligned} \tag{6.17}$$

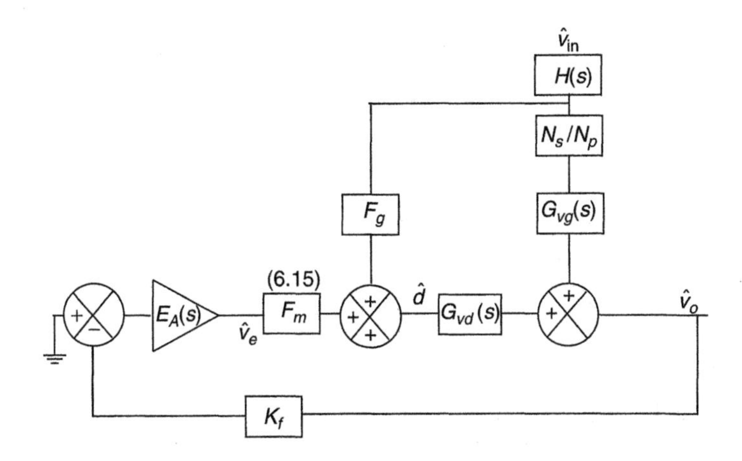

Figure 6.12: AC block diagram for peak current-mode in DCM

Regrouping (6.17), we get conducted susceptibility:

$$C_s(s) = \frac{\hat{v}_o}{\hat{v}_{in}} = \frac{\left[F_g \bullet G_{vd}(s) + G_{vg}(s) \bullet \dfrac{N_s}{N_p}\right] \bullet H(s)}{1 + K_f \bullet E_A(s) \bullet F_m \bullet G_{vd}(s)} \tag{6.18}$$

6.10 Peak Current-Mode CCM Closed Loop

Under a higher load or having a large energy storage inductor, the CCM operation may commence. The switch current in this case changes shape to what was shown in Figure 6.3. The steady-state block diagram is changed accordingly. First, the analytical form for the switch current needs to be modified:

$$i_p = \frac{N_s \bullet i_B}{N_p} + \frac{V_{in}}{L_p} t \tag{6.19}$$

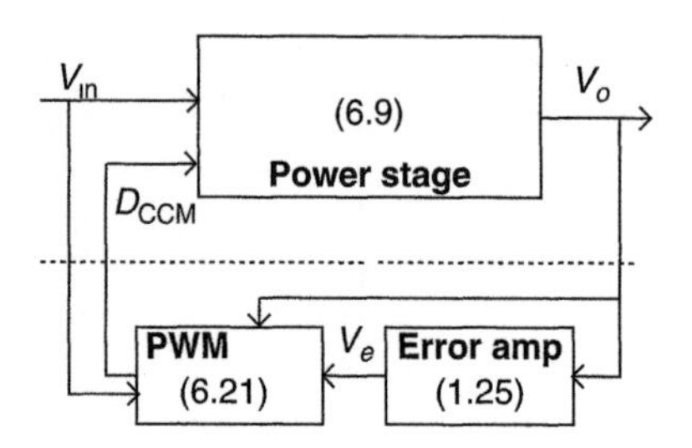

Figure 6.13: Block diagram of peak current-mode in CCM

where i_B is from (6.6). By the same logic mentioned in Section 6.8, the steady-state duty cycle under CCM operation is derived:

$$\frac{N_s \bullet i_A \bullet R_s}{N_p \bullet n_i} = \frac{N_s \bullet R_s}{N_p \bullet n_i}\left[\frac{V_o}{(1-D_{\text{CCM}})R_L} + \frac{(V_o + V_D)(1-D_{CCM})}{2 \bullet L_s \bullet f_s}\right] = V_e \quad (6.20)$$

We can also rewrite (6.20):

$$\frac{N_s \bullet R_s}{N_p \bullet n_i}\left[\frac{V_o \bullet N_p \bullet (V_o + V_D)}{R_L \bullet N_s \bullet V_{\text{in}} \bullet D_{\text{CCM}}} + \frac{N_s \bullet V_{\text{in}} \bullet D_{\text{CCM}}}{2 \bullet L_s \bullet f_s \bullet N_p}\right] = V_e \quad (6.21)$$

However, unlike (6.14), D_{CCM} cannot be neatly expressed as an explicit function of other circuit components and parameters. We do not attempt to solve it. But we still have a block diagram for this subject in the steady state (Figure 6.13).

We notice that one more interesting effect is added: The PWM action is also under the direct influence of the output. Is there some advantage, or disadvantage, in terms of performance? It is not clear.

6.11 Peak Current-Mode CCM Small-Signal Stability

By the same argument stated in section 6.9 and the fact of (6.21), the PWM gain factors for this section are much more sophisticated. Explicit, symbolic derivatives are out of the question. We again resort to the implicit function approach. We first define

$$f(V_e, V_{\text{in}}, V_o, D_{\text{CCM}}, \ldots) = \frac{N_s \cdot R_s}{N_p \cdot n_t} \cdot \left[\frac{V_o \cdot N_p \cdot (V_o + V_D)}{R_L \cdot N_s \cdot V_{\text{in}} \cdot D_{\text{CCM}}} + \frac{N_s \cdot V_{\text{in}} \cdot D_{\text{CCM}}}{2 \cdot L_s \cdot f_s \cdot N_p}\right] - V_e = 0 \quad (6.22)$$

We then derive all three modulation gains by

$$dD_{\text{DCM}} = F_m \cdot dV_e + F_g \cdot dV_{\text{in}} + F_v \cdot dV_o = -\frac{(\partial f/\partial V_e)}{(\partial f/\partial D_{\text{CCM}})} dV_e - \frac{(\partial f/\partial V_{\text{in}})}{(\partial f/\partial D_{\text{CCM}})} dV_{\text{in}} - \frac{(\partial f/\partial V_o)}{(\partial f/\partial D_{\text{CCM}})} dV_o \quad (6.23)$$

Subsequently, the overall block diagram (Figure 6.14) is given.

The loop gain under the condition of constant input is

$$T(s) = -K_f \cdot E_A(s) \cdot F_m \cdot \frac{G_{vpd}(s) \cdot \dfrac{N_s}{N_p \cdot M'} \cdot H_e(s)}{1 + F_v \cdot G_{vpd}(s) \cdot \dfrac{N_s}{N_p \cdot M'} \cdot H_e(s)} \quad (6.24)$$

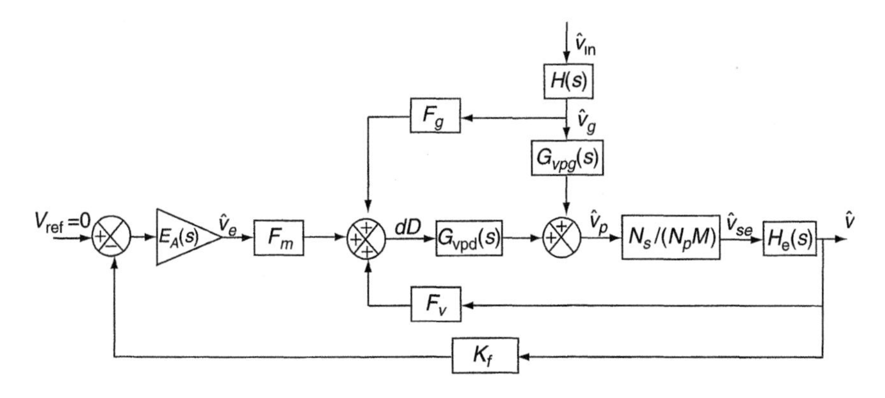

Figure 6.14: AC block diagram for peak current-mode in CCM

where $E_A(s)$ is not signed. And the conducted susceptibility is

$$C_s(s) = \frac{H(s) \bullet [G_{vpd}(s) \bullet F_g + G_{vpg}(s)] \dfrac{N_s}{N_p \bullet M'} \bullet H_e(s)}{1 + [K_f \bullet E_A(s) \bullet F_m - F_v] \bullet G_{vpd}(s) \bullet \dfrac{N_s}{N_p \bullet M'} \bullet H_e(s)} \tag{6.25}$$

6.12 Output Capacitor

Referring to Figure 6.1 or Figure 6.10, we here find the proper size for the output capacitor such that it meets the output ripple voltage requirement, δv. We first find the one for DCM, then that for CCM.

Under the DCM, and if we ignore the input filter losses R_f, the secondary current is given by (referring to Figure 6.2):

$$i_s(t) = I_{sp} - \frac{V_o + V_D}{L_s} t = \frac{N_p}{N_s} \sqrt{\frac{2 \bullet V_o \bullet (V_o + V_D)}{L_p \bullet f_s \bullet R_L}} - \frac{V_o + V_D}{L_s} t \tag{6.26}$$

Since the capacitor does not pass DC current, the AC current is then given as

$$i_c(t) = i_s(t) - \frac{V_o}{R_L} = \frac{N_p}{N_s} \sqrt{\frac{2 \bullet V_o \bullet (V_o + V_D)}{L_p \bullet f_s \bullet R_L}} - \frac{V_o + V_D}{L_s} t - \frac{V_o}{R_L} \tag{6.27}$$

We are used to the differential equation $C \bullet dv/dt = i_c$ for capacitor current. We also understand from calculus that, if the behavior of a variable x is governed by a simple differential equation, $a \bullet dx/dt = b$, the variable's extreme (maximum or minimum) value occurs at $b = 0$. By the same token, we say the output capacitor voltage peaks at zero current crossing. We next find the current zero-crossing time, t_0:

$$i_c(t_0) = \frac{N_p}{N_s} \sqrt{\frac{2 \bullet V_o \bullet (V_o + V_D)}{L_p \bullet f_s \bullet R_L}} - \frac{V_o + V_D}{L_s} t_0 - \frac{V_o}{R_L} = 0$$

$$t_0 = \frac{\dfrac{N_p}{N_s} \sqrt{\dfrac{2 \bullet V_o \bullet (V_o + V_D)}{L_p \bullet f_s \bullet R_L}} - \dfrac{V_o}{R_L}}{\dfrac{V_o + V_D}{L_s}} \tag{6.28}$$

The capacitor accumulates a net charge

$$\delta Q = \frac{\left[\frac{N_p}{N_s}\sqrt{\frac{2 \cdot V_o \cdot (V_o + V_D)}{L_p \cdot f_s \cdot R_L}} - \frac{V_o}{R_L}\right] t_0}{2}$$

$$= \frac{1}{2} \frac{\left[\frac{N_p}{N_s}\sqrt{\frac{2 \cdot V_o \cdot (V_o + V_D)}{L_p \cdot f_s \cdot R_L}} - \frac{V_o}{R_L}\right]^2}{\frac{V_o + V_D}{L_s}} = C \cdot \delta v \qquad (6.29)$$

The DCM output capacitor size is

$$C_{\mathrm{DCM}} = \frac{\left[\frac{N_p}{N_s}\sqrt{\frac{2 \cdot V_o \cdot (V_o + V_D)}{L_p \cdot f_s \cdot R_L}} - \frac{V_o}{R_L}\right]^2 \cdot \frac{L_s}{V_o + V_D}}{2 \cdot \delta v} \qquad (6.30)$$

For the case of CCM, we refer to (6.6) and Figure 6.3. Based on the concept of charge balance, the net charge the output capacitor acquires during the switch turn-off time equals what it releases during the turn-on time; that is,

$$\delta Q = \frac{V_o}{R_L} \cdot \frac{D_{\mathrm{CCM}}}{f_s} = \frac{V_o \cdot \frac{N_p}{N_s}(V_o + V_D)}{R_L \cdot f_s \cdot \left[V_{\mathrm{in}} + \frac{N_p}{N_s}(V_o + V_D)\right]} = C \cdot \delta v$$

$$C_{\mathrm{CCM}} = \frac{V_o \cdot \frac{N_p}{N_s}(V_o + V_D)}{R_L \cdot f_s \cdot \left[V_{\mathrm{in}} + \frac{N_p}{N_s}(V_o + V_D)\right] \delta v} \qquad (6.31)$$

6.13 Accelerated Steady-State Output

On many occasions, it is desirable to get some idea about the output ripple profile in the time domain instead of the ripple magnitude alone. We again invoke the procedure of accelerated steady-state analysis and treat

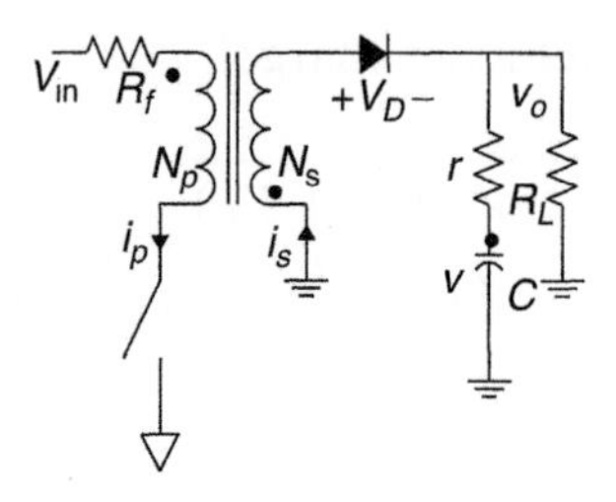

Figure 6.15: Output circuit

the output circuit including capacitor equivalent series resistance (Figure 6.15).

The secondary current considered as a current source feeds the output load, RC, in parallel. The current source can be written analytically as (6.25)

$$i_s(t) = \begin{cases} \frac{N_p}{N_s}\sqrt{\frac{2 \cdot V_o \cdot (V_o + V_D)}{L_p \cdot f_s \cdot R_L}} - \frac{V_o + V_D}{L_s} t = a - b \cdot t & \text{during } D_2 T_s \\ 0 \quad \text{other times} \end{cases} \tag{6.32}$$

At the capacitor node and the output node, one main equation and an auxiliary equation can be written:

$$C\frac{dv}{dt} = \frac{v_O - v}{r}$$

$$v_O = \frac{r \cdot R}{r + R} \cdot i_s + \frac{R}{r + R} \cdot v = R_p \cdot i_s + k \cdot v$$

Consolidating both, we get

$$\frac{dv}{dt} + \frac{1-k}{r \cdot C} v = \frac{R_p}{r \cdot C} i_s \tag{6.33}$$

Taking a Laplace transform with an unknown starting state, V_0, we get the capacitor voltage in the transformed domain as

$$V(s) = \frac{V_0}{\left(s + \frac{1-k}{r \cdot C}\right)} + \frac{R_p}{r \cdot C} \frac{I_s(s)}{\left(s + \frac{1-k}{r \cdot C}\right)} \tag{6.34}$$

where

$$I_s(s) = \begin{cases} \frac{a}{s} - \frac{b}{s^2} & 0 < t < D_2 \bullet T_s \\ 0 & D_2 \bullet T_s < t < T_s \end{cases} \tag{6.35}$$

As a result, we have, for the two time segments,

$$\begin{aligned} V_2(s) &= \frac{V_{02}}{\left(s + \frac{1-k}{r \bullet C}\right)} + \frac{R_p}{r \bullet C} \frac{a}{s\left(s + \frac{1-k}{r \bullet C}\right)} - \frac{R_p}{r \bullet C} \frac{b}{s^2\left(s + \frac{1-k}{r \bullet C}\right)} \\ V_1(s) &= \frac{V_{01}}{\left(s + \frac{1-k}{r \bullet C}\right)} \mathrm{e}^{-D_2 \bullet T_s \bullet s} \end{aligned} \tag{6.36}$$

Inverse Laplace transformation gives us the ideal capacitor voltage:

$$\begin{aligned} v_2(t) &= \left[V_{02} \bullet f_1(t) + \frac{R_p \bullet a}{r \bullet C} \bullet f_2(t) - \frac{R_p \bullet b}{r \bullet C} \bullet f_3(t)\right][u(t) - u(t - D_2 T_s)] \\ v_1(t) &= [V_{01} \bullet f_1(t - D_2 \bullet T_s)][u(t - D_2 T_s) - u(t - T_s)] \end{aligned} \tag{6.37}$$

Continuity of states at the two time boundaries, $D_2 T_s$ and T_s, yields

$$\begin{aligned} V_{01} - V_{02} \bullet f_1(D_2 \bullet T_s) &= \frac{R_p \bullet a}{r \bullet C} \bullet f_2(D_2 \bullet T_s) - \frac{R_p \bullet b}{r \bullet C} \bullet f_3(D_2 \bullet T_s) \\ V_{01} \bullet f_1(T_s - D_2 \bullet T_s) - V_{02} &= 0 \end{aligned} \tag{6.38}$$

Equation (6.38) enables us to solve the two unknown, cyclic starting states, V_{01} and V_{02}. With this knowledge, the main output is given:

$$v_o(t) = \begin{cases} R_p\left[\frac{N_p}{N_s} \bullet \frac{V_{\text{in}} \bullet D_{\text{DCM}}}{L_p \bullet f_s} - \frac{V_o + V_D}{L_s} t\right][u(t) - u(t - D_2 T_s)] + k \bullet v_2(t) \\ k \bullet v_1(t) \end{cases} \tag{6.39}$$

Both equations (6.37) and (6.39) and the estimate of equivalent series resistance, r, also give us the ripple current, and consequently power dissipation, for the capacitor:

$$i_c(t) = \begin{cases} \frac{1}{r}\left\{R_p\left[\frac{N_p}{N_s} \cdot \frac{V_{\text{in}} \cdot D_{\text{DCM}}}{L_p \cdot f_s} - \frac{V_o + V_D}{L_s} t\right] \right. \\ \left. [u(t) - u(t - D_2 T_s)] + (k-1) \cdot v_2(t)\right\} \\ \frac{1}{r}(k-1) \cdot v_1(t) \end{cases} \tag{6.40}$$

6.14 A Complete DCM Example

In this section, we give a complete example based on Figure 6.16. First, the closed-loop steady state is developed, then the AC small-signal loop gain.

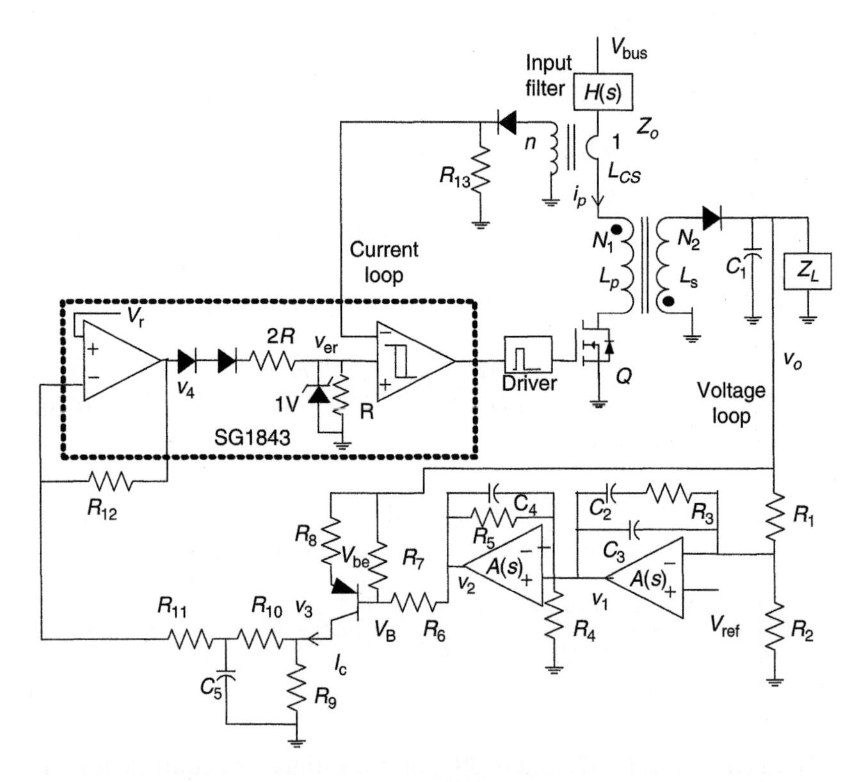

Figure 6.16: Flyback converter in DCM with current-mode control

A. Steady-State Closed-Form Output Equation

The operation of the converter can be briefly described as follows. At the initiation of an internal clock, f_s, residing in the PWM integrated circuit, SG1843 (silicon general semiconductor), the power switch, Q, is turned on. This action generates a ramping up current, i_p, at the primary winding with a slope determined by the associated primary inductance, L_p, and the effective input voltage, V_b. Meanwhile, the output voltage, v_o, is processed by the feedback loop and eventually transformed into a control (error) voltage, v_{er}. Both the control voltage and the scaled version of the ramp current are fed to a hysteretic comparator. At the instant the sensed current in voltage form intercepts the control voltage, the power switch ceases conduction and remains off until the next clock-on cycle.

At the steady state, that is, under a constant load and constant line input, the intermediate control voltage v_2 is expressed as

$$v_2 = M_1 - M_2 \cdot v_o \tag{6.41}$$

where

$$M_1 = \left(1 + \frac{R_5}{R_4}\right) A \cdot V_{\text{ref}}$$

$$M_2 = \left(1 + \frac{R_5}{R_4}\right) A \cdot \frac{R_2}{R_1 + R_2}$$

Subsequent stages of transconductance amplifier and operational amplifier develop the intermediate control voltage further and generate

$$\begin{aligned} I_c &= M_3 \cdot v_2 + M_4 \cdot v_o + M_5 \\ v_3 &= M_6 + M_7 \cdot I_c \\ v_{er} &= M_8 + M_9 \cdot v_3 \end{aligned} \tag{6.42}$$

where

$$M_3 = h_{\mathrm{FE}} \frac{R_7}{-[R_7 \cdot R_6 + (1 + h_{\mathrm{FE}})R_8(R_7 + R_6)]}$$

$$M_4 = -M_3$$

$$M_5 = h_{\mathrm{FE}} \frac{(R_7 + R_6)V_{be}}{-[R_7 \cdot R_6 + (1 + h_{\mathrm{FE}})R_8(R_7 + R_6)]}$$

$$M_6 = \frac{R_9}{R_9 + R_{10} + R_{11}} V_r$$

$$M_7 = \frac{R_9(R_{11} + R_{10})}{R_9 + R_{10} + R_{11}}$$

$$M_8 = \frac{1}{3}\left[\left(1 + \frac{R_{12}}{R_{11} + R_{10}}\right) V_r - 2 \cdot V_D\right]$$

$$M_9 = \frac{1}{3} \cdot \frac{R_{12}}{R_{11} + R_{10}}$$

The effective error voltage, v_{er}, in turn, generates a steady-state duty cycle:

$$D = \frac{n \cdot L_p \cdot f_s}{V_b \cdot R_{13}} \cdot v_{\mathrm{er}} = M_{10} \cdot v_{\mathrm{er}} \tag{6.43}$$

and the output

$$v_o = D \cdot \frac{N_s \cdot V_{\mathrm{bus}}}{N_p} \sqrt{\frac{R_L}{2 \cdot L_s \cdot f_s}} = D \cdot M_{11} \tag{6.44}$$

where the power stage gain, M_{11}, is given in section 6.1 in approximation. The preceding set of equations can then be back substituted starting from (6.44) using all M, and the process yields the closed-form, closed-loop output:

$$v_o = \frac{M_{10} \cdot M_{11} \cdot \{M_8 + M_9[M_6 + M_7(M_3 \cdot M_1 + M_5)]\}}{1 + M_{10} \cdot M_{11} \cdot M_9 \cdot M_7(M_3 \cdot M_2 - M_4)} \tag{6.45}$$

B. Output Sensitivity

The closed-form solution gives designers the ability to evaluate many performance merits of a design analytically and numerically. Among them, load regulation, line regulation, and component sensitivities are the three most sought after figures. For instance, the load regulation sensitivity can be expressed as

$$S_R = \frac{\partial v_o}{\partial M_{11}} \frac{\partial M_{11}}{\partial R_L} \tag{6.46}$$

C. AC Loop Gain

To study the small-signal behavior of the converter, the schematic of Figure 6.16 is transformed into Figure 6.17, in which the transfer functions of individual blocks are identified and derived as follows.

Figure 6.18(b) gives the corresponding small-signal equivalent circuit of the first stage of error amplifier. Based on the equivalent circuit and considering the nonideal gain, $A(s)$, of the operational amplifier, the input-to-output transfer function is obtained:

$$EA_1(s) = \frac{v_1}{v_o} = -\frac{R_2}{R_1 + R_2} \bullet \frac{A(s)}{R_p\left[\dfrac{1}{R_p} + \dfrac{1 + A(s)}{Z_{f1}(s)}\right]} \tag{6.47}$$

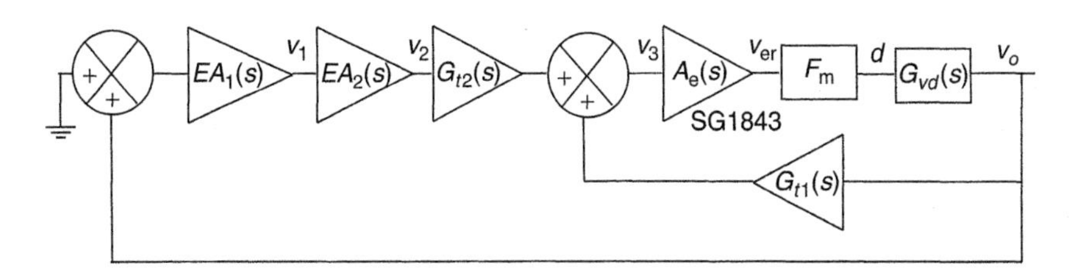

Figure 6.17: Small-signal block diagram (constant input form)

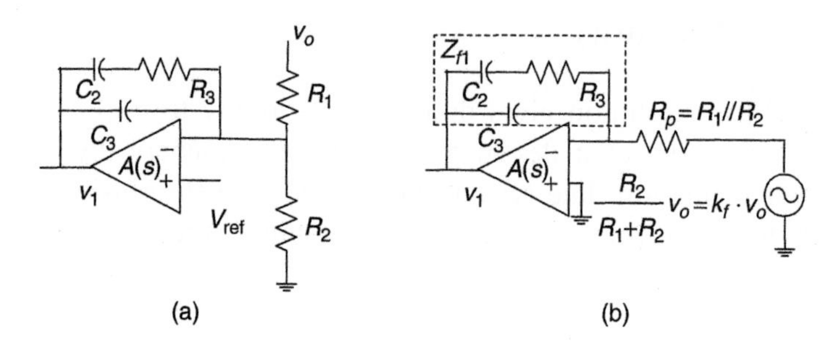

Figure 6.18: Error amplifier: (a) DC, (b) AC

AQ1

where $R_p = R_1/R_2$ and

$$A(s) = \frac{A}{\left(\frac{s}{2 \cdot \pi \cdot f_p} + 1\right)}$$

represents the open-loop gain of the operational amplifier with single-pole gain-roll-off at frequency f_p and DC gain A, which is also used in M_1 and M_2.

By exactly the same procedure, the second-stage error amplifier gives

$$EA_2(s) = \frac{v_2}{v_1} = A(s)\left\{1 - \frac{A(s)}{Z_{f2}(s) \cdot \left[\frac{1 + A(s)}{Z_{f2}(s)} + \frac{1}{R_4}\right]}\right\} \tag{6.48}$$

For the transistor voltage-to-current converter, more steps are involved. Figure 6.19 gives both the DC and AC circuits of the circuit. Three nodal equations can be written for Figure 6.19(b):

$$\begin{gathered}\left(\frac{1}{R_6} + \frac{1}{R_7} + \frac{1}{h_i}\right) v_B - \frac{1}{h_i} v_x = \frac{v_o}{R_7} + \frac{v_2}{R_6} \\ \frac{1 + h_f}{h_i} v_B - \left(\frac{1}{R_8} + \frac{1 + h_f}{h_i}\right) v_x = -\frac{v_o}{R_8} \\ \frac{h_f}{h_i} v_B - \frac{h_f}{h_i} v_x + \frac{1}{Z_c(s)} v_c = 0\end{gathered} \tag{6.49}$$

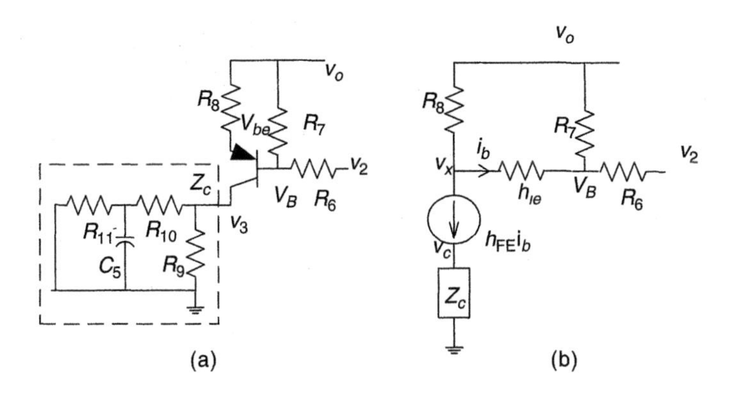

Figure 6.19: Transistor buffer: (a) DC, (b) AC

The equation set yields, at note v_c,

$$v_c = \frac{\begin{vmatrix} \frac{1}{R_6}+\frac{1}{R_7}+\frac{1}{h_\imath} & -\frac{1}{h_i} & \frac{v_o}{R_7}+\frac{v_2}{R_6} \\ \frac{1+h_f}{h_i} & -\left(\frac{1}{R_8}+\frac{1+h_f}{h_i}\right) & -\frac{v_o}{R_8} \\ \frac{h_f}{h_i} & -\frac{h_f}{h_i} & 0 \end{vmatrix}}{D_e} \tag{6.50}$$

and the transistor stage gains

$$G_{t1}(s) = \frac{\partial v_c}{\partial v_o}$$

$$= \frac{\begin{vmatrix} \frac{1+h_f}{h_i} & -\left(\frac{1}{R_8}+\frac{1+h_t}{h_t}\right) \\ \frac{h_f}{h_i} & -\frac{h_f}{h_i} \end{vmatrix}\frac{1}{R_7} + \begin{vmatrix} \frac{1}{R_6}+\frac{1}{R_7}+\frac{1}{h_1} & \frac{-11}{h_i} \\ \frac{h_f}{h_i} & -\frac{h_f}{h_i} \end{vmatrix}\frac{1}{R_8}}{D_e} \tag{6.51}$$

$$G_{t2}(s) = \frac{\partial v_c}{\partial v_2} = \frac{\begin{vmatrix} \frac{1+h_f}{h_\imath} & -\left(\frac{1}{R_8}+\frac{1+h_f}{h_i}\right) \\ \frac{h_f}{h_i} & -\frac{h_f}{h_i} \end{vmatrix}\frac{1}{R_6}}{D_e}$$

where

$$D_e = \begin{vmatrix} \frac{1}{R_6}+\frac{1}{R_7}+\frac{1}{h_i} & \frac{-1}{h_i} & 0 \\ \frac{1+h_f}{h_i} & -\left(\frac{1}{R_8}+\frac{1+h_f}{h_i}\right) & 0 \\ \frac{h_f}{h_i} & \frac{-h_f}{h_i} & \frac{1}{Z_c(s)} \end{vmatrix}$$

Following the voltage-to-current converter, the internal error amplifier of the PWM IC yields one more transfer function, based on Figure 6.20; that is,

$$A_e(s) = \frac{v_{\text{er}}}{v_3} = \frac{v_{\text{er}}}{v_{\text{eq}}}\frac{v_{\text{eq}}}{v_3}$$
$$= -\frac{1}{3} \bullet \frac{R_{12}}{R_{11} + \left(\frac{1}{R_{10}} + C_5 \bullet s\right)^{-1}} \bullet \frac{(C_5 \bullet s)^{-1}}{R_{10} + (C_5 \bullet s)^{-1}} \tag{6.52}$$

With both the effective error voltage and the current feedback available at the input terminals of the comparator, the steady-state duty cycle is established, given by (6.43). The AC, small-signal perturbation of (6.43) produces

$$\delta D = F_m \bullet \delta v_{\text{er}} + F_g \bullet \delta V_{\text{bus}} \tag{6.53}$$

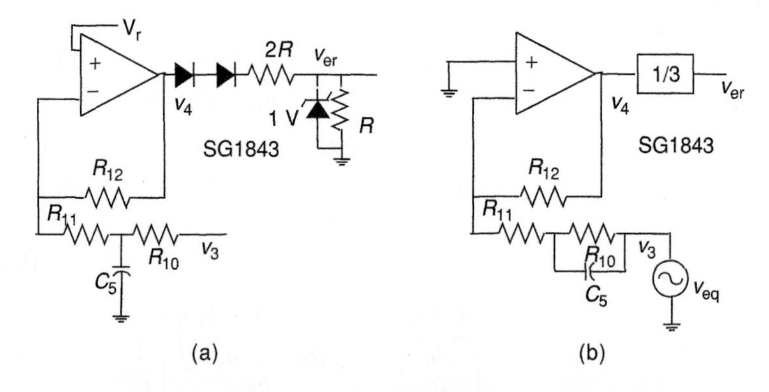

Figure 6.20: PWM internal error amplifier: (a) DC, (b) AC

where

$$F_m = \frac{\partial D}{\partial v_{\text{er}}} = \frac{L_p \bullet f_s \bullet n}{R_s \bullet V_{\text{bus}}}, \quad F_g = \frac{\partial D}{\partial V_{\text{bus}}} = \frac{-L_p \bullet f_s \bullet n \bullet V_{\text{er}}}{R_s \bullet V_{\text{bus}}^2}$$

The last, but not the least important, block that needs to be developed is the power stage. The development of power stage transfer function is accomplished by invoking the canonical model and modifying it such that the magnetic isolation and the line input filter are included and reflected to the secondary side as given in (2.22) section 2.2. The power stage duty cycle-to-output transfer function can be shown to be (Z_o, the output impedance of input filter, $g_1 = 0$)

$$G_{vd}(s) = \left[1 - g_2 \frac{j_1}{j_2} \frac{r_1 \bullet Z_s(s)}{r_1 + Z_s(s)}\right] \frac{j_2}{\frac{1}{Z_L(s)} + \frac{1}{r_2} + C_1 \bullet s} \tag{6.54}$$

where

$$Z_s(s) = \left(\frac{N_2}{N_1}\right)^2 Z_o(s)$$

Once all blocks have been treated for AC analysis, the base diagram of Figure 6.17 can be further simplified to result in Figure 6.21. This step is taken because a loop-gain study must be conducted under a steady state, that is, with a constant input and constant load. It is also pointed out that, due to the transistor voltage-to-current converter, multiple loops exist. However, the outer loop is considered the main loop, since this is where the command reference is applied. With this selection of

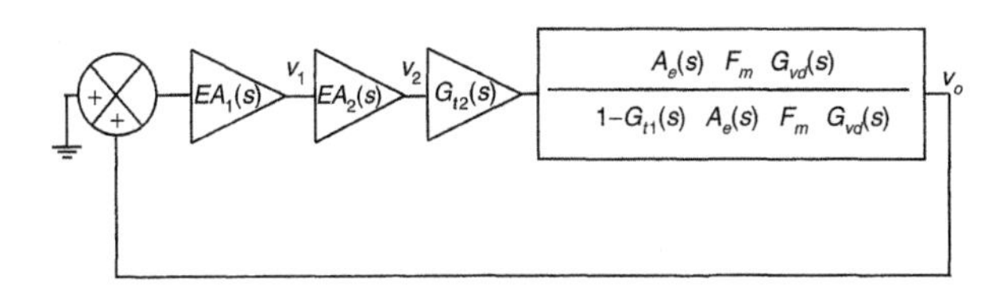

Figure 6.21: Block diagram for loop gain

loop definition and by absorbing the inner loop as shown in Figure 6.21, the loop gain is finally given as

$$T(s) = \frac{EA_1(s) \bullet EA_2 \bullet G_{t2}(s) \bullet A_e(s) \bullet F_m \bullet G_{vd}(s)}{1 - G_{t1}(s) \bullet A_e(s) \bullet F_m \bullet G_{vd}(s)} \tag{6.55}$$

To verify the validity of the loop-gain formulation process, the end result (6.55) is computed for an actual design. The theoretical results, shown in Figure 6.22, compare extremely well against the actual measurement of Figure 6.23. The comparison hence gives credence to (6.55).

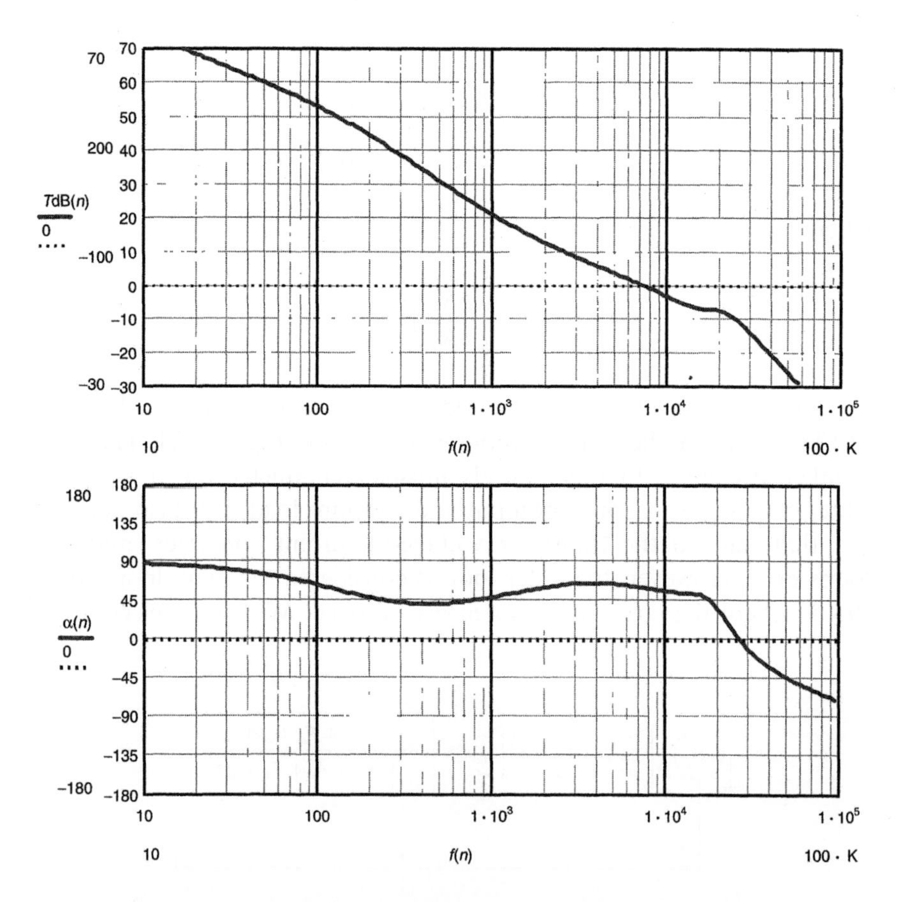

Figure 6.22: Theoretical loop gain: Gain 10dB/div; Phase 45°/div

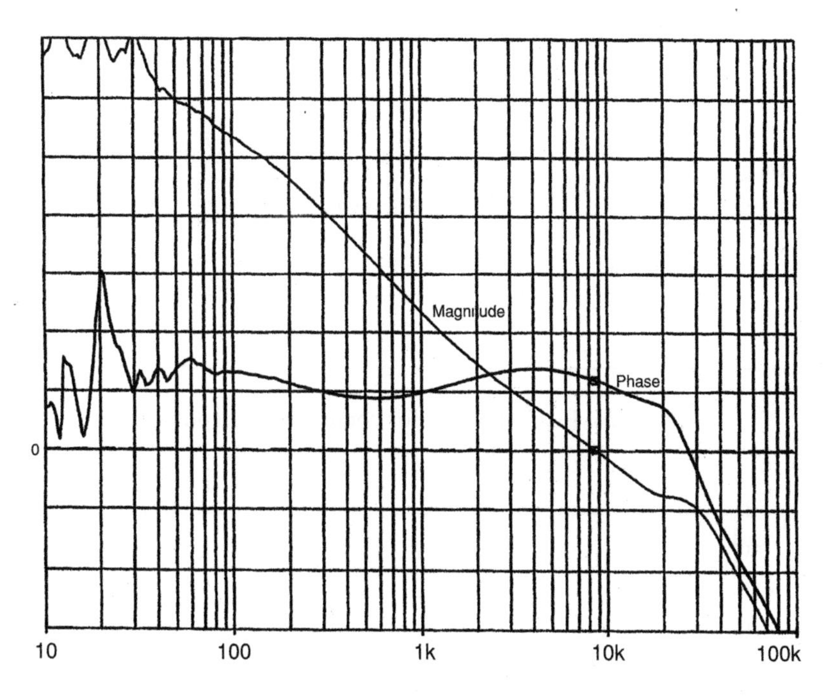

Figure 6.23: Actual measurement: Magnitude 10dB/div; Phase 45°/div

D. Conducted Susceptibility

Conducted susceptibility is a measure of how well the converter rejects input perturbation δV_{bus}. This performance parameter is made particularly necessary for specifying supplies powering RF equipment, since poor supply rejection exhibited by the equipment can cause contamination of RF transmissions. It is therefore very desirable to estimate in advance the parameter figure for a converter. This task can be performed by reconfiguring and modifying Figure 6.17, which results in Figure 6.24. Here, caution should be exercised in formulating the control loop-gain, because two feedback paths coexist. Moreover, a slight complication also exists in the input chain; that is, the presence of feedforward effect, F_g, which plays a part in perturbing the duty cycle, too. However, as shown in (6.15), the feedforward coefficient is inversely proportional to the square of the input, V_{bus}. Its magnitude is, in effect, much smaller than

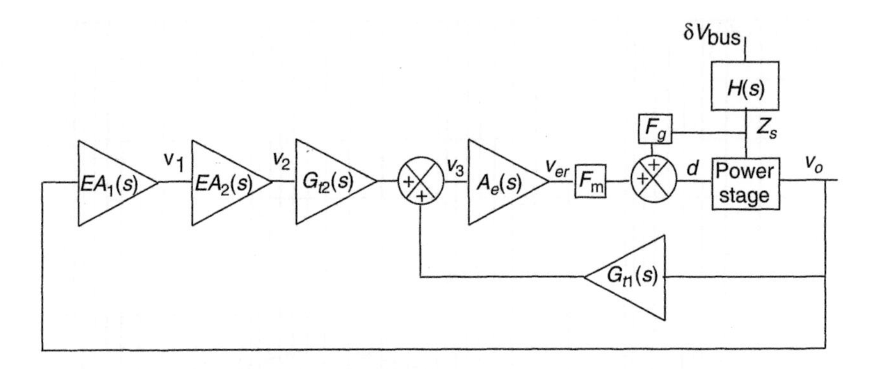

Figure 6.24: Block diagram for conducted susceptibility

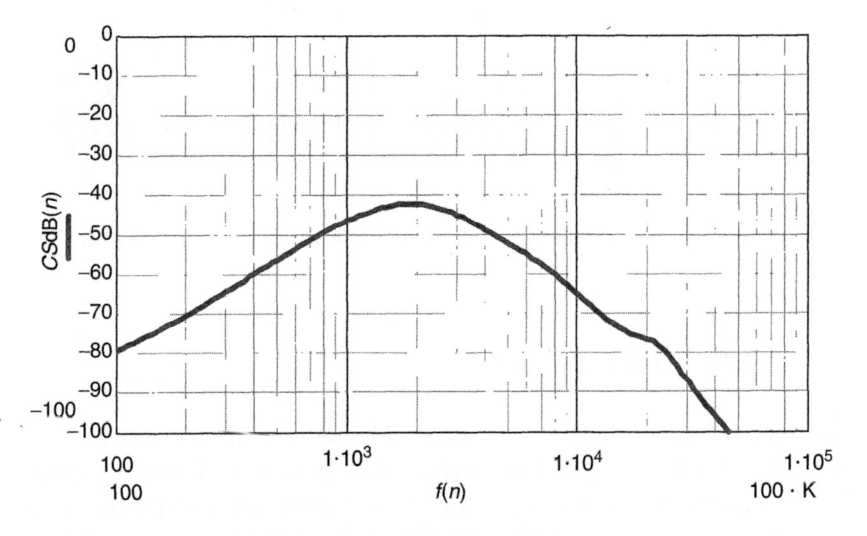

Figure 6.25: Conducted susceptibility, theoretical: 10dB/div

the feedback coefficient, F_m, which is inversely proportional to V_{bus}. The feedforward effect can be included and lead to a reformulated loop-gain equation:

$$T_{CS}(s) = [EA_1(s) \cdot EA_2(s) \cdot G_{t2}(s) + G_{t1}(s)] \cdot A_e(s) \cdot F_m \cdot G_{vd}(s) \qquad (6.56)$$

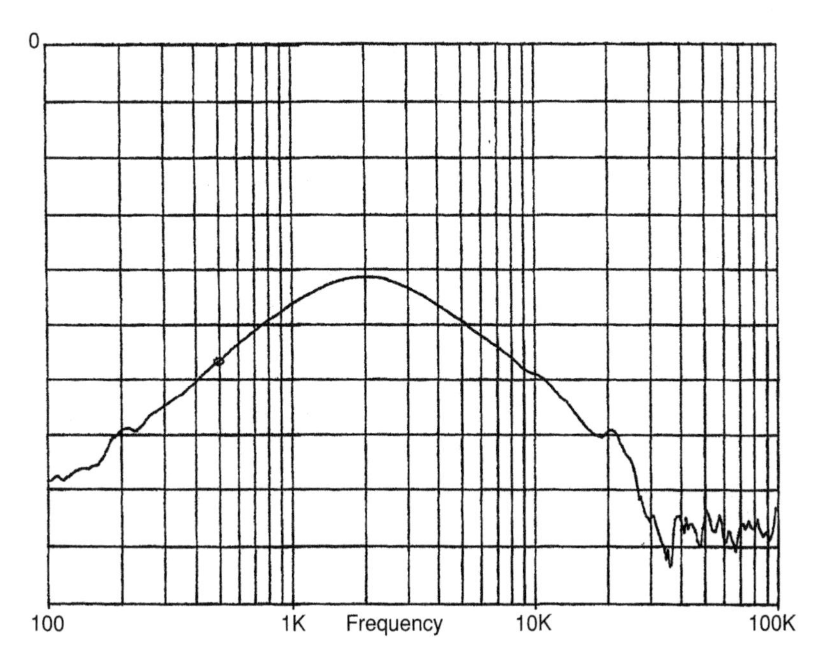

Figure 6.26: Conducted susceptibility, measurement; −10dB/div

The resultant conducted susceptibility is then expressed as

$$CS(s) = \frac{\dfrac{N_2}{N_1} H(s) \left[\dfrac{r_1}{Z_s(s) + r_1} \dfrac{g_2}{\dfrac{1}{Z_L(s)} + \dfrac{1}{r_2} + C_1 \bullet s} + F_g \bullet G_{vd}(s) \right]}{1 - T_{CS}(s)} \tag{6.57}$$

Again, the analytical computation of (6.57) yields Figure 6.25, which also matches the actual measurement, Figure 6.26, extremely well.

Chapter 7

Nonisolated Boost Converter

We shall depart from the presentation flow that has been the hallmark for the past six. Since current-mode control enjoys obvious advantages over voltage-mode control, we can omit the latter without disrupting the key aspects of the chapter. Because of that, this will be a short chapter. Nevertheless, we cover duty-cycle determination, critical inductance, and the steady state under closed loop, loop gain, and the like. All analyses focus on a single schematic, Figure 7.1, which is derived from Figure 6.10 by replacing the power stage with the nonisolated boost stage.

7.1 Duty-Cycle Determination

Figure 7.1 also shows major current waveforms for DCM and CCM operations. Given that, the procedure for determining the duty cycle for both cases is basically the same as that for the flyback converters. The case for CCM was covered in Wu [2]. We consider only DCM in this section.

We start with the requirement of volt–second, flux balance across the boost inductor L:

$$V_{\text{eff}} \cdot D_{\text{DCM}} + [V_{\text{eff}} - (V_D + V_o)] \cdot D_2 = 0 \tag{7.1}$$

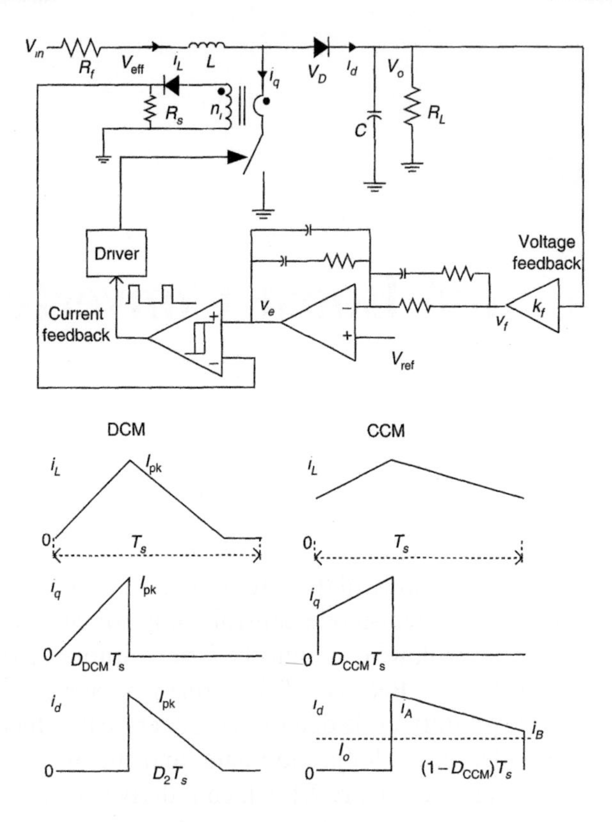

Figure 7.1: Nonisolated boost converter

The switch peak current is

$$I_{pk} = \frac{V_{eff}}{L \cdot f_s} \cdot D_{DCM} \tag{7.2}$$

The input line has a DC current

$$I_{in} = \frac{1}{2}(D_{DCM} + D_2)I_{pk} = \frac{1}{2}(D_{DCM} + D_2)\frac{V_{eff}}{L \cdot f_s} \cdot D_{DCM} \tag{7.3}$$

Considering the input losses R_f, the effective input is

$$V_{\text{eff}} = V_{\text{in}} - I_{\text{in}} R_f = V_{\text{in}} - \frac{1}{2}\left(D_{\text{DCM}} + \frac{V_{\text{eff}} \bullet D_{\text{DCM}}}{V_o + V_D - V_{\text{eff}}}\right) \frac{V_{\text{eff}}}{L \bullet f_s} \bullet D_{\text{DCM}} \bullet R_f$$

$$V_{\text{eff}} = V_{\text{in}} - \frac{1}{2}\left(1 + \frac{V_{\text{eff}}}{V_o + V_D - V_{\text{eff}}}\right) \frac{V_{\text{eff}}}{L \bullet f_s} \bullet D_{\text{DCM}}^2 \bullet R_f \tag{7.4}$$

The DC load current is given as

$$\frac{V_o}{R_L} = \frac{1}{2} D_2 \bullet I_{\text{pk}} = \frac{V_{\text{eff}}^2 \bullet D_{\text{DCM}}^2}{2 \bullet L \bullet f_s (V_o + V_D - V_{\text{eff}})} \tag{7.5}$$

The open-loop DCM duty cycle is embedded in (7.4) and (7.5).

7.2 Critical Inductance

From the CCM current waveform, we again understand that the critical inductance can be decided by the boundary condition $i_B = 0$:

$$\begin{aligned} i_B &= \frac{I_o}{1 - D_{\text{DCM}}} - \frac{(V_o + V_D - V_{\text{in}}) \bullet (1 - D_{\text{CCM}})}{2 \bullet L_{\text{cri}} \bullet f_s} = 0 \\ L_{\text{cri}} &= \frac{(V_o + V_D - V_{\text{in}}) V_{\text{in}}^2}{2 \bullet I_o \bullet f_s \bullet (V_o + V_D)^2} \end{aligned} \tag{7.6}$$

7.3 Peak Current-Mode Closed-Loop Steady State in CCM

Based on the current waveform given in Figure 7.1, the main power switch terminates conduction when the current feedback signal in voltage form crosses the control error voltage generated by the voltage feedback loop; that is,

$$V_e = \frac{R_s}{n_i} i_A = \frac{R_s}{n_i}\left[\frac{V_o}{(1 - D_{\text{CCM}}) R_L} + \frac{V_{\text{in}} \bullet D_{\text{CCM}}}{2 \bullet L \bullet f_s}\right] \tag{7.7}$$

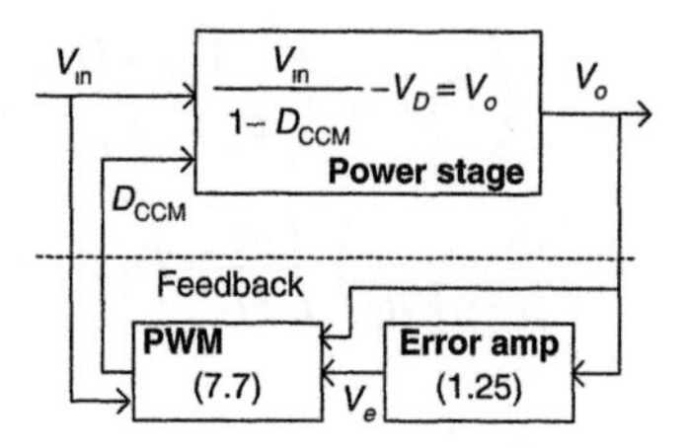

Figure 7.2: Nonisolated boost converter in the steady state, CCM

With (7.7) available, we are ready to close the loop. This leads to Figure 7.2.

If so desired, readers can replace the approximation for the power stage with the more accurate one that accounts for losses and was given in Wu [2].

7.4 Peak Current-Mode Small-Signal Stability in CCM

Following the same steps for other converters, we derive the small-signal gains for the PWM block by defining an implicit function:

$$f(V_e, V_o, V_{\text{in}}, D_{\text{CCM}}) = V_e - \frac{R_s}{n_i}\left[\frac{V_o}{(1 - D_{\text{CCM}})R_L} + \frac{V_{\text{in}} \cdot D_{\text{CCM}}}{2 \cdot L \cdot f_s}\right] = 0 \tag{7.8}$$

The small-signal gains are then

$$\begin{aligned} dD_{\text{CCM}} &= F_m \cdot \delta V_e + F_v \cdot \delta V_o + F_g \cdot \delta V_{\text{in}} \\ &= -\frac{\partial f/\partial V_e}{\partial f/\partial D_{\text{CCM}}}\delta V_e - \frac{\partial f/\partial V_o}{\partial f/\partial D_{\text{CCM}}}\delta V_o - \frac{\partial f/\partial V_{\text{in}}}{\partial f/\partial D_{\text{CCM}}}\delta V_{\text{in}} \end{aligned} \tag{7.9}$$

This is shown in Figure 7.3.

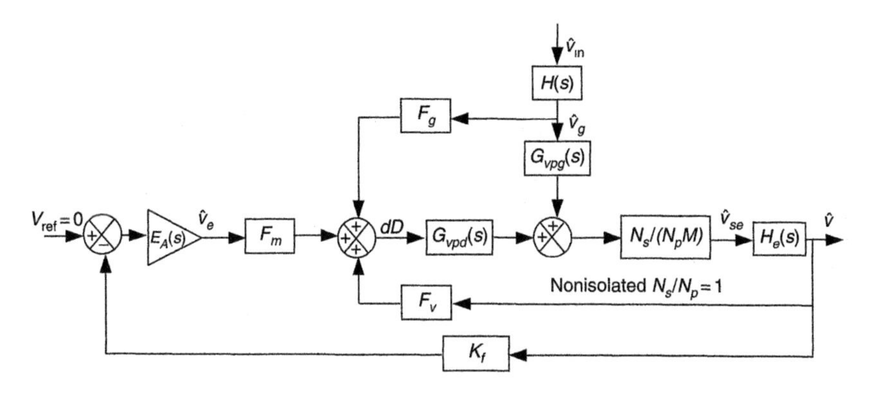

Figure 7.3: AC block diagram for current-mode boost converter in CCM

The loop gain is given by

$$T(s) = -K_f \bullet E_A(s) \bullet F_m \bullet \frac{G_{vpd}(s) \bullet \dfrac{1}{M(D)} \bullet H_e(s)}{1 - F_v \bullet G_{vpd}(s) \bullet \dfrac{1}{M(D)} \bullet H_e(s)} \tag{7.10}$$

7.5 Peak Current-Mode Closed-Loop Steady State in DCM

The main power switch ceases conduction when the current feedback signal in voltage form crosses the control error voltage generated by the voltage feedback loop:

$$V_e = \frac{R_s}{n_i} \frac{V_{\text{in}} \bullet D_{\text{DCM}}}{L \bullet f_s}, \quad D_{\text{DCM}} = \frac{n_i \bullet L \bullet f_s \bullet V_e}{R_s \bullet V_{\text{in}}} \tag{7.11}$$

The steady-state closed-loop block diagram becomes Figure 7.4.

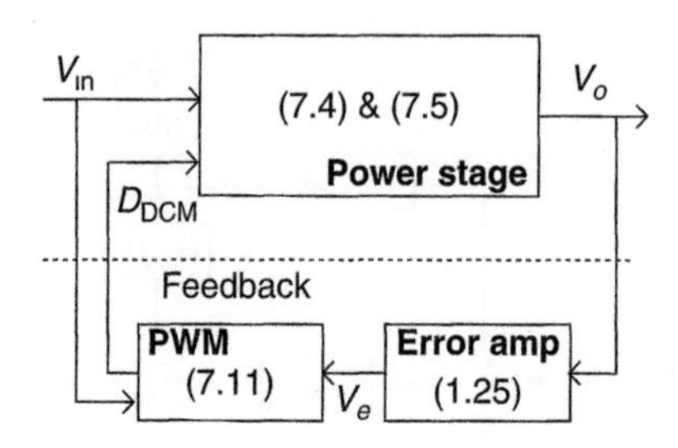

Figure 7.4: Nonisolated boost converter in the steady state, DCM

7.6 Peak Current-Mode Small-Signal Stability in DCM

Equation (7.12) gives the small-signal PWM gains:

$$\begin{aligned} dD_{\mathrm{DCM}} &= F_m \bullet \delta V_e + F_g \bullet \delta V_{\mathrm{in}} \\ &= \frac{n_i \bullet L \bullet f_s}{R_s \bullet V_{\mathrm{in}}} \delta V_e - \frac{n_i \bullet L \bullet f_s \bullet V_e}{R_s \bullet V_{\mathrm{in}}^2} \delta V_{\mathrm{in}} \end{aligned} \tag{7.12}$$

And the overall block diagram is in Figure 7.5.

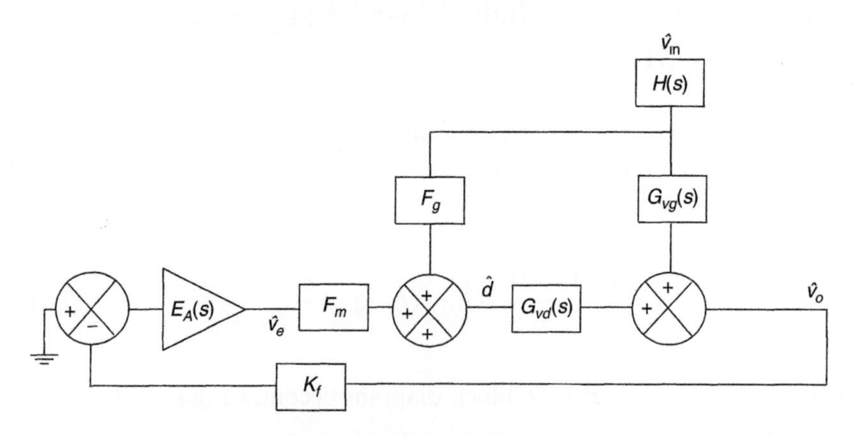

Figure 7.5: AC block diagram for current-mode boost converter in DCM

The control loop gain is

$$T(s) = -K_f \bullet E_A(s) \bullet F_m \bullet G_{vd}(s) \tag{7.13}$$

7.7 DCM Output Capacitor Size

For DCM, the pulsating output current, i_d, is given as

$$i_d(t) = I_{\text{pk}} - \frac{V_o + V_D - V_{\text{in}}}{L} t = \frac{V_{\text{eff}} \bullet D_{\text{DCM}}}{L \bullet f_s} - \frac{V_o + V_D - V_{\text{in}}}{L} t \tag{7.14}$$

For engineering purposes, the line input series loss can be ignored. In that case, (7.14) becomes

$$i_d(t) = \sqrt{\frac{2(V_o + V_D - V_{\text{in}})V_o}{L \bullet f_s \bullet R_L}} - \frac{V_o + V_D - V_{\text{in}}}{L} t \tag{7.15}$$

The output capacitor current is then given as

$$i_c(t) = \sqrt{\frac{2(V_o + V_D - V_{\text{in}})V_o}{L \bullet f_s \bullet R_L}} - \frac{V_o}{R_L} - \frac{V_o + V_D - V_{\text{in}}}{L} t \tag{7.16}$$

It crosses zero at

$$t_0 = \frac{\left(\sqrt{\frac{2(V_o + V_D - V_{\text{in}})V_o}{L \bullet f_s \bullet R_L}} - \frac{V_o}{R_L}\right) L}{V_o + V_D - V_{\text{in}}} \tag{7.17}$$

The output capacitor acquires a net charge increase of

$$\delta Q = \frac{I_{\text{pk}} - \frac{V_o}{R_L}}{2} t_0 = \frac{\left[\sqrt{\frac{2(V_o + V_D - V_{\text{in}})V_o}{L \bullet f_s \bullet R_L}} - \frac{V_o}{R_L}\right]^2 L}{2(V_o + V_D - V_{\text{in}})} = C \bullet \delta v \tag{7.18}$$

Given an output ripple voltage requirement, v, the capacitor value is

$$C = \frac{\left[\sqrt{\frac{2(V_o + V_D - V_{\text{in}})V_o}{L \cdot f_s \cdot R_L}} - \frac{V_o}{R_L}\right]^2 L}{2(V_o + V_D - V_{\text{in}}) \cdot \delta v} \tag{7.19}$$

7.8 CCM Output Capacitor Size

For CCM operation, the output capacitor acquires, or loses, a net charge

$$\delta Q = I_o \cdot \frac{D_{\text{CCM}}}{f_s} = \frac{V_o(V_o + V_D - V_{\text{in}})}{R_L(V_o + V_D)f_s} = C \cdot \delta v \tag{7.20}$$

The output capacitor value is easily obtained, given a specified v.

If desired, a more accurate estimate for the capacitor needed can be obtained using the same procedure and equations given in Section 6.13 but with the duty cycle determined in Wu [2] for CCM or in Section 7.1 for DCM.

Chapter 8

Quasi-Resonant Converters

Chapter 9 of Wu [2] shows that power switching devices encounter severe electrical stresses, consume significant power, and generate heat that must be properly displaced. Excessive electrical stresses in the form of repetitive voltages or currents and heat, as well, eventually shorten the operating life of those devices. Figure 9.10 of Wu [2] indicates that the major power dissipation for those devices takes place at the switching edges, when neither the device voltage across nor the current through equals zero. This is true for both the main switch on the primary side and the rectifiers on the secondary side. It is therefore imperative to create, by some means, either a zero-voltage or zero-current state for at least one, or better both, at the time of transition.

It was also long understood that an inductor in series with a switch can slow down the time rate of current through the switch while a capacitor in parallel is capable of slowing down the rise of voltage across. This being given, the first question is where to incorporate both components and the second is what value to use. In 1983, both questions were answered by a device (patent 4,415,959). We examine the circuit revealed in the patent in the next section.

8.1 How Does It Work?

Sometimes, one feels that nature seems deceptively simple. This sentiment also holds for the answer to the section's question. In Figure 8.1, a capacitor, C_r, in parallel with a diode and the leakage inductance L_r in series with a second switch are all it takes to do the job. At first sight, its resemblance to the forward converter, Figure 1.1, strikes one as impossible, if not "why didn't I think of it?" Well, creativity is akin to a lightning strike. One unexpectedly conceives a superb solution out of a hidden function buried layers deep in concept. Anyway, the similarity between Figure 1.1 and Figure 8.1 ends at their appearance. Their behaviors are oceans apart.

The circuit's dynamic behavior at a steady state may be described by four distinctive phases with unique topological structures, as shown in Figure 8.2. The circuit passes through all phases in succession for each

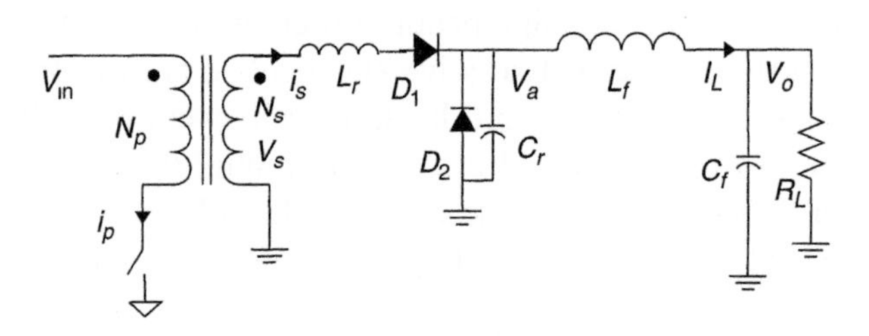

Figure 8.1: Quasi-resonant converter, QRC, $V_s = (N_s/N_p)V_{in}$

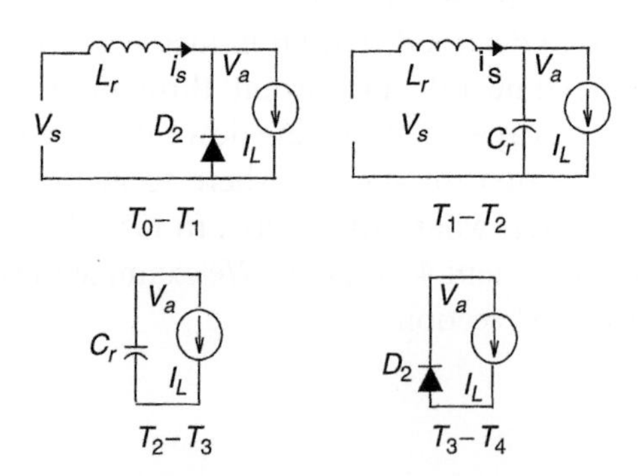

Figure 8.2: Quasi-resonant converter transitions

(clock) switching cycle. During the first phase $T_0 - T_1$, current, i_s, builds up linearly from zero through the leakage inductance L_r. Meanwhile, the balance of load current, $I_L - i_s$, is supported by the output filter choke, L_f, and flows through the diode D_2. This linear phase ends when i_s ramps up to equaling I_L. At that time D_2 ceases conduction and the second phase follows.

During the second phase, $T_1 - T_2$, the leakage inductor and the shunt capacitor form a resonant tank. The excess current, $i_s - I_L$, charges the tank capacitor starting from zero volts. This resonant phase ends when the resonant current reaches zero and heads negative. However, diode D_1 does not permit current reversal. The circuit therefore enters the third phase.

During the third phase, $T_2 - T_3$, the capacitor discharges at constant current and supports the full load. This phase ends when the capacitor voltage crosses zero volts and is slightly negative. At that time, diode D_2 snaps into conduction and the circuit enters the last phase $T_3 - T_4$.

During the final phase, the output filter choke supports the load. The load current circulates through D_2. This phase ends when a new clock cycle is issued.

8.2 Mathematical Analysis

During phase 1, $0 < t < T_1$, the input loop in Figure 8.2 is governed by a simple equation, $L_r \bullet di_s/dt = V_s$ and a zero-current initial condition. Together, it gives the circuit state as

$$i_s(0) = 0, \quad i_s(t) = \frac{V_s}{L_r} t, \quad v_C(t) = 0 \tag{8.1}$$

This phase lasts until $i_s(T_1) = I_L$. The duration for this phase is given by

$$T_1 = \frac{L_r \bullet I_L}{V_s} \tag{8.2}$$

During phase 2, $T_1 < t < T_2$, two equations describe the circuit function:

$$L_r \frac{di_s}{dt} = V_s - v_C, \quad C_r \frac{dv_C}{dt} = i_s - I_L \tag{8.3}$$

The initial conditions are $i_s(T_1) = I_L$ and $v_C(T_1) = 0$. The solutions are

$$i_s(t) = I_L + \frac{V_s}{Z_n} \bullet \sin\left[\omega(t - T_1)\right], \quad \omega = \frac{1}{\sqrt{L_r \bullet C_r}}, \quad Z_n = \sqrt{\frac{L_r}{C_r}},$$
$$v_C(t) = V_s \bullet \{1 - \cos\left[\omega(t - T_1)\right]\} \tag{8.4}$$

The second phase lasts until $i_s(T_2) = 0$; that is,

$$T_2 = T_1 + \frac{\pi + \sin^{-1}\left(\frac{Z_n \bullet I_L}{V_s}\right)}{\omega} \tag{8.5}$$

During phase 3, $T_2 < t < T_3$, the resonant capacitor discharges according to $C_r \bullet dv_C/dt = -I_L$ and initial condition $v_C(T_2) \neq 0$. The solution for this phase is

$$v_C(t) = v_C(T_2) - \frac{I_L}{C_r}(t - T_2) \tag{8.6}$$

The resonant capacitor ceases discharge when $v_C(T_3) = 0$; that is,

$$T_3 = T_2 + \frac{C_r \bullet V_s}{I_L}\left\{1 - \cos\left[\pi + \sin^{-1}\left(\frac{Z_n \bullet I_L}{V_s}\right)\right]\right\} \tag{8.7}$$

During the last phase, both the inductor current and the capacitor voltage remain zero. The duration of this phase is determined by the control mechanism external to the resonant tank and associated output filter.

Furthermore, we can obtain a better visual understanding by displaying all four phases in continuity using an example (Figure 8.3). As shown, the pulsating current and voltage exhibited by the conventional forward converters are replaced by resonant current and voltage waveforms without sharp edges. Both waveforms start at zero and end at zero. By so doing, both the switching losses and noise are markedly reduced.

The energy transferred in one switching cycle, T_s, may also be given as

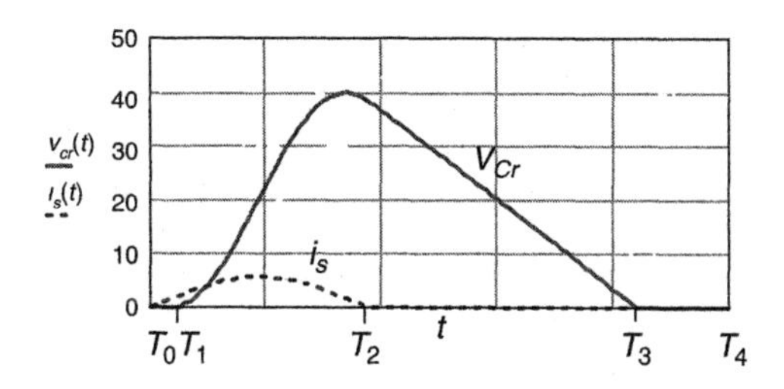

Figure 8.3: Quasi-resonant converter waveforms

$$E_t/\text{pulse} = V_s\left(\int_0^{T_1}\frac{V_s}{L_r}t\cdot dt + \int_{T_1}^{T_2}\left\{I_L + \frac{V_s}{\sqrt{\frac{L_r}{C_r}}}\sin[\omega(t-T_1)]\right\}dt\right)$$
$$= V_o \cdot I_L \cdot T_s \tag{8.8}$$

where the quantity in the large parentheses is the total charge delivered to the resonant tank and all circuits that follow by the source V_s. The equality is established based on the physical principle that energy equals the product of voltage and charge. With further manipulation, the output voltage can be expressed as

$$V_o = \frac{V_s}{T_s}\left(\frac{L_r \cdot I_L}{2V_s} + \frac{\pi + \sin^{-1}\left(\frac{Z_n \cdot I_L}{V_s}\right)}{\omega} + \frac{C_r \cdot V_s \cdot \left\{1 - \cos\left[\pi + \sin^{-1}\left(\frac{Z_n \cdot I_L}{V_s}\right)\right]\right\}}{I_L}\right) \tag{8.9}$$

However, readers are cautioned that the expression does not give the output voltage as an explicit function, since the load current, I_L, on the right-hand side actually equals V_o/R_L; R_L is the load resistance. Another factor that makes it prohibitive for symbolic processing is the arcsine function.

In the preceding treatment, an approximation that enables us to proceed with the analysis was made early on. That is, the filter inductance and capacitance is fairly large, so that it presents a constant current sink

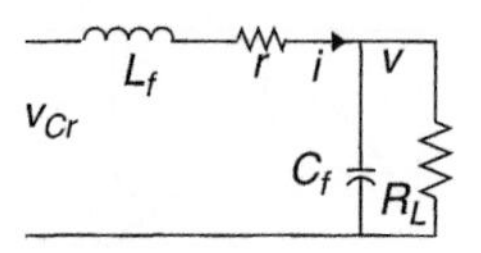

Figure 8.4: QRC output circuit

for the resonant tank. However, for analyzing the steady-state output ripple, we remove the approximation, include the filter (Figure 8.4), and shift the time origin to T_1. In essence, we are treating the resonant capacitor voltage as a source driving the output filter.

Based on Figure 8.4, the resonant tank loop and the output node give

$$\begin{aligned} L_f \frac{di}{dt} + r \bullet i &= v_{Cr} - v \\ i &= C_f \frac{dv}{dt} + \frac{v}{R_L} \end{aligned} \tag{8.10}$$

In Laplace transformed forms, these are

$$\begin{aligned} I(s) &= F_1(s) \bullet I_{0n} + F_2(s) \bullet V_{0n} + \frac{s + \frac{1}{\tau_C}}{L_f \bullet D(s)} \bullet V_{Cr}(s) \\ V(s) &= F_3(s) \bullet I_{0n} + F_4(s) \bullet V_{0n} + \frac{1}{L_f \bullet C_f \bullet D(s)} \bullet V_{Cr}(s) \end{aligned} \tag{8.11}$$

where I_{0n} and V_{0n} are cyclic starting conditions and

$$\tau_L = \frac{L_f}{r}, \quad \tau_C = R_L \bullet C_f, \quad D(s) = s^2 + \left(\frac{1}{\tau_L} + \frac{1}{\tau_C}\right)s + \left(\frac{1}{\tau_L \bullet \tau_C} + \frac{1}{L_f \bullet C_f}\right),$$

$$F_1(s) = \frac{s + \frac{1}{\tau_C}}{D(s)}, \quad F_2(s) = -\frac{1}{L_f \bullet D(s)}, \quad F_3(s) = \frac{1}{C_f \bullet D(s)}, \quad F_4(s) = \frac{s + \frac{1}{\tau_L}}{D(s)},$$

$$V_{Cr}(s) = \begin{cases} V_s\left(\frac{1}{s} - \frac{s}{s^2 + \omega^2}\right), & \text{Duration I } T_1 < t < T_2, \\ \frac{V_a}{s} - \frac{I_L}{C} \bullet \frac{1}{s^2}, \; V_a = V_s\left[1 - \cos\left(\pi + \sin^{-1}\frac{Z_n \bullet I_L}{V_s}\right)\right], \\ \text{Duration II } T_2 < t < T_3 \quad 0 \quad \text{Duration III} \end{cases} \tag{8.12}$$

Consequently, in duration I,

$$I(s) = F_1(s) \bullet I_{01} + F_2(s) \bullet V_{01} + \frac{V_s}{L_f} \bullet [F_5(s) - F_6(s)]$$

$$V(s) = F_3(s) \bullet I_{01} + F_4(s) \bullet V_{01} + \frac{V_s}{L_f \bullet C_f} \bullet [F_7(s) - F_8(s)] \quad (8.13)$$

where

$$F_5(s) = \frac{s + \frac{1}{\tau_C}}{s \bullet D(s)}, \quad F_6(s) = \frac{s \bullet \left(s + \frac{1}{\tau_C}\right)}{D(s) \bullet (s^2 + \omega^2)},$$

$$F_7(s) = \frac{1}{s \bullet D(s)}, \quad F_8(s) = \frac{s}{D(s) \bullet (s^2 + \omega^2)} \quad (8.14)$$

In duration II (delay factor $\exp(-T_2 s)$ implied),

$$I(s) = F_1(s) \bullet I_{02} + F_2(s) \bullet V_{02} + \frac{V_a}{L_f} \bullet F_5(s) - \frac{I_L}{L_f \bullet C} F_9(s)$$

$$V(s) = F_3(s) \bullet I_{02} + F_4(s) \bullet V_{02} + \frac{V_a}{L_f \bullet C_f} \bullet F_7(s) - \frac{I_L}{L_f \bullet C_f \bullet C} F_{10}(s) \quad (8.15)$$

where

$$F_9(s) = \frac{s + \frac{1}{\tau_C}}{s^2 \bullet D(s)}, \quad F_{10}(s) = \frac{1}{s^2 \bullet D(s)} \quad (8.16)$$

In duration III (delay factor $\exp(-T_3 s)$ implied)

$$I(s) = F_1(s) \bullet I_{03} + F_2(s) \bullet V_{03}$$
$$V(s) = F_3(s) \bullet I_{03} + F_4(s) \bullet V_{03} \quad (8.17)$$

By taking the inverse Laplace transformation and as a result of continuity of state at the topological transition boundaries, the following are established:

$$X_1 = \begin{bmatrix} I_{01} \\ V_{01} \end{bmatrix}, \quad X_2 = \begin{bmatrix} I_{02} \\ V_{02} \end{bmatrix}, \quad X_3 = \begin{bmatrix} I_{03} \\ V_{03} \end{bmatrix},$$

$$A_1 = \begin{bmatrix} f_1(T_1) & f_2(T_1) \\ f_3(T_1) & f_4(T_1) \end{bmatrix}, \quad B_1 = \begin{bmatrix} \frac{V_s}{L_f}[f_5(T_1) - f_6(T_1)] \\ \frac{V_s}{L_f \bullet C_f}[f_7(T_1) - f_8(T_1)] \end{bmatrix} \tag{8.18}$$

$$A_2 = \begin{bmatrix} f_1(T_2 - T_1) & f_2(T_2 - T_1) \\ f_3(T_2 - T_1) & f_4(T_2 - T_1) \end{bmatrix},$$

$$B_2 = \begin{bmatrix} \frac{V_a}{L_f} f_5(T_2 - T_1) - \frac{I_L}{L_f \bullet C}[f_9(T_2 - T_1)] \\ \frac{V_a}{L_f \bullet C_f} f_7(T_2 - T_1) - \frac{I_L}{L_f \bullet C_f \bullet C}[f_{10}(T_2 - T_1)] \end{bmatrix}, \tag{8.19}$$

$$A_3 = \begin{bmatrix} f_1(T_3 - T_2) & f_2(T_3 - T_2) \\ f_3(T_3 - T_2) & f_4(T_3 - T_2) \end{bmatrix}$$

Therefore,

$$\begin{aligned} A_1 \bullet X_1 + B_1 &= X_2 \\ A_2 \bullet X_2 + B_2 &= X_3 \\ A_3 \bullet X_3 &= X_1 \\ X_1 &= (I - A_3 \bullet A_2 \bullet A_1)^{-1}(A_3 \bullet A_2 \bullet B_1 + A_3 \bullet B_2) \end{aligned} \tag{8.20}$$

The steady-state output is then given by

$$\begin{aligned} v(t) = {} & \left\{ X_1^T \bullet \begin{bmatrix} f_3(t) \\ f_4(t) \end{bmatrix} + \frac{V_s}{L_f \bullet C_f}[f_7(t) - f_8(t)] \right\} [u(t) - u(t - T_1)] \\ & + \left\{ X_2^T \bullet \begin{bmatrix} f_3(t - T_1) \\ f_4(t - T_1) \end{bmatrix} + \frac{V_a}{L_f \bullet C_f} f_7(t - T_1) - \frac{I_L}{L_f \bullet C_f \bullet C} f_{10}(t - T_1) \right\} \\ & [u(t - T_1) - u(t - T_2)] + X_3^T \bullet \begin{bmatrix} f_3(t - T_2) \\ f_4(t - T_2) \end{bmatrix} \\ & [u(t - T_2) - u(t - T_s)] \end{aligned} \tag{8.21}$$

8.3 Steady-State Closed Loop and Stability

A typical control loop for quasi-resonant converter output is shown in Figure 8.5. We can translate the circuit diagram into a block diagram (Figure 8.6).

We can describe the error voltage with (1.25), which is more precise, or with the approximation

$$V_e = A(V_{\text{ref}} - K_f \bullet V_o) \tag{8.22}$$

Other circuit blocks include the feedback divider K_f, the error amplifier A, and the voltage-controlled oscillator, VCO. Equation (8.22) and the VCO equation give the open-loop feedback operation, while equation (8.9) gives the open-loop power stage implicit function. By combining all

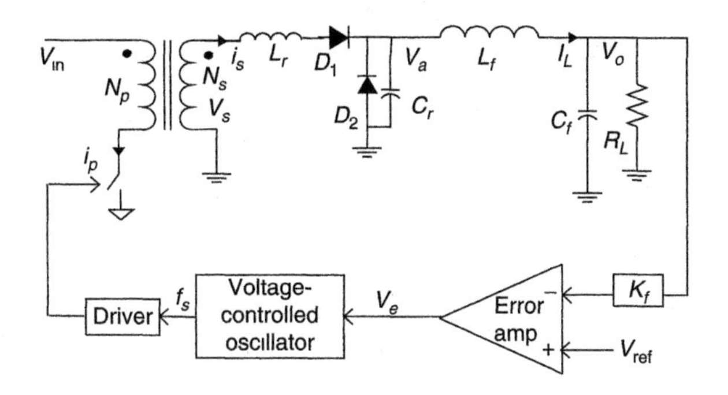

Figure 8.5: QRC control loop

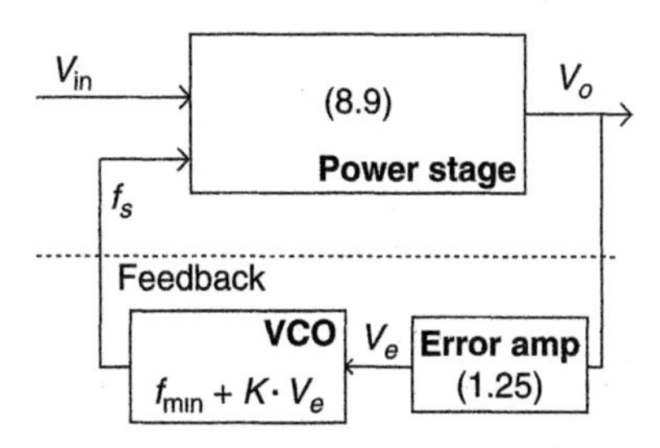

Figure 8.6: QRC control loop block diagram

three equations, the following represents the closed-loop operation for QRC in the steady state:

$$g(R_L, V_o, V_{ref}, \ldots) = V_o - V_s \bullet [f_{\min} + K \bullet A \bullet (V_{\text{ref}} - K_f \bullet V_o)]$$

$$\left[\begin{array}{l} \dfrac{L \bullet \dfrac{V_o}{R_L}}{2 \bullet V_s} + \dfrac{\pi + \sin^{-1} \dfrac{V_o/R_L}{V_s/Z_n}}{\omega} \\ + \dfrac{C \bullet V_s \bullet \left[1 - \cos\left(\pi + \sin^{-1} \dfrac{V_o/R_L}{V_s/Z_n} \right) \right]}{\dfrac{V_o}{R_L}} \end{array} \right] = 0 \tag{8.23}$$

In this implicit form, it prevents us from performing direct graphical analysis. However, by employing a Jacobian determinant in a single dimension, the output sensitivity figure (for instance, load regulation) can still be obtained; that is,

$$S_{RL} = -\frac{\partial g(R_L, V_o, V_{\text{ref}}, \ldots)/\partial R_L}{\partial g(R_L, V_o, V_{\text{ref}}, \ldots)/\partial V_o} \tag{8.24}$$

Next we perform the quintessential AC small-signal study, the open-loop gain in particular. For this task, Figure 8.5 is converted into Figure 8.7, in which the power-stage small-signal gain is no doubt

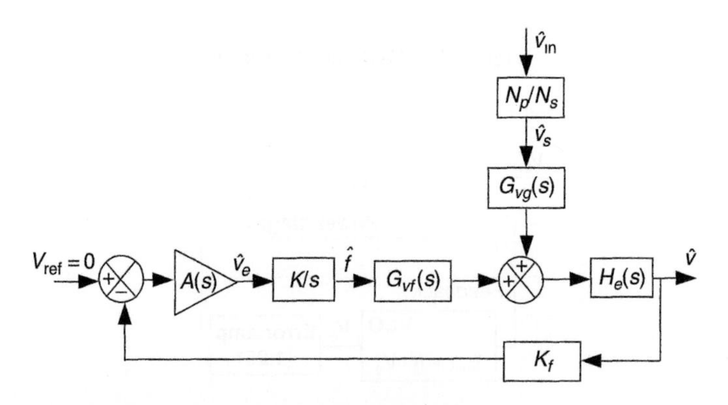

Figure 8.7: QRC small-signal stability block diagram

the core. This core gain is derived by first defining a function from equation (8.9):

$$f(V_o, V_s, f_s, \ldots) = V_o - V_s \bullet f_s \bullet \left\{ \frac{L \bullet V_o}{2 \bullet V_s \bullet R_L} + \frac{\pi + \sin^{-1} \frac{V_o/R_L}{V_s/Z_n}}{\omega} + \frac{C \bullet V_s \bullet \left[1 - \cos\left(\pi + \sin^{-1} \frac{V_o/R_L}{V_s/Z_n} \right) \right]}{V_o/R_L} \right\} = 0 \quad (8.25)$$

Then, the frequency-to-output modulation gain is

$$G_{vf} = -\frac{\partial f(V_o, V_s, f_s, \ldots)/\partial f_s}{\partial f(V_o, V_s, f_s, \ldots)/\partial V_o} \quad (8.26)$$

And the source-to-output modulation gain is

$$G_{vg} = -\frac{\partial f(V_o, V_s, f_s, \ldots)/\partial V_s}{\partial f(V_o, V_s, f_s, \ldots)/\partial V_o} \quad (8.27)$$

The total loop gain is then given by

$$T(s) = -K_f \bullet A(s) \bullet \frac{K}{s} \bullet G_{vf} \bullet \frac{\left(\frac{1}{R_L} + C_f \bullet s \right)^{-1}}{L_f \bullet s + \left(\frac{1}{R_L} + C_f \bullet s \right)^{-1}} \quad (8.28)$$

where *K/s* is the VCO gain.

8.4 Design Issues

By virtue of operating in resonant mode, the QRC design steps are distinctively different from those for conventional forward, boost, or flyback converters. Two fundamental constraints must be met to operate the resonant tank properly and take advantage of zero-current and zero-voltage switching. Zero-current switching demands that the resonant tank current start from zero and end at zero. This requirement can be satisfied only by setting the maximum load current to a value that is less than the sinusoidal peak current amplitude attainable in the resonant tank; otherwise, the argument for arcsine is larger than 1. In other words,

$$\frac{V_s}{Z_n} \geq I_{L,\ \max} \tag{8.29}$$

The other requirement is that the duration sum for all operation topologies equals the switching period. However, the free-wheeling time is not bounded by a particular circuit behavior and may be set to a few percent, say, 5%, of the switching period. Therefore, this second requirement can be formulated as

$$\frac{0.95}{f_{s,\ \max}} = \frac{L_r \bullet I_{L,\ \max}}{V_s} + \frac{\pi + \sin^{-1}\dfrac{I_{L,\ \max}}{V_s/Z_n}}{1/\sqrt{L_r \bullet C_r}} + \frac{C_r \bullet V_s \bullet \left[1 - \cos\left(\pi + \sin^{-1}\dfrac{I_{L,\ \max}}{V_s/Z_n}\right)\right]}{I_{L,\ \max}} \tag{8.30}$$

Given the maximum load, $I_{L,\ \max}$; transferred supply, $V_s = (V_{\text{in}}/N_p)N_s$; and switching frequency, f_s, the resonant tank elements, L_r and C_r, can be solved numerically using both equations.

8.5 Example and Dilemma

In Lui, Orugnati, and Lee [5], an example was given with $L_r = 1.6\,\mu\text{H}$, $C_r = 0.064\,\mu\text{F}$, $L_f = 100\,\mu\text{H}$, $C_f = 10\,\mu\text{F}$, $V_s = 20$ V, $f_s = 150$ KHz. Theoretical computation based on the procedure of section 8.2 gives the results of Figure 8.8. The filter inductor current is also computed. Its result is given in Figure 8.9.

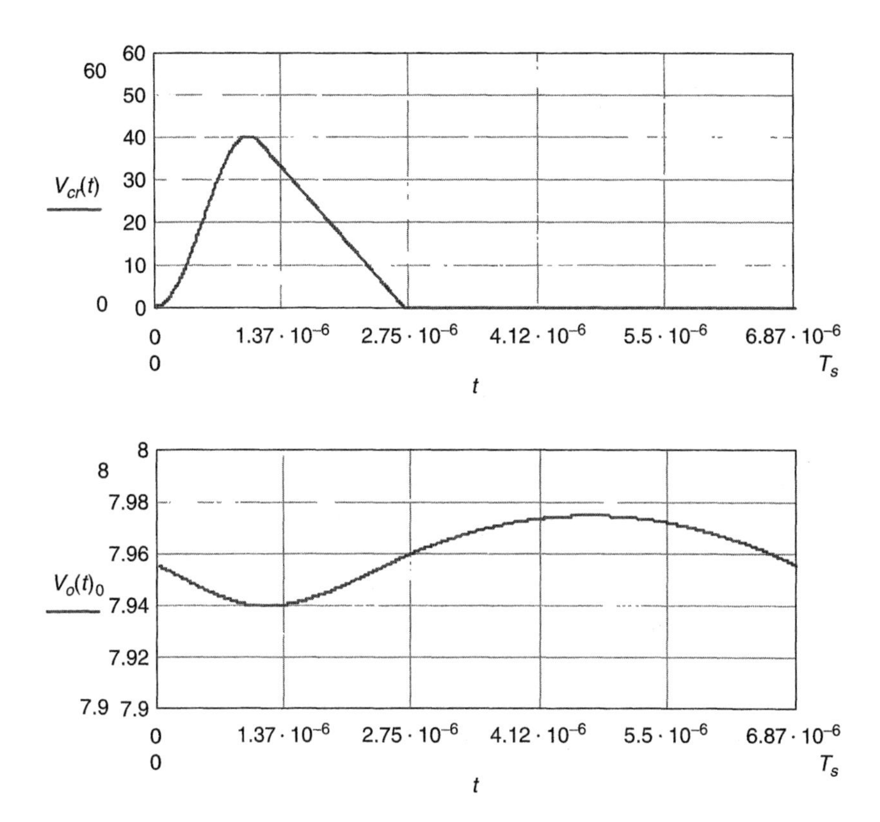

Figure 8.8: Example output; resonant capacitor and filtered output

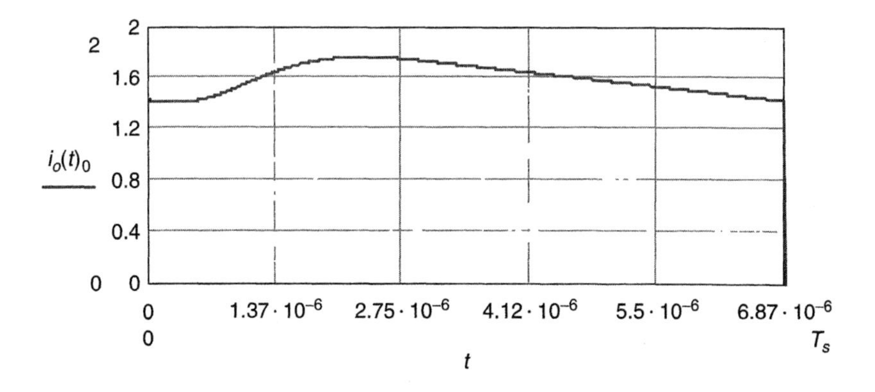

Figure 8.9: Example output filter inductor current

However, we create a dilemma. The filter inductor current is not really a constant, as we took the liberty to assume in section 8.2. Worst yet, the current ripple is not insignificant.

Chapter 9

Class-E Resonant Converter

In the course of the previous presentation, it was shown that the main power train has numerous dissipative elements that degrade converter efficiency. Among these dissipative factors, some are physical in nature—for instance, winding resistance—and cannot be eliminated. Others originating from the functional mechanism, for example, the switching losses of power switches, may be minimized or even totally eliminated.

The switching losses arise, as shown in Wu's Figure 9.10 [2], because either the switch voltage rises too quickly ahead of the decreasing current at turn-off or the switch current surges in advance of the diminishing switch voltage at turn-on. In both cases, power pulses are created where the switch voltage and current profiles cross each other. The switching losses were considered an insignificant nuisance in years past, when most switching converters operated at only tens of kilohertz. However, with the march toward higher switching frequencies of 500 KHz and beyond, the switching losses overtake other dissipations in importance and demand attention. As a result, the concepts of zero-voltage switching (ZVS) and zero-current switching (ZCS) were born.

Zero-voltage switching in general is implemented by placing a capacitor across the power switch (Figure 9.1). The motivation for taking this approach is based simply on the recognition that the capacitor's voltages are continuous. In contrast, zero-current switching is implemented by connecting an inductor in series with the power switch, Figure 9.2. In this case, the property of continuity of inductive current is employed.

Furthermore, both techniques can be enhanced by taking the advantage of resonant circuits; that is, by adding an inductor to Figure 9.1 or a capacitor to Figure 9.2, so a series resonant circuit, Figure 9.3, is formed. Based on the basic form, it is only a matter of adding several more components to yield complete power stages of the various forms. For instance, Figure 9.4 is the most basic and popular topology of ZVS. It was often used as a high-efficiency power amplifier and given a name,

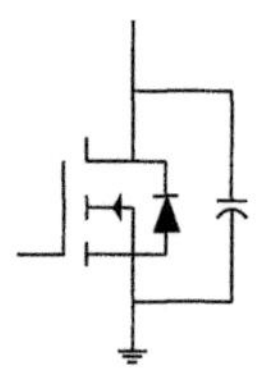

Figure 9.1: Zero-voltage switch

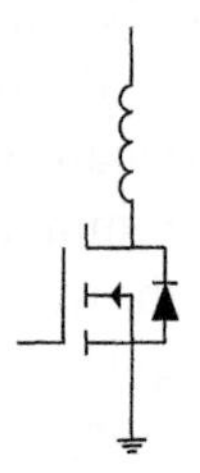

Figure 9.2: Zero-current switch

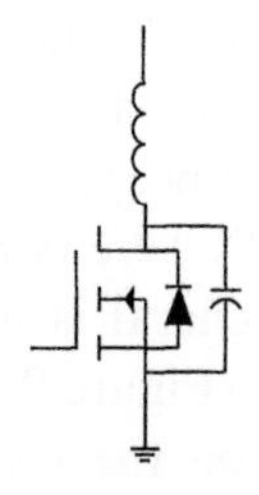

Figure 9.3: Resonant switch

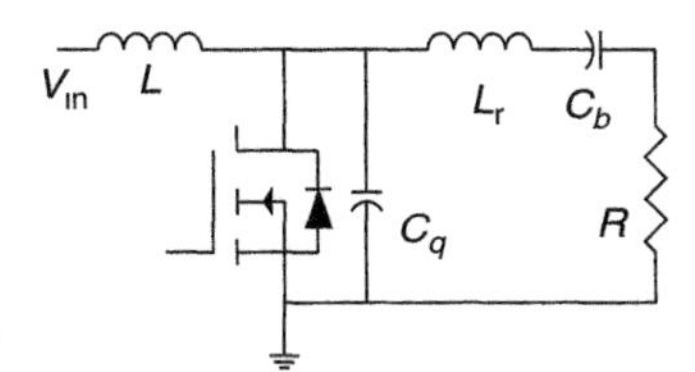

Figure 9.4: Class-E power amplifier

Class E. However, it has also been given new life as a power converter. In its new role, the class-E power converter, as shown in Figure 9.5(a), has been studied extensively by Kazimierczuk [6 and 7].

However, since its inception [8] in 1975, class-E power amplifiers and their derivatives persistently present a major challenge, or perhaps difficulty, for circuit designers (see Sokal and Sokal [8]). This is the daunting task of reaching a more exact analytical solution without making some self-conflicting assumptions. For instance, when analyzing the basic form of the class-E power stage (Figure 9.5(a)), Kazimierczuk [6] started the process by assuming a very large input inductance and treated the source current as constant. In addition, a high-Q condition that enables the analysis to be carried out was also assumed. This practice, even to this date, not only has been largely inherited in many reports, but also has inspired other assumptions. For example, Jozwik and Kazimierczuk [9] also assumed a pure sinusoidal branch current, i_b, in the resonant tank of Figure 9.5(a). In all, assumptions of this nature attain expedience at the expense of accuracy. The situation is made even worse with the switching frequency being lately pushed toward the megahertz range, in which inductance values less than 100 microhenry are freely used.

In other words, those assumptions invoked for the purpose of easing the analytical procedures are being seriously challenged by the advance of art. Therefore, to avoid the unnecessary introduction of inaccuracy, on the one hand, and improve the understanding of resonant DC–DC power converters, on the other, this chapter removes all assumptions regarding the input inductor value, the circuit Q factor, and the resonant branch current. Furthermore, the goal of this writing is to obtain a more accurate and complete solution in closed form for the converter power stage in a steady state. It is accomplished by deriving, in

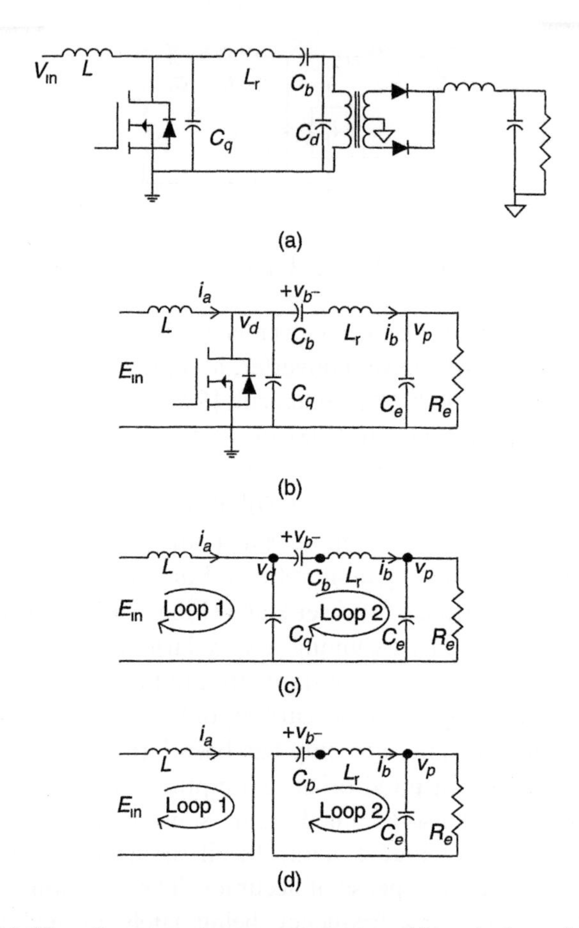

Figure 9.5: (a) Class-E resonant power converter; (b) equivalent circuit; (c) off-time equivalent; (d) on-time equivalent

section 9.1, the starting state of a steady-state switching cycle. Section 9.2 then expresses all steady-state node voltages, branch voltages, and branch currents in compact, closed forms over one cycle. Further studies are carried out in section 9.3 by discussing the development of DC output formulation. Section 9.4 presents AC loop gain. The remaining sections give the type II error amplifier, damping effects of rectifiers, and a numerical example with the assistance of math software capable of performing inverse Laplace transformation. A good match between the theoretical results and the actual measurements attests to the validity of the analytical process. The last section concludes the chapter with discussion regarding one critical assumption that is retained in the analysis.

9.1 Starting States of the Steady State

As suggested by Kazmierczuk [6], the power stage of a class-E resonant converter can be transformed into an equivalent circuit (Figure 9.5(b)), which delivers the same average load power as the original circuit of Figure 9.5(a). Based on the equivalent circuit, two loop and three nodal equations can be written respectively for the switch-off time, t_{off}, in terms of five state variables, i_a, v_d, v_b, i_b, and v_p; that is, during t_{off} (Figure 9.5(c)) with a designated subscript of 1, loop 1 gives

$$\frac{di_{a1}}{dt} + \frac{1}{L} \bullet v_{d1} = \frac{E_{\text{in}}}{L} \tag{9.1}$$

while node v_d, node C_b, loop 2, and node v_p give

$$\begin{aligned}
\frac{1}{C_q} \bullet i_{a1} - \frac{dv_{d1}}{dt} - \frac{1}{C_q} \bullet i_{b1} &= 0 \\
\frac{dv_{b1}}{dt} - \frac{1}{C_b} \bullet i_{b1} &= 0 \\
-\frac{1}{L_r} \bullet v_{d1} + \frac{1}{L_r} \bullet v_{b1} + \frac{di_{b1}}{dt} + \frac{1}{L_r} \bullet v_{p1} &= 0 \\
-\frac{1}{C_e} \bullet i_{b1} + \frac{dv_{p1}}{dt} + \frac{1}{R_e \bullet C_e} \bullet v_{p1} &= 0
\end{aligned} \tag{9.2}$$

Under the condition of a steady state and at the beginning of switch-off time, each of the five state variables bears a yet unknown starting condition: i_{a10}, v_{d10}, v_{b10}, and v_{p10}. However, because C_q was completely discharged in the previous switch-on time, the starting voltage, v_{d10}, of the switch has a null value. This reduces the number of unknowns by one, and by taking the Laplace transformation of (9.1) and (9.2), a new equation set is generated:

$$
\begin{aligned}
s \cdot I_{a1} + \frac{1}{L} \cdot V_{d1} + 0 + 0 + 0 &= i_{a10} + \frac{E_{\text{in}}}{L \cdot s} \\
\frac{1}{C_q} \cdot I_{a1} - s \cdot V_{d1} + 0 - \frac{1}{C_q} \cdot I_{b1} + 0 &= 0 \\
0 + 0 + s \cdot V_{b1} - \frac{1}{C_b} \cdot I_{b1} + 0 &= v_{b10} \\
0 - \frac{1}{L_r} \cdot V_{d1} + \frac{1}{L_r} \cdot V_{b1} + s \cdot I_{b1} + \frac{1}{L_r} \cdot V_{p1} &= i_{b10} \\
0 + 0 + 0 - \frac{1}{C_e} \cdot I_{b1} + \left(s + \frac{1}{R_e \cdot C_e}\right) \cdot V_{p1} &= v_{p10}
\end{aligned}
\tag{9.3}
$$

This equation set, in matrix form, yields

$$
I_{a1}(s) = \frac{\begin{vmatrix} i_{a10} + \dfrac{E_{\text{in}}}{L \cdot s} & \dfrac{1}{L} & 0 & 0 & 0 \\ 0 & -s & 0 & -\dfrac{1}{C_q} & 0 \\ v_{b10} & 0 & s & -\dfrac{1}{C_b} & 0 \\ i_{b10} & -\dfrac{1}{L_r} & \dfrac{1}{L_r} & s & \dfrac{1}{L_r} \\ v_{p10} & 0 & 0 & -\dfrac{1}{C_e} & s + \dfrac{1}{R_e \cdot C_e} \end{vmatrix}}{\begin{vmatrix} s & \dfrac{1}{L} & 0 & 0 & 0 \\ \dfrac{1}{C_q} & -s & 0 & -\dfrac{1}{C_q} & 0 \\ 0 & 0 & s & -\dfrac{1}{C_b} & 0 \\ 0 & -\dfrac{1}{L_r} & \dfrac{1}{L_r} & s & \dfrac{1}{L_r} \\ 0 & 0 & 0 & -\dfrac{1}{C_e} & s + \dfrac{1}{R_e \cdot C_e} \end{vmatrix}}
\tag{9.4}
$$

$$V_{d1}(s) = \frac{\begin{vmatrix} s & i_{a10} + \frac{E_{\text{in}}}{L \bullet s} & 0 & 0 & 0 \\ \frac{1}{C_q} & 0 & 0 & -\frac{1}{C_q} & 0 \\ 0 & v_{b10} & s & -\frac{1}{C_b} & 0 \\ 0 & i_{b10} & \frac{1}{L_r} & s & \frac{1}{L_r} \\ 0 & v_{p10} & 0 & -\frac{1}{C_e} & s + \frac{1}{R_e \bullet C_e} \end{vmatrix}}{D_1(s)} \tag{9.5}$$

$$V_{b1}(s) = \frac{\begin{vmatrix} s & \frac{1}{L} & i_{a10} + \frac{E_{\text{in}}}{L \bullet s} & 0 & 0 \\ \frac{1}{C_q} & -s & 0 & -\frac{1}{C_q} & 0 \\ 0 & 0 & v_{b10} & -\frac{1}{C_b} & 0 \\ 0 & -\frac{1}{L_r} & i_{b10} & s & \frac{1}{L_r} \\ 0 & 0 & v_{p10} & -\frac{1}{C_e} & s + \frac{1}{R_e \bullet C_e} \end{vmatrix}}{D_1(s)} \tag{9.6}$$

$$I_{b1}(s) = \frac{\begin{vmatrix} s & \frac{1}{L} & 0 & i_{a10} + \frac{E_{\text{in}}}{L \bullet s} & 0 \\ \frac{1}{C_q} & -s & 0 & 0 & 0 \\ 0 & 0 & s & v_{b10} & 0 \\ 0 & -\frac{1}{L_r} & \frac{1}{L_r} & i_{b10} & \frac{1}{L_r} \\ 0 & 0 & 0 & v_{p10} & s + \frac{1}{R_e \bullet C_e} \end{vmatrix}}{D_1(s)} \tag{9.7}$$

$$V_{p1}(s) = \frac{\begin{vmatrix} s & \frac{1}{L} & 0 & 0 & i_{a10} + \frac{E_{\text{in}}}{L \cdot s} \\ \frac{1}{C_q} & -s & 0 & -\frac{1}{C_q} & 0 \\ 0 & 0 & s & -\frac{1}{C_b} & v_{b10} \\ 0 & -\frac{1}{L_r} & \frac{1}{L_r} & s & i_{b10} \\ 0 & 0 & 0 & -\frac{1}{C_e} & v_{p10} \end{vmatrix}}{D_1(s)} \tag{9.8}$$

The equations in Laplace transformation form can be further processed to produce time-domain functions. For instance, the source current (9.4) in the frequency domain can be expanded and rewritten as

$$\begin{aligned} I_{a1}(s) = {} & i_{a10} \cdot \frac{H_{1n}(s)}{D_1(s)} + v_{b10} \cdot \frac{H_{2n}(s)}{D_1(s)} + i_{b10} \cdot \frac{H_{3n}(s)}{D_1(s)} + v_{p10} \cdot \frac{H_{4n}(s)}{D_1(s)} \\ & + E_{\text{in}} \cdot \frac{H_{5n}(s)}{D_1(s)} \end{aligned} \tag{9.9}$$

In theory, this can be inverse transformed, which brings forth

$$\begin{aligned} i_{a1}(t) = {} & i_{a10} \cdot h_1(t) + v_{b10} \cdot h_2(t) + i_{b10} \cdot h_3(t) + v_{p10} \cdot h_4(t) \\ & + E_{\text{in}} \cdot h_5(t) \end{aligned} \tag{9.10}$$

Therefore, at the end of switch-off time, t_{off},

$$\begin{aligned} i_{a1}(t_{\text{off}}) = {} & i_{a10} \cdot h_1(t_{\text{off}}) + v_{b10} \cdot h_2(t_{\text{off}}) + i_{b10} \cdot h_3(t_{\text{off}}) \\ & + v_{p10} \cdot h_4(t_{\text{off}}) + E_{\text{in}} \cdot h_5(t_{\text{off}}) \end{aligned} \tag{9.11}$$

By applying the same procedure against (9.5) through (9.8), the following is generated:

$$\begin{aligned} v_{b1}(t_{\text{off}}) = {} & i_{a10} \cdot f_1(t_{\text{off}}) + v_{b10} \cdot f_2(t_{\text{off}}) + i_{b10} \cdot f_3(t_{\text{off}}) + v_{p10} \cdot f_4(t_{\text{off}}) \\ & + E_{in} \cdot f_5(t_{\text{off}}) \end{aligned} \tag{9.12}$$

$$i_{b1}(t_{\text{off}}) = i_{a10} \bullet g_1(t_{\text{off}}) + v_{b10} \bullet g_2(t_{\text{off}}) + i_{b10} \bullet g_3(t_{\text{off}}) + v_{p10} \bullet g_4(t_{\text{off}}) + E_{\text{in}} \bullet g_5(t_{\text{off}})$$

$$v_{p1}(t_{\text{off}}) = i_{a10} \bullet p_1(t_{\text{off}}) + v_{b10} \bullet p_2(t_{\text{off}}) + i_{b10} \bullet p_3(t_{\text{off}}) + v_{p10} \bullet p_4(t_{\text{off}}) + E_{\text{in}} \bullet p_5(t_{\text{off}})$$

These equations, of course, can be placed in a matrix form

$$X_2 = A_1 \bullet X_1 + B_1 \bullet E_{\text{in}} \tag{9.13}$$

where

$$X_1 = \begin{bmatrix} i_{a10} \\ v_{b10} \\ i_{b10} \\ v_{p10} \end{bmatrix}, \; X_2 = \begin{bmatrix} i_{a1}(t_{\text{off}}) \\ v_{b1}(t_{\text{off}}) \\ i_{b1}(t_{\text{off}}) \\ v_{p1}(t_{\text{off}}) \end{bmatrix}, \; B_1 = \begin{bmatrix} h_5(t_{\text{off}}) \\ f_5(t_{\text{off}}) \\ g_5(t_{\text{off}}) \\ p_5(t_{\text{off}}) \end{bmatrix},$$

$$A_1 = \begin{bmatrix} h_1(t_{\text{off}}) & h_2(t_{\text{off}}) & h_3(t_{\text{off}}) & h_4(t_{\text{off}}) \\ f_1(t_{\text{off}}) & f_2(t_{\text{off}}) & f_3(t_{\text{off}}) & f_4(t_{\text{off}}) \\ g_1(t_{\text{off}}) & g_2(t_{\text{off}}) & g_3(t_{\text{off}}) & g_4(t_{\text{off}}) \\ p_1(t_{\text{off}}) & p_2(t_{\text{off}}) & p_3(t_{\text{off}}) & p_4(t_{\text{off}}) \end{bmatrix}$$

In other words, in the course of switch-off time, the starting-state X_1 evolves into X_2 at time t_{off}.

Following the termination of switch-off and when the capacitor, C_q, voltage rings back to zero, the transistor switch is turned on due to the implementation of zero-voltage switching. The action alters the circuit topology and the switch-on dynamics commencing with X_2 serves as the starting state for the switch-on time, $t_{\text{on}} = T_s - t_{\text{off}}$.

During the time interval of t_{on}, Figure 9.5(d), loop 1, node C_b, loop 2, and v_p give, with subscript 2,

$$\begin{aligned} \frac{di_{a2}}{dt} &= \frac{E_{\text{in}}}{L} \\ \frac{dv_{b2}}{dt} - \frac{1}{C_b} \bullet i_{b2} &= 0 \\ \frac{di_{b2}}{dt} + \frac{1}{L_r} \bullet v_{b2} + \frac{1}{L_r} \bullet v_{p2} &= 0 \\ \frac{dv_{p2}}{dt} - \frac{1}{C_e} \bullet i_{b2} + \frac{1}{R_e \bullet C_e} \bullet v_{p2} &= 0 \end{aligned} \tag{9.14}$$

Assuming the starting state for the switch-on time, $T_s - t_{\text{off}}$, as $(i_{a20}\ v_{b20}\ i_{b20}\ v_{p20})^T$, the first equation of (9.14) is solved separately:

$$i_{a2}(t) = \left[i_{a20} + \frac{E_{\text{in}}}{L} \bullet (t - t_{\text{off}})\right] \bullet u(t - t_{\text{off}}) \tag{9.15}$$

while the rest gives

$$V_{b2}(s) = \frac{\begin{vmatrix} v_{b20} & -\dfrac{1}{C_b} & 0 \\ i_{b20} & s & \dfrac{1}{L_r} \\ v_{p20} & -\dfrac{1}{C_e} & s + \dfrac{1}{R_e \bullet C_e} \end{vmatrix} \bullet \mathrm{e}^{-t_{\text{off}} \bullet s}}{\begin{vmatrix} s & -\dfrac{1}{C_b} & 0 \\ \dfrac{1}{L_r} & s & \dfrac{1}{L_r} \\ 0 & -\dfrac{1}{C_e} & s + \dfrac{1}{R_e \bullet C_e} \end{vmatrix}} \tag{9.16}$$

$$I_{b2}(s) = \frac{\begin{vmatrix} s & v_{b20} & 0 \\ \dfrac{1}{L_r} & i_{b20} & \dfrac{1}{L_r} \\ 0 & v_{p20} & s + \dfrac{1}{R_e \bullet C_e} \end{vmatrix} \bullet \mathrm{e}^{-t_{\text{off}} \bullet s}}{D_2(s)} \tag{9.17}$$

$$V_{p2}(s) = \frac{\begin{vmatrix} s & -\dfrac{1}{C_b} & v_{b20} \\ \dfrac{1}{L_r} & s & i_{b20} \\ 0 & -\dfrac{1}{C_e} & v_{p20} \end{vmatrix} \bullet \mathrm{e}^{-t_{\text{off}} \bullet s}}{D_2(s)} \tag{9.18}$$

Equations (9.16), (9.17), and (9.18) again can be solved along the same procedure as outlined for $i_{a1}(t)$, (9.10). This step yields

$$\begin{aligned}
v_{b2}(t) &= [v_{b20} \bullet q_1(t - t_{\text{off}}) + i_{b20} \bullet q_2(t - t_{\text{off}}) \\
&\quad + v_{p20} \bullet q_3(t - t_{\text{off}})] \bullet u(t - t_{\text{off}}) \\
i_{b2}(t) &= [v_{b20} \bullet r_1(t - t_{\text{off}}) + i_{b20} \bullet r_2(t - t_{\text{off}}) \\
&\quad + v_{p20} \bullet r_3(t - t_{\text{off}})] \bullet u(t - t_{\text{off}}) \\
v_{p2}(t) &= [v_{b20} \bullet w_1(t - t_{\text{off}}) + i_{b20} \bullet w_2(t - t_{\text{off}}) \\
&\quad + v_{p20} \bullet w_3(t - t_{\text{off}})] \bullet u(t - t_{\text{off}})
\end{aligned} \tag{9.19}$$

Consequently, at the end of switch-on time and because of the requirement of continuity of states, the following equality is established:

$$A_2 \bullet X_2 + B_2 \bullet E_{\text{in}} = X_1 \tag{9.20}$$

where

$$A_2 = \begin{bmatrix} 1 & 0 & 0 & 0 \\ 0 & q_1(T_s - t_{\text{off}}) & q_2(T_s - t_{\text{off}}) & q_3(T_s - t_{\text{off}}) \\ 0 & r_1(T_s - t_{\text{off}}) & r_2(T_s - t_{\text{off}}) & r_3(T_s - t_{\text{off}}) \\ 0 & w_1(T_s - t_{\text{off}}) & w_2(T_s - t_{\text{off}}) & w_3(T_s - t_{\text{off}}) \end{bmatrix}$$

$$B_2 = \begin{bmatrix} \dfrac{T_s - t_{\text{off}}}{L} \\ 0 \\ 0 \\ 0 \end{bmatrix}$$

Finally, (9.13) and (9.20) give the unknown starting state under the steady-state condition as

$$X_1 = (I - A_2 \bullet A_1)^{-1} \bullet (A_2 \bullet B_1 + B_2) \bullet E_{\text{in}} \tag{9.21}$$

and X_2 is given by (9.13) once X_1 is obtained.

9.2 Time-Domain Steady-State Solutions

With both starting states for the two distinctive time intervals available, all branch currents and node voltages under the steady-state condition can be expressed for just one switching cycle as

$$
\begin{aligned}
i_a(t) &= \left\{ X_1^T \bullet \begin{bmatrix} h_1(t) \\ h_2(t) \\ h_3(t) \\ h_4(t) \end{bmatrix} + E_{\text{in}} \bullet h_5(t) \right\} \bullet [u(t) - u(t - t_{\text{off}})] \\
&+ \left\{ X_2^T \bullet \begin{bmatrix} 1 \\ 0 \\ 0 \\ 0 \end{bmatrix} + \frac{E_{\text{in}}}{L} \bullet (t - t_{\text{off}}) \right\} \bullet [u(t - t_{\text{off}}) - u(t - T_s)] \\
v_d(t) &= \left\{ X_1^T \bullet \begin{bmatrix} x_{d1}(t) \\ x_{d2}(t) \\ x_{d3}(t) \\ x_{d4}(t) \end{bmatrix} + E_{\text{in}} \bullet x_{d5}(t) \right\} \bullet [u(t) - u(t - t_{\text{off}})] \\
v_b(t) &= \left\{ X_1^T \bullet \begin{bmatrix} f_1(t) \\ f_2(t) \\ f_3(t) \\ f_4(t) \end{bmatrix} + E_{\text{in}} \bullet f_5(t) \right\} \bullet [u(t) - u(t - t_{\text{off}})] \\
&+ \left\{ X_2^T \bullet \begin{bmatrix} 0 \\ q_1(t - t_{\text{off}}) \\ q_2(t - t_{\text{off}}) \\ q_3(t - t_{\text{off}}) \end{bmatrix} + \frac{E_{\text{in}}}{L} \bullet (t - t_{\text{off}}) \right\} \bullet [u(t - t_{\text{off}}) - u(t - T_s)] \\
i_b(t) &= \left\{ X_1^T \bullet \begin{bmatrix} g_1(t) \\ g_2(t) \\ g_3(t) \\ g_4(t) \end{bmatrix} + E_{\text{in}} \bullet g_5(t) \right\} \bullet [u(t) - u(t - t_{\text{off}})] \\
&+ \left\{ X_2^T \bullet \begin{bmatrix} 0 \\ r_1(t - t_{\text{off}}) \\ r_2(t - t_{\text{off}}) \\ r_3(t - t_{\text{off}}) \end{bmatrix} + \frac{E_{\text{in}}}{L} \bullet (t - t_{\text{off}}) \right\} \bullet [u(t - t_{\text{off}}) - u(t - T_s)]
\end{aligned}
\tag{9.22}
$$

$$v_p(t) = \left\{ X_1^T \bullet \begin{bmatrix} p_1(t) \\ p_2(t) \\ p_3(t) \\ p_4(t) \end{bmatrix} + E_{\text{in}} \bullet p_5(t) \right\} \bullet [u(t) - u(t - t_{\text{off}})]$$

$$+ \left\{ X_2^T \bullet \begin{bmatrix} 0 \\ w_1(t - t_{\text{off}}) \\ w_2(t - t_{\text{off}}) \\ w_3(t - t_{\text{off}}) \end{bmatrix} + \frac{E_{\text{in}}}{L} \bullet (t - t_{\text{off}}) \right\} \bullet [u(t - t_{\text{off}}) - u(t - T_s)]$$

$$v_{Lr}(t) = v_d(t) - v_b(t) - v_p(t)$$

Other derived component current, voltage, or power consumption values can also be obtained. For instance, the current through C_q is given as

$$i_{Cq}(t) = C_q \bullet \frac{dv_d(t)}{dt} \tag{9.23}$$

Current conservation requirements then lead to the switch current:

$$i_{\text{sw}}(t) = i_a(t) - i_b(t) - i_{Cq}(t) \tag{9.24}$$

and the average switch power

$$P_q = f_s \bullet \int_0^{1/f_s} v_d(t) \bullet i_{\text{sw}}(t) \bullet dt \tag{9.25}$$

In theory, a similar process can be applied against all dissipative components, namely, series resistance. However, most equivalent series resistance of inductors and capacitors are ignored for simplicity in this study.

More performance figures of interest to designers may also be evaluated, for example, $i_{\text{sw}}(t)$ vs $v_d(t)$ trajectories with time, as a parameter gives a better view of the cyclic stress to which the transistor switch is subject. This is left for the reader to investigate.

9.3 Closed-Loop DC Analysis

The resonant DC–DC converter studied in this chapter is designed to have a constant switch-off time, t_{off}. However, for output regulation, the switch-on time is modulated. This control strategy implies a frequency-modulated power stage and a control loop, as shown in Figure 9.6, where after being scaled and processed by a constant feedback ratio, K_f, and the error amplifier, the output generates an error voltage, V_{er}. The error voltage, in turn, feeds a voltage-controlled oscillator, VCO, which produces a variable switching frequency driving the driver and the power stage.

By slightly altering the switching frequency, the output voltage is regulated against load and line variation. Here, since frequency modulation is involved, it is desirable to reformulate the primary voltage, $v_p(t)$, as a function of both switching frequency, f_s, and time, t:

$$\begin{aligned} v_p(t) &= F_1(f_s, t) \cdot [u(t) - u(t - t_{\text{off}})] \\ &\quad + F_2(f_s, t) \cdot [u(t - t_{\text{off}}) - u(t - \frac{1}{f_s})] \end{aligned} \tag{9.26}$$

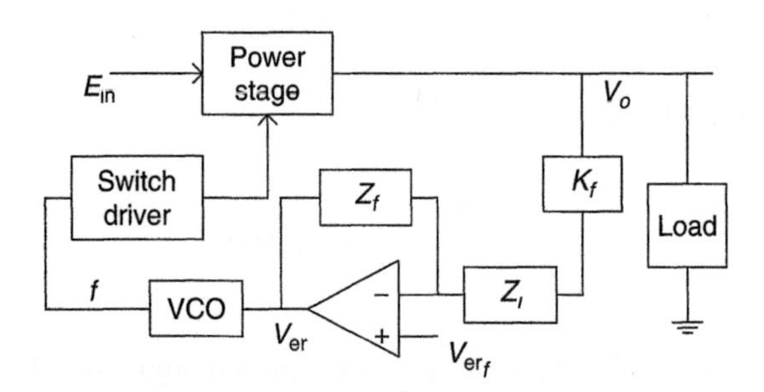

Figure 9.6: Steady-state block diagram

where

$$F_1(f_s,t)=\left\{[I-Y(f_s,t_{\text{off}})\cdot A_1]^{-1}\cdot\left[Y(f_s,t_{\text{off}})\cdot B_1+\begin{bmatrix}\dfrac{\frac{1}{f_s}-t_{\text{off}}}{L}\\0\\0\\0\end{bmatrix}\right]\cdot E_{\text{in}}\right\}^T$$

$$\cdot\begin{bmatrix}p_1(t)\\p_2(t)\\p_3(t)\\p_4(t)\end{bmatrix}+E_{\text{in}}\cdot p_5(t)$$

$$F_2(f_s,t)=\left\{A_1\cdot[I-Y(f_s,t_{\text{off}})\cdot A_1]^{-1}\cdot\left[Y(f_s,t_{\text{off}})\cdot B_1+\begin{bmatrix}\dfrac{\frac{1}{f_s}-t_{\text{off}}}{L}\\0\\0\\0\end{bmatrix}\right]\right.$$

$$\left.\cdot E_{\text{in}}+B_1\cdot E_{\text{in}}\right\}^T\cdot\begin{bmatrix}0\\w_1(t-t_{\text{off}})\\w_2(t-t_{\text{off}})\\w_3(t-t_{\text{off}})\end{bmatrix}$$

$$Y(f_s,t_{\text{off}})=\begin{bmatrix}1&0&0&0\\0&q_1\left(\frac{1}{f_s}-t_{\text{off}}\right)&q_2\left(\frac{1}{f_s}-t_{\text{off}}\right)&q_3\left(\frac{1}{f_s}-t_{\text{off}}\right)\\0&r_1\left(\frac{1}{f_s}-t_{\text{off}}\right)&r_2\left(\frac{1}{f_s}-t_{\text{off}}\right)&r_3\left(\frac{1}{f_s}-t_{\text{off}}\right)\\0&w_1\left(\frac{1}{f_s}-t_{\text{off}}\right)&w_2\left(\frac{1}{f_s}-t_{\text{off}}\right)&w_3\left(\frac{1}{f_s}-t_{\text{off}}\right)\end{bmatrix}\tag{9.27}$$

Through the transformer turn ratio, n, and by including rectifier drop, V_{RC}, the power stage gives a DC, open-loop output voltage as a function of switching frequency, f_s:

$$V_{\mathrm{DC}}(f_s) = f_s \cdot \int_0^{\frac{1}{f_s}} (|\frac{1}{n} \cdot v_p(f_s, t)| - V_{RC}) \cdot dt \tag{9.28}$$

or

$$V_{\mathrm{DC}}(f_s) = \frac{f_s}{n} \cdot \left[\int_0^{t_{\mathrm{off}}} |F_1(f_s, t)| \cdot dt + \int_{t_{\mathrm{off}}}^{\frac{1}{f_s}} |F_2(f_s, t)| \cdot dt \right] - V_{RC} \tag{9.29}$$

Based on the block diagram of Figure 9.6, the closed-loop DC equation can then be written as

$$\begin{aligned} &\frac{A(V_{\mathrm{ref}} - K \cdot V_{\mathrm{DC}})G_{\mathrm{VCO}} + f_o}{n} \left[\int_0^{t_{\mathrm{off}}} |F_1(A, V_{\mathrm{ref}}, K, V_{\mathrm{DC}}, G_{\mathrm{VCO}}, f_o, t)| dt \right. \\ &\quad \left. + \int_{t_{\mathrm{off}}}^{\frac{1}{A(V_{\mathrm{ref}} - K \cdot V_{\mathrm{DC}})G_{\mathrm{VCO}} + f_o}} |F_2(A, V_{\mathrm{ref}}, K, V_{\mathrm{DC}}, G_{\mathrm{VCO}}, f_o, t)| dt \right] \\ &\quad - V_{RC} - V_{\mathrm{DC}} = 0 \end{aligned} \tag{9.30}$$

where A stands for the open-loop gain of the error amplifier, V_{ref} the command reference voltage, K_f the DC feedback factor, G_{VCO} the voltage-controlled-oscillator gain, and f_o the minimum VCO output frequency.

As can be seen from equation (9.30), the closed-loop DC output cannot be easily expressed explicitly in terms of circuit components and parameters, since symbol V_{DC} is embedded extremely deep in the second integral and in each component function of $F_2(f_s, t)$. It is therefore left in this implicit form. However, even in these undesirable forms, more studies can be conducted. For instance, by sweeping the switching frequency, f_s, the open-loop DC output equation (9.29) gives the output sensitivity against the switching frequency. By the same token, (9.30) can be employed for closed-loop output sensitivity studies. For example, although very tedious, the output sensitivity against the VCO gain, G_{VCO}, can be derived as follows by first assigning the left-hand side of (9.30) a function symbol f_{DC}. The sensitivity figure is then given by taking the Jacobian

$$\frac{\partial V_{\mathrm{DC}}}{\partial G_{\mathrm{VCO}}} = -\frac{\frac{\partial f_{\mathrm{DC}}}{\partial G_{\mathrm{VCO}}}}{\frac{\partial f_{\mathrm{DC}}}{\partial V_{\mathrm{DC}}}} \tag{9.31}$$

Similar studies, of course, can be performed for other parameters and give the designers a better perception as to the direction where improvement can be achieved and how to implement it.

9.4 Closed-Loop AC Analysis

In addition to the DC studies given in the previous section, AC small-signal loop response can also be evaluated employing equation (9.29). For this latter study, the small-signal gain of the power stage at low-perturbation frequency can be described as

$$G_p(f_s) = \frac{dV_{\mathrm{DC}}(f_s)}{df_s} \tag{9.32}$$

The power stage gain can then be incorporated into the overall AC loop gain formulation. This last step results in the control loop gain function as

$$T(s) = -K_f \bullet A(s) \bullet G_{\mathrm{VCO}} \bullet G_p(f_s) \bullet H_p(s) \tag{9.33}$$

Among the five factors contributing to the loop gain, all but $H_p(s)$ have been, or can be, dealt with easily. Here, K_f is the feedback divider ratio, $A(s)$ the error amplifier, G_{VCO} the VCO linear gain, and so forth. For the last item, and from the viewpoint of practical application, it can be considered to be dominated by the output filter with series damping. The damping effect is thought to come from the dynamic resistance of the full-wave rectifier. With this in mind, the power stage transfer function $H_p(s)$ is given as

$$H_p(s) = \frac{\left(\frac{1}{R_L} + C_o \bullet s\right)^{-1}}{L_o \bullet s + \left(\frac{1}{r_s} + C_s \bullet s\right)^{-1} + \left(\frac{1}{R_L} + C_o \bullet s\right)^{-1}} \tag{9.34}$$

where, in reference to Figure 9.5(a), r_s stands for the series damping resistance, consisting of the rectifier dynamic series resistance and the series resistance of the output filter choke; C_s for parasitic capacitance across the damping resistor; L_o the output filter choke; C_o the output filter capacitor; and R_L the load.

The damping resistance of output rectifiers should be examined for two possible cases. If the regular junction diodes are used, the dynamic resistance can be derived by the following set of equations, representing the junction diodes operating in the first quadrant of the I–V plane:

$$I(V_d) = I_s \cdot (e^{q \cdot V_d / k \cdot T} - 1)$$
$$r_s = \frac{1}{\partial I(V_d)/\partial V_d} \tag{9.35}$$

Here, the forward current of rectifiers is expressed as a function of the forward voltage and other fundamental constants of physics.

By the same token, if MOSFET synchronous rectifiers are used, the damping resistance is obtained by the following two equations:

$$I(V_{\mathrm{ds}}) = \beta \cdot (V_{\mathrm{gs}} - V_t)^2 \cdot (1 - \lambda \cdot V_{\mathrm{ds}}) \cdot \tan\mathrm{h}(\alpha \cdot V_{\mathrm{ds}})$$
$$r_s = \frac{1}{\partial I(V_{\mathrm{ds}})/\partial V_{\mathrm{ds}}} \tag{9.36}$$

where the drain-to-source current is expressed in terms of drain-to-source voltage, V_{ds}, gate-to-source voltage, V_{gs}, and other curve-fitting parameters, depending on the specific device.

On the surface and in mathematical form, both appear to be similar. However, once both dynamic resistances are plotted as a function of their operating points (Figures 9.7 and 9.8), a critical difference clearly

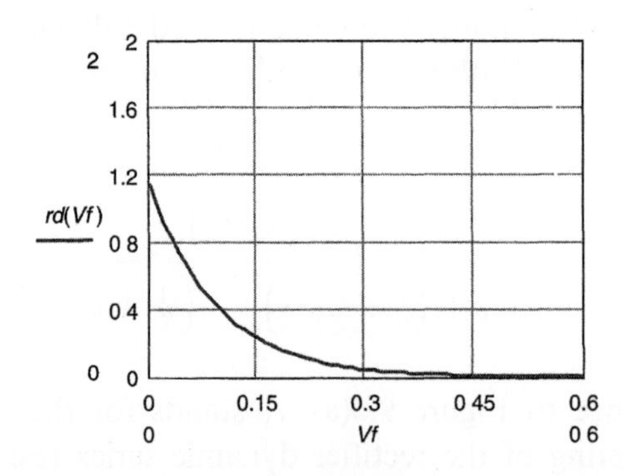

Figure 9.7: Dynamic impedance of a diode

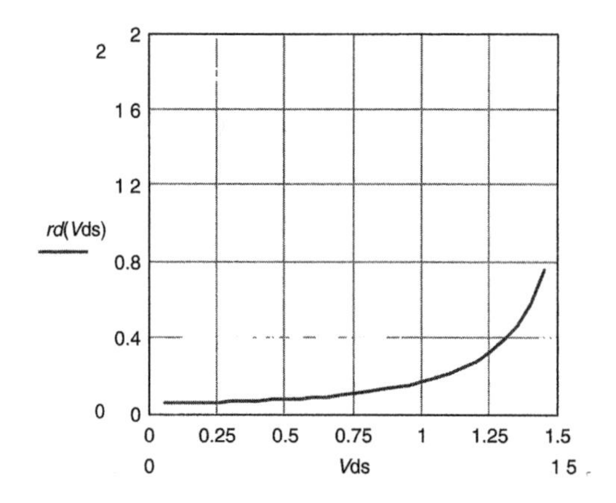

Figure 9.8: Dynamic impedance of a MOSFET

emerges. What the plot indicates is that both devices are not necessarily interchangeable. Designers hoping to do so should exercise caution.

9.5 Type II Amplifier

Up to this point, all factors but K_f and $A(s)$ have been treated. In general, the DC feedback ratio K_f is implemented easily with resistive components and requires no further analysis. The compensation amplifier, $A(s)$, however deserves more attention.

To give the compensation amplifier an in-depth treatment, (9.33) is partitioned into two groups and presented as

$$T(s) = M(s) \cdot A(s) \tag{9.37}$$

where

$$M(s) = -K \cdot G_{\text{VCO}} \cdot G_p(f_s) \cdot H_p(s) \tag{9.38}$$

The modulator group, $M(s)$, can be analyzed separately and gives a magnitude and phase profile, as shown in Figure 9.9. Based on the modulator gain (magnitude) curve and assuming the overall loop unity-gain

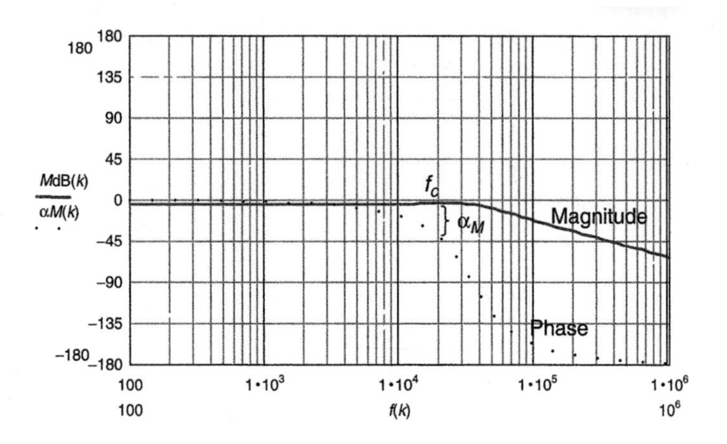

Figure 9.9: Modulator gain and phase

crossover occurs at the selected frequency, f_c, the modulator has a gain of $-G$ db and phase α_M. A modulator of this nature requires at least a type II compensation amplifier (Figure 9.10) to make the complete loop stable. The steps leading to the design of a type II amplifier capable of offering a maximum phase boost of 90° can be summarized as in the figures.

At the selected crossover frequency, the phase boost, α_b, needed for a desired phase margin, α_m, can be given as

$$\alpha_b = \alpha_m - \alpha_M - 90 \tag{9.39}$$

The phase boost angle is consequently translated to a pole/zero separation factor:

$$k_s = \tan\left[\frac{\pi}{180} \cdot \left(\frac{\alpha_b}{2} + 45\right)\right] \tag{9.40}$$

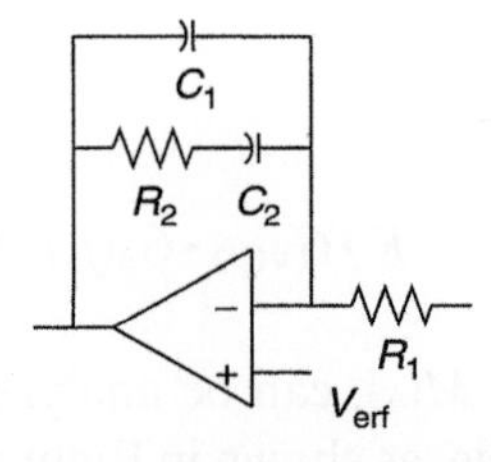

Figure 9.10: Type II amplifier

With a preselected input resistor, $R1$, for the type II amplifier (Figure 9.10), the rest of the component values are given as

$$C_1 = \frac{1}{2 \cdot \pi \cdot R_1 \cdot k_s \cdot f_c \cdot 10^{\frac{G}{20} - \log k_s}}$$
$$C_2 = (k_s^2 - 1) \cdot C_1 \tag{9.41}$$
$$R_2 = \frac{k_s}{2 \cdot \pi \cdot f_c \cdot C_2}$$

Subtle differences exist between the type II and type III amplifiers presented in Figure 6.19 of Wu [2] and Figures 3.8 through 3.15 of Wu [10]. Readers are encouraged to examine the difference. Chapter 11 gives more details regarding type II and type III amplifiers.

9.6 Example

To verify the validity of the preceding analytical procedure, a numerical example is given. In the example, the following component values, referring to Figures 9.5(a) and 9.5(b), and operating parameters are used:

$$L = 160\,\mu\text{H}$$
$$C_q = 850\,\text{pF}$$
$$C_b = 0.022\,\mu\text{F}$$
$$L_r = 43\,\mu\text{H}$$
$$C_d = 2420\,\text{pF}$$
$$R_L = 1.2\,\text{ohm}$$
$$C_e = (\pi^2/8) \cdot C_d$$
$$R_e = 0.81 \cdot n^2 \cdot R_L,\ n = 9$$
$$t_{\text{off}} = 600\,\text{ns}$$
$$f_s = 545\,\text{KHz}$$

The system shown in Figure 9.5(c), however, is a fifth-order system; that is, the system characteristic equation in the frequency domain, $D_1(s)$, is a fifth-degree polynomial in Laplace operator, s. Furthermore, most coefficients of the polynomial are complex combinations of circuit components:

$$D_1(s) = -\left\{ s^5 + \frac{1}{R_e \cdot C_e} \cdot s^4 + \left[\frac{1}{L_r} \cdot \left(\frac{1}{C_e} + \frac{1}{C_b} \right) + \left(\frac{1}{L} + \frac{1}{L_r} \right) \cdot \frac{1}{C_q} \right] \cdot s^3 \right.$$
$$+ \left[\frac{1}{L_r} \cdot \left(\frac{1}{C_q} + \frac{1}{C_b} \right) + \frac{1}{L \cdot C_q} \right] \cdot \frac{1}{R_e \cdot C_e} \cdot s^2 + \frac{1}{L \cdot L_r \cdot C_q} \cdot \left(\frac{1}{C_e} + \frac{1}{C_b} \right) \cdot s$$
$$\left. + \frac{1}{L \cdot L_r \cdot C_q \cdot C_b \cdot R_e \cdot C_e} \right\} \tag{9.42}$$

Polynomials in this form simply cannot be placed in the conventional factored forms needed to perform inverse Laplace transformation by hand. Therefore, all transfer functions, a total of 29 in this study, are left in their primitive forms and MathCAD V.7 (Mathsoft, Inc.), capable of inverse transformation, is invoked to obtain the corresponding time-domain functions. Once this is done, the all-state transition matrix A_1, B_1, A_2, and B_2 are hence available and eventually lead to

$$X_1 = \begin{bmatrix} 0.799 \\ 57.329 \\ -0.961 \\ -68.631 \end{bmatrix}, \quad X_2 = \begin{bmatrix} 0.259 \\ 72.247 \\ 1.415 \\ 75.809 \end{bmatrix}$$

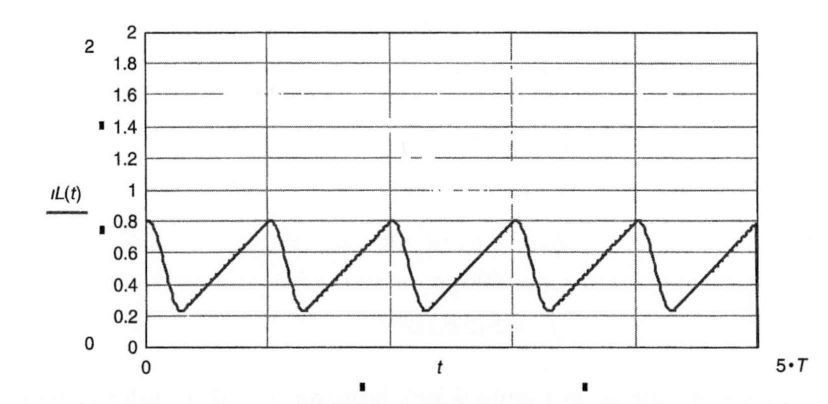

Figure 9.11: Input current

These starting states are then plugged into (9.22) to yield the input source current $i_a(t)$ (Figure 9.11); the switch voltage $v_d(t)$ (Figure 9.12); the blocking capacitor voltage $v_b(t)$ (Figure 9.13); the resonant tank inductor current $i_b(t)$ (Figure 9.14); the primary voltage $v_p(t)$ (Figure 9.15); and the resonant inductor voltage $vL_r(t)$ (Figure 9.16). In comparison with the actual measurements, all theoretical predictions match very well.

In addition, steady-state solutions shown as time-domain waveforms, (9.29) gives the DC output voltage of the power stage as a function of the switching frequency (Figure 9.17). Differentiation of (9.29) against the switching frequency gives the small-signal gain $G_p(f_s)$, (9.32), and is

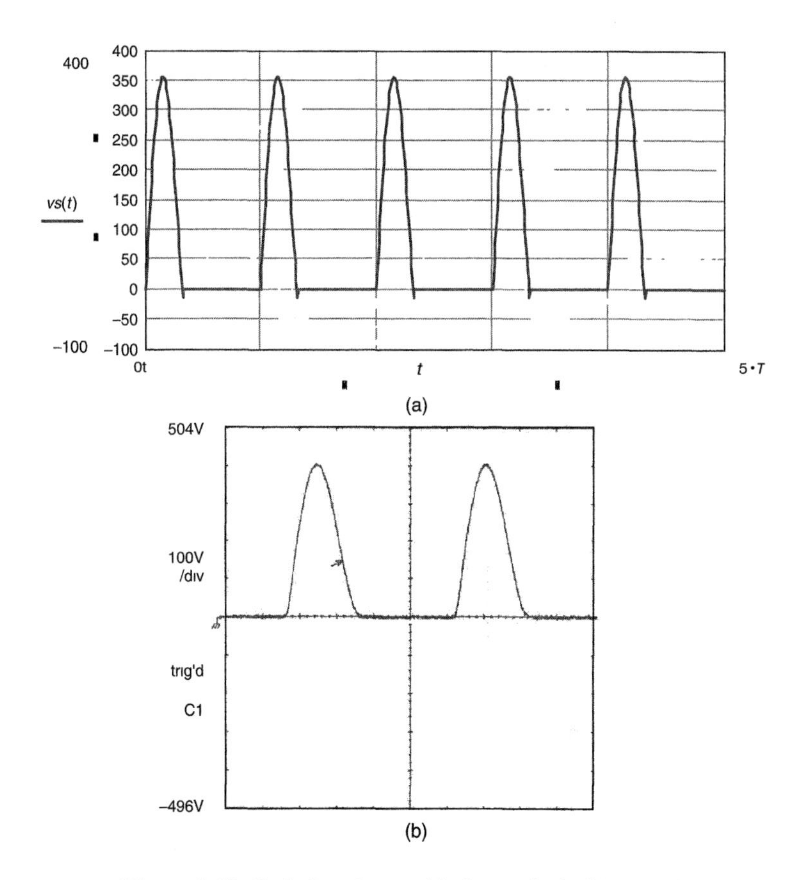

Figure 9.12: Switch voltage, (a) theoretical, (b) actual

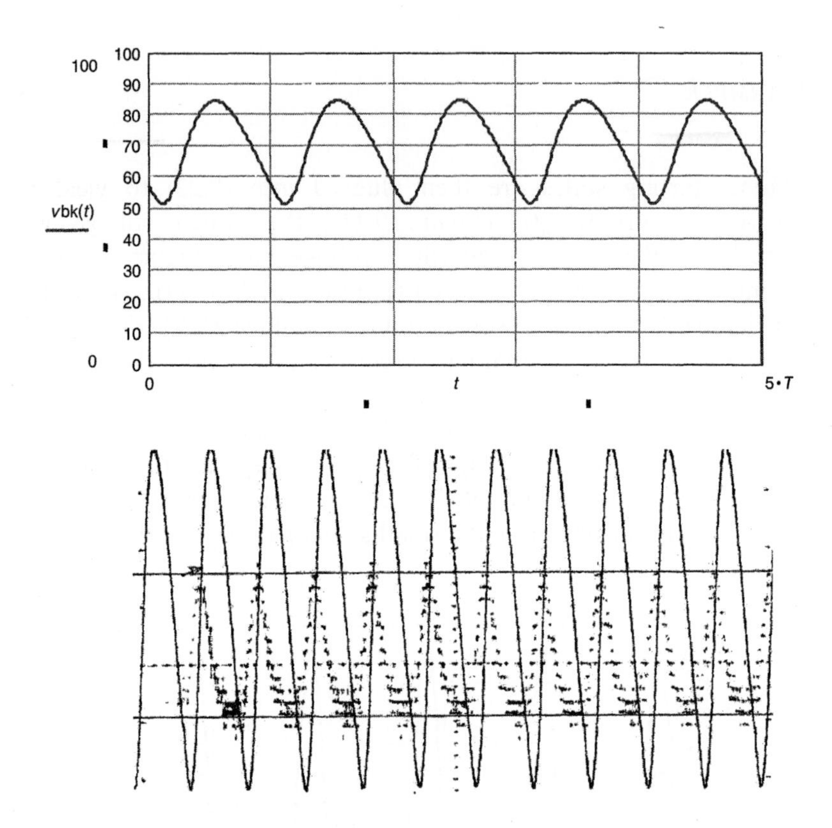

Figure 9.13: Blocking capacitor voltage, theoretical and actual

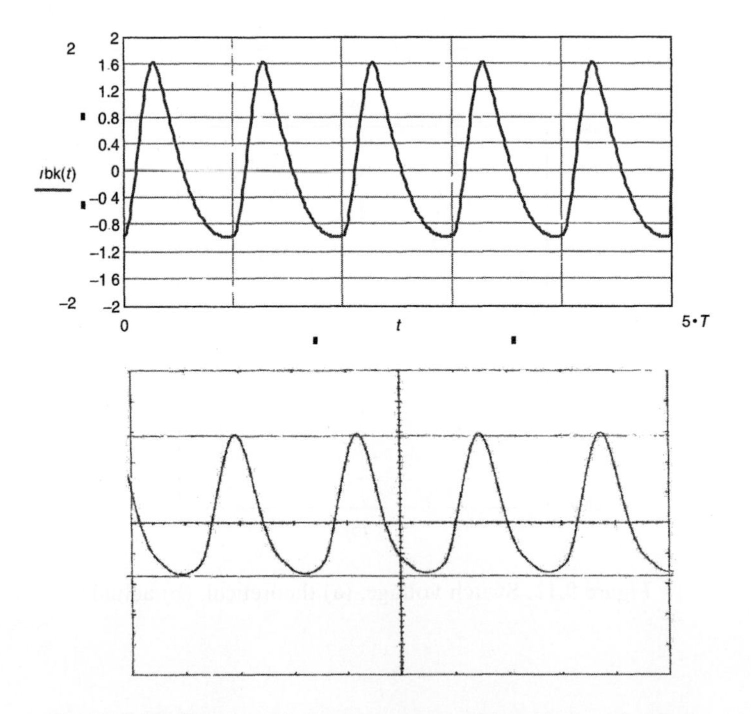

Figure 9.14: Resonant inductor current, theoretical and actual

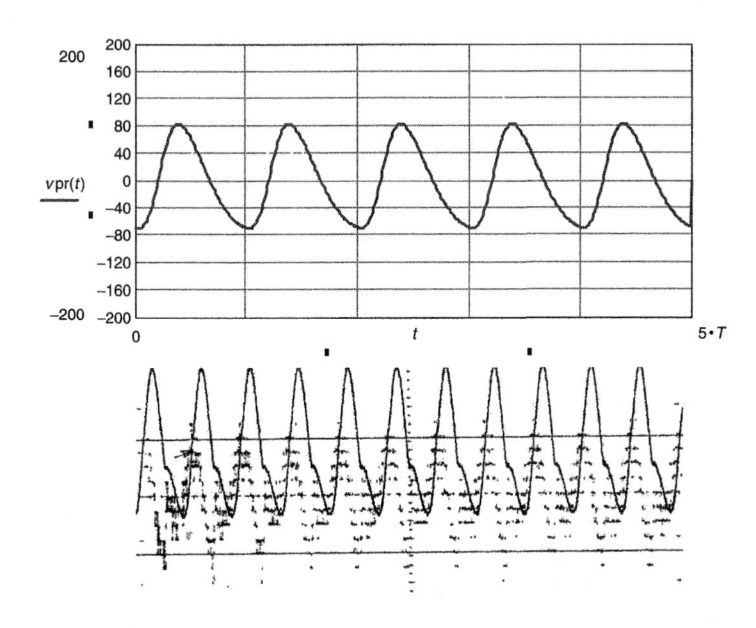

Figure 9.15: Primary (resonant capacitor) voltage, theoretical and actual

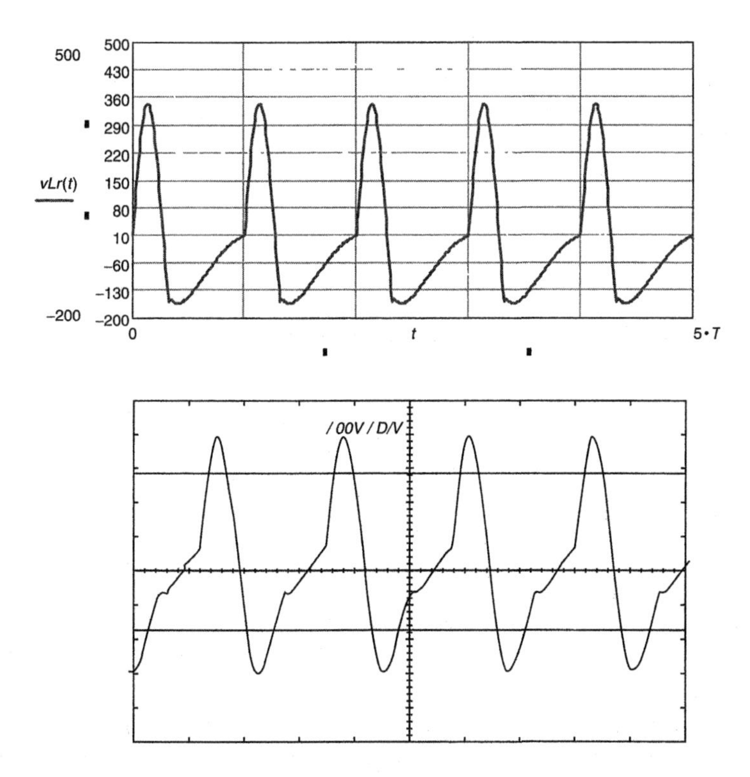

Figure 9.16: Resonant inductor voltage, theoretical and actual

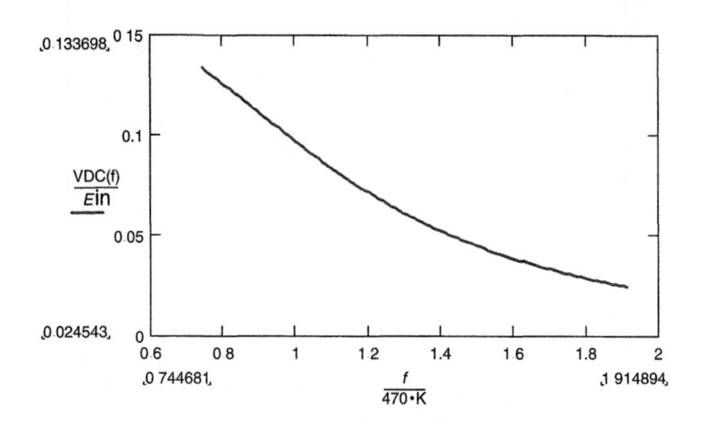

Figure 9.17: DC output as function of switching frequency

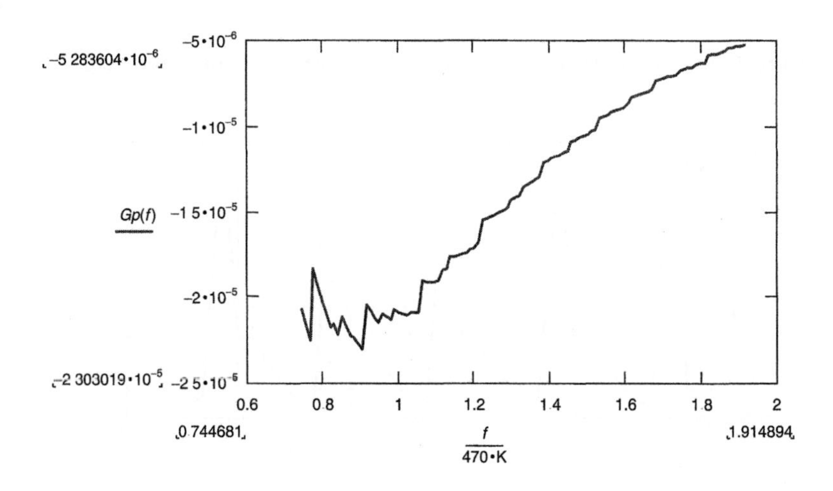

Figure 9.18: Power stage gain curve

plotted in Figure 9.18. This leads further to the closed-loop loop-gain shown in Figure 9.19, which compares well with Figure 9.20, the actual measurement.

More time-domain waveforms are worth investigating. The rectified secondary voltage comes to mind first. It is expressed as $[|v_{pr}(\mathrm{i})/n| - 0.5]$

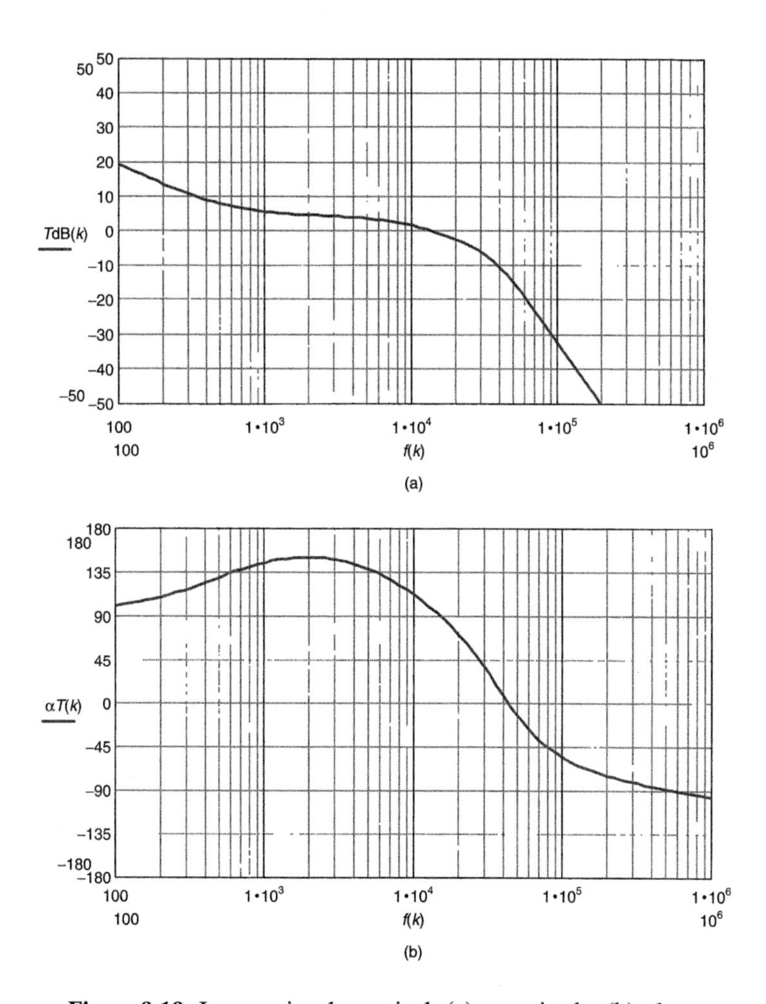

Figure 9.19: Loop gain, theoretical: (a) magnitude, (b) phase

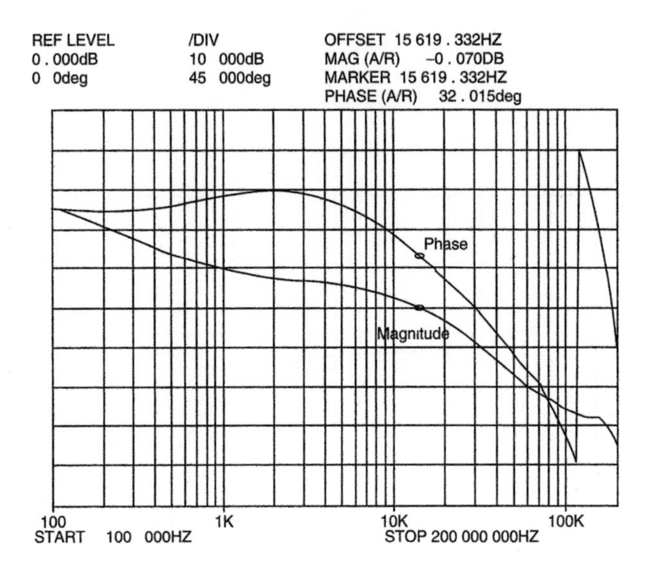

Figure 9.20: Loop gain, actual

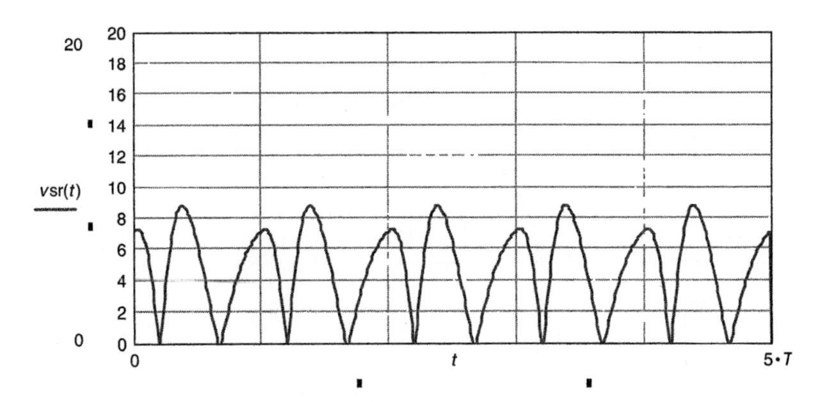

Figure 9.21: Rectified secondary voltage

(Figure 9.21), where the transformer turn ratio and Schottky diode drop are included. The current through capacitor, C_q, in parallel with the switch, is expressed as $C_q[dv_d(t)/dt]$ and its relationship with the input current and the resonant current is given in Figure 9.22.

Last, the main switch and its associated body diode current is expressed as the input current subtracted by the resonant current and the

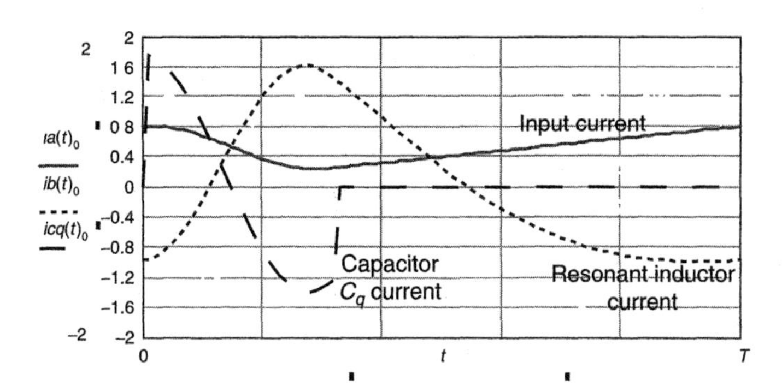

Figure 9.22: Relationship of C_q and the switch with the input current and the resonant current

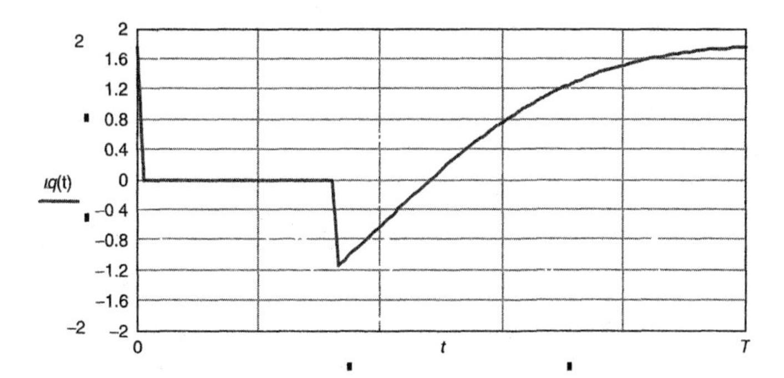

Figure 9.23: Switch and body diode current

current through C_q. In addition, we may calculate the power; for instance, the switch power losses equal the product of switch voltage (Figure 9.12) and switch current (Figure 9.23).

9.7 Discussion

As mentioned in the introduction, the analytical procedure associated with resonant converters often employs many assumptions because of its complexity. Among those assumptions, some are legitimate and some not. Assumptions considered questionable include the ideal sinusoidal branch current, ideal sinusoidal node voltage, constant source current, and so on. Those assumptions are removed in this study. However, one critical assumption and its derivations are retained: the assumption of effective power transfer consisting of only the fundamental component at the resonant frequency, as per Kazimierczuk [6]. This assumption leads to the establishment of the equivalent circuit of Figure 9.5(b) and the rule of load reflection giving the effective load used in the analysis.

However, this study shows that none of the branch currents or node voltages approaches an ideal sinusoidal waveform. This implies that multiple harmonics are present and calls into question the validity of power transfer at the fundamental frequency. As a result, the rule of load reflection may also be called into question and needs to be reexamined.

By assuming an ideal sinusoidal current through the resonant inductor, L_r, and the power transfer at the same frequency, Kazimierczuk [6] reached the end result of the load reflection rule that gives $R_e = 0.81n^2R_L$, R_L being the actual load.

One then may ask, given Kazimierczuk's [6] approach based on the existence of just one frequency, if the load reflection rule is still applicable for a periodic source current consisting of multiple harmonics. The answer to this question is yes, and its justification stems from the recognition that the load reflection rule holds for each harmonic standing alone and consequently for all harmonics superimposed (summed). This is the reason why the equivalent circuit of Figure 9.5(b) is used, even though the resonant branch current is eventually known to be not sinusoidal.

Another matter that needs to be discussed is the validity of the equivalent circuit shown in Figure 9.5(b). As is shown, the circuit is valid to a large extent, but it can be improved further. The improvement comes in the form of adding the magnetizing inductance of the transformer primary winding to the equivalent circuit. This step changes Figure 9.5 to Figure 9.24. By so doing, the system order increases by one and makes $D_1(s)$ a 6° polynomial while $D_2(s)$ becomes a 4° equation. Without question, the analytical procedure associated with this improved circuit is significantly more demanding than what has been obtained. Moreover, several dissipative components can be added

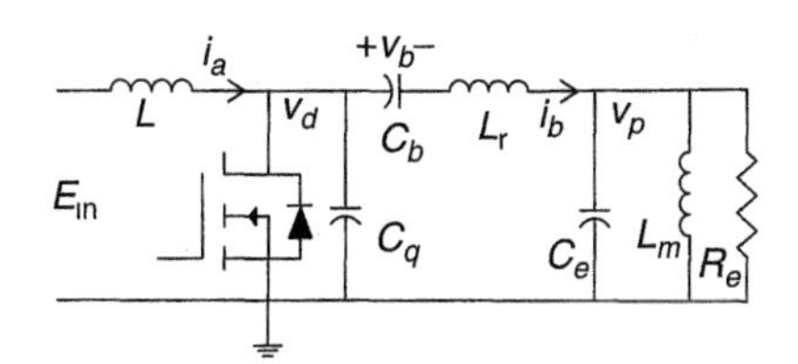

Figure 9.24: Adding magnetizing inductance

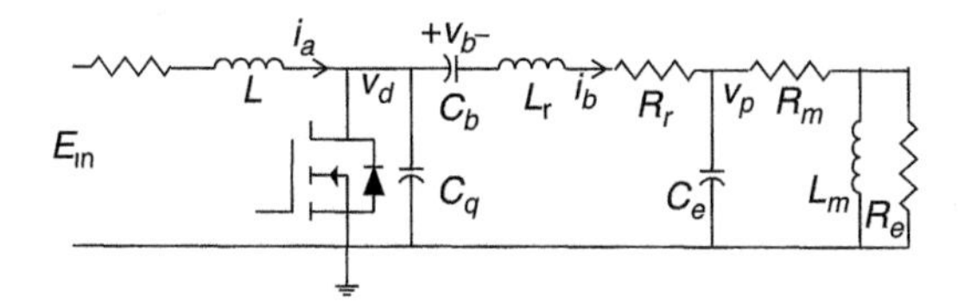

Figure 9.25: Adding dissipative components

to Figure 9.24, resulting in Figure 9.25. This latter step does not change the system order but it does add damping properties to the system. In theory, this improvement makes the analysis results better represent the actual hardware performance. Although extremely cumbersome, the procedure may yield results that far outweigh the additional effort. More studies are therefore possible and suggested in this regard.

Chapter 10

AC–DC Power-Factor Correction Supplies

Conventional single-phase AC–DC rectification (Figure 10.1) was treated in Wu [10], who showed in that analysis, as was well understood in the past, that the input line current is pulsating and nonsinusoidal. This implies high harmonic contents and a low power factor. The former produces electromagnetic interferences and the latter gives rise to low-power transfer efficiency. It is therefore imperative to reshape, by some means, the input current such that not only is it in phase with the rectified source voltage but also sinusoidal in its time-domain profile with low harmonic distortion. That is the essence of power-factor correction, PFC.

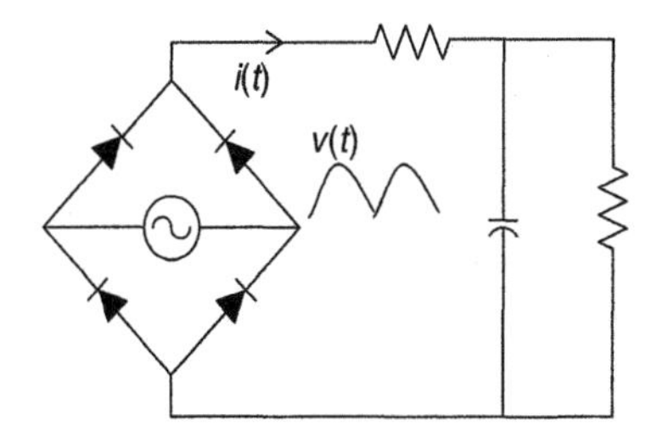

Figure 10.1: AC–DC rectifier and filter

In the old days, the objective was met by passive means, adding huge capacitor banks to the AC line. In contrast, modern techniques approach it through active means, employing a switch-mode converter and forcing line current to match the line voltage. We attempt to cover the topic by briefly going over some definition issues, then examine the fundamental theory for boosting the PFC, the output capacitor size, the boost inductor selection, high-power parallel operation, current sharing issues, startup considerations, output short-circuit protection, control issues, and three-phase PFC, in that order.

10.1 Fundamental Definition

Given a circuit (Figure 10.2) that contains passive components and active switching devices and is driven by a single-frequency sinusoid, the instantaneous input power is given as

$$p(t) = v(t) \cdot i(t) = V \cdot \sin(\omega t) \cdot \sum_n I_n \cdot \sin(n \cdot \omega \cdot t + \theta_n), \quad n = 1,2,\ldots \tag{10.1}$$

The input current is expressed as a Fourier series because of the understanding that it is cyclic but nonsinusoidal. We can expand the expression, which leads to

$$p(t) = V \sum_n (I_n \cdot \cos\theta_n \cdot \sin\omega t \cdot \sin n\omega t + I_n \cdot \sin\theta_n \cdot \sin\omega t \cdot \cos n\omega t) \tag{10.2}$$

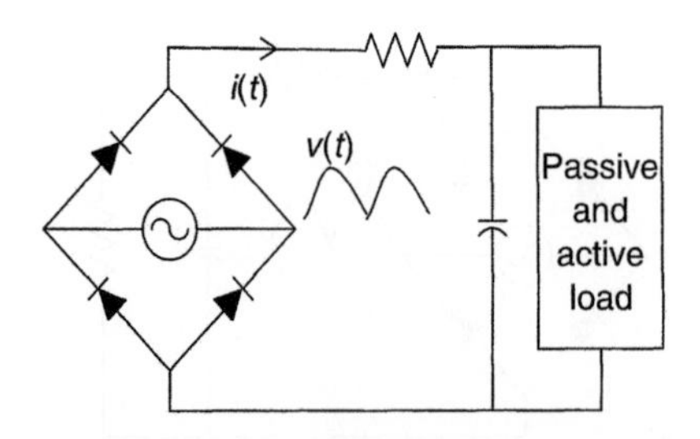

Figure 10.2: AC–DC rectifier with a nonlinear load

We then take the average of (10.2):

$$< p(t) >= V\sum_n (I_n \bullet \cos\theta_n \bullet < \sin\omega t \bullet \sin n\omega t > + I_n \bullet \sin\theta_n \bullet \\ < \sin\omega t \bullet \cos n\omega t >) \tag{10.3}$$

From the theory of orthogonal functions,

$$< \sin\omega t \bullet \sin n\omega t > = \begin{cases} ½ & n = 1 \\ 0 & n = 2,3,\ldots \end{cases} \\ < \sin\omega t \bullet \cos n\omega t > = 0 \quad n = 1,2,3,\ldots \tag{10.4}$$

Therefore,

$$< p(t) >= V\frac{I_1 \bullet \cos\theta_1}{2} = \frac{V}{\sqrt{2}}\frac{I_1}{\sqrt{2}}\cos\theta_1 = V_{\text{rms}} \bullet I_{1\_\text{rms}}\cos\theta_1 \tag{10.5}$$

This equation is the basis for the power factor, PF, definition, which is defined as

$$PF = \frac{< p(t) >}{V_{\text{rms}} \bullet i_{\text{rms}}} = \frac{I_{1\_\text{rms}}}{i_{\text{rms}}}\cos\theta_1 = k_d \bullet k_\theta \tag{10.6}$$

where k_θ is attributed to the angular displacement of the fundamental current referred to the input voltage source, while k_d represents current shape distortion. For instance, given a single-phase AC source of 120 Vrms, 60 Hz, line resistance 0.5 ohm, filter capacitor 200 μF, and expected 140 VDC output loaded at 57 ohm, the steady-state output voltage and the rectified input line current are as shown in Figure 10.3. The RMS (Root Mean Squared) input current is 4.45 A. That is translated to an input apparent power $V_{\text{rms}} \bullet i_{\text{rms}} = 534$ W. The average power is 351 W. The power factor is therefore 0.658.

The input line current has a $\sin\omega t$ component of 4.138 A and a $\cos\omega t$ component of 2.158 A. It gives an RMS current at the fundamental frequency of 3.3 A and a phase angle of $\theta_1 = 27.543°$. The displacement factor is therefore $k_\theta = \cos\theta_1 = 0.887$, while the distortion factor $k_d = 0.742$. The power factor is $k_d \bullet k_\theta = 0.658$. Both computations agree.

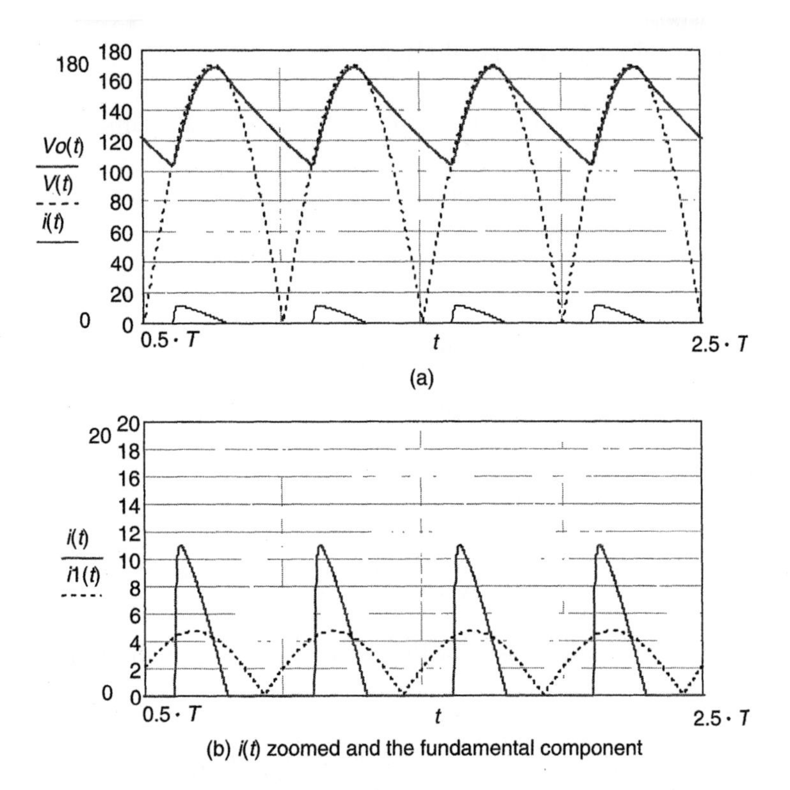

Figure 10.3: Example AC–DC rectifier and filtered output

10.2 Single-Phase Single-Stage Nonisolated Boost PFC

Among the three basic topologies, only the boost and the flyback are suitable for power-factor correction applications. The buck (step-down) configuration is excluded because its energy storage inductor is not at the input side. Without an input-side inductor, the input current pulsates. Of the remaining two, the boost configuration is preferred, for the flyback (buck boost) does not perform well over the span of a complete rectification cycle, in particular, the transition, boost to buck or buck to boost, that takes place at some point in time on the rising, or falling, half cycle. The boost topology is not without its problems at the low point (cusp) of a rectified sinusoid. However, this can be taken care of passively with the

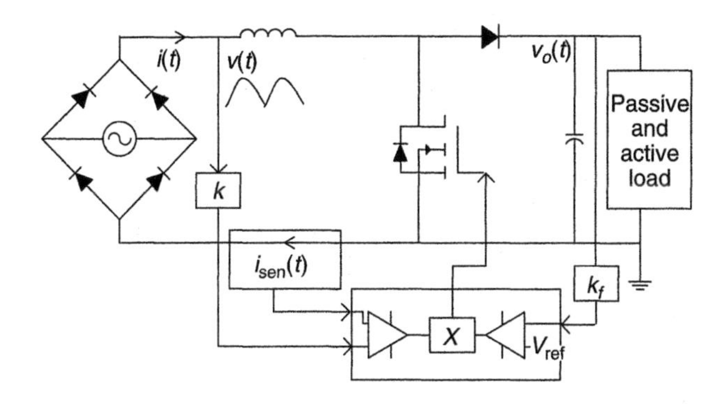

Figure 10.4: Boost PFC

output capacitor. In contrast, it is rather difficult to assure a smooth transition between boost and buck twice every rectified cycle. It is a control issue. It is also an EMI (Electro-Magnetic Intereference) issue. For those considerations, we focus on the boost PFC alone.

The boost PFC typically comes in the form of Figure 10.4, in which the boosted output, $v_o(t)$ is fed back and sampled. The sampled output is compared with a reference. Meanwhile, the switched line current is also sampled, $i_{sen}(t)$ and compared against a reduced line voltage. Both the voltage error and the current error are multiplied. The product controls the power switch.

By so doing, the line current is forced to follow the line voltage. However, since a product voltage commands the power switch, one may ask just which error is the primary control factor. In general, the current loop has a wider bandwidth (fast), because after all, current shaping is the goal. The voltage loop is therefore low in bandwidth (slow), and the boosted output voltage is expected to carry significant ripple at twice the line frequency. In the next section, we attempt to size the output capacitor.

10.3 Output Capacitor Size

To formulate and understand the mechanism of a boost PFC, Figure 10.4 is simplified to Figure 10.5. In the figure, the input current is

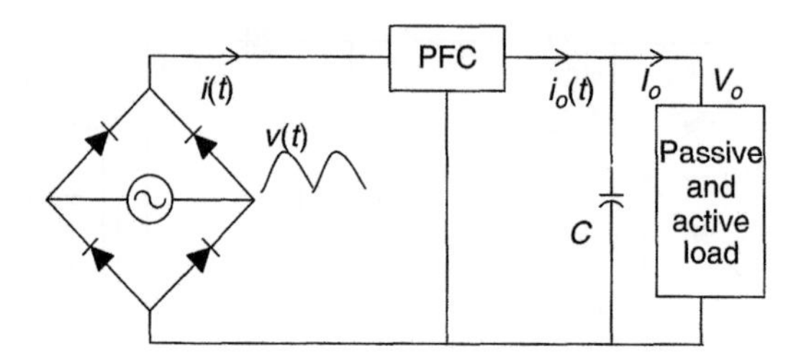

Figure 10.5: Simplified boost PFC

assumed to be well controlled and exhibits a sinusoid shape in phase with the line voltage. The input power is therefore given as

$$p_{\text{in}}(t) = \sqrt{2}V_{\text{rms}}|\sin\omega t| \cdot \sqrt{2}I_{\text{rms}}|\sin\omega t| = V_{\text{rms}}I_{\text{rms}}(1 - \cos 2\omega t) \quad (10.7)$$

If the PFC is assumed to have a conversion efficiency η, the output power is given by

$$p_o(t) = i_o(t) \cdot V_o = \eta \cdot V_{\text{rms}}I_{\text{rms}}(1 - \cos 2\omega t) \quad (10.8)$$

or

$$i_o(t) = \frac{\eta \cdot V_{\text{rms}}I_{\text{rms}}(1 - \cos 2\omega t)}{V_o} = I_o - \frac{\eta \cdot V_{\text{rms}}I_{\text{rms}}}{V_o}\cos 2\omega t \quad (10.9)$$

Clearly, the PFC output current consists of two components, the load current I_o and the capacitor current. The latter is seen to run at twice the line frequency. As a matter of fact, (10.8) yields an average power $< p_o(t) >= \eta \cdot V_{\text{rms}} \cdot I_{\text{rms}} =$ load power. The term running at twice the line frequency yields a zero average. We can integrate the capacitor current, the second term on the right-hand side of (10.9), and obtain the PFC output ripple voltage as

$$v_o(t) = \frac{1}{C}\int \frac{\eta \cdot V_{\text{rms}}I_{\text{rms}}}{V_o}(\cos 2\omega t)dt = \frac{\eta \cdot V_{\text{rms}}I_{\text{rms}}}{2 \cdot \omega \cdot C \cdot V_o}\sin(2\omega t + \pi) \quad (10.10)$$

Given a requirement on the ripple magnitude, δv_o, the output capacitor size can be estimated using (10.10).

$$C = \frac{\eta \cdot V_{\mathrm{rms}} I_{\mathrm{rms}}}{2 \cdot \omega \cdot \delta v_o \cdot V_o} \tag{10.11}$$

In all, the key waveforms for a PFC power stage can be summarized as the following, given $V_{\mathrm{rms}} = 120$ V, output power 2000 W, and output voltage 210 V. Figures 10.6 through 10.9 present a graphic demonstration of this.

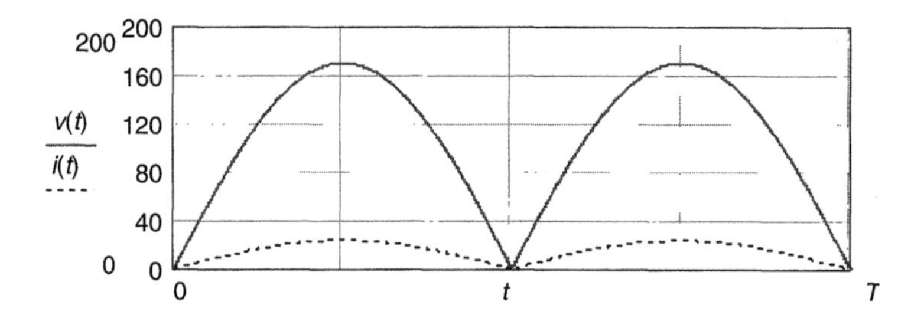

Figure 10.6: Input line voltage and current

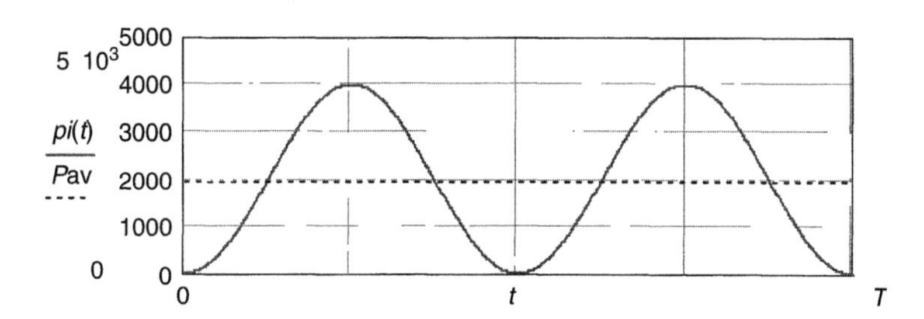

Figure 10.7: Input instantaneous power and average power

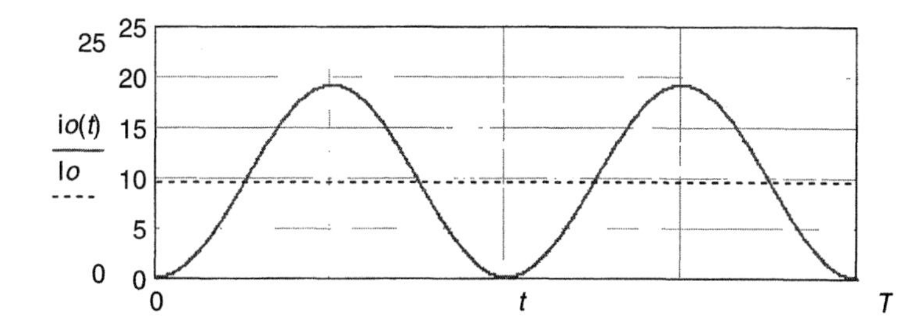

Figure 10.8: Output current and load current

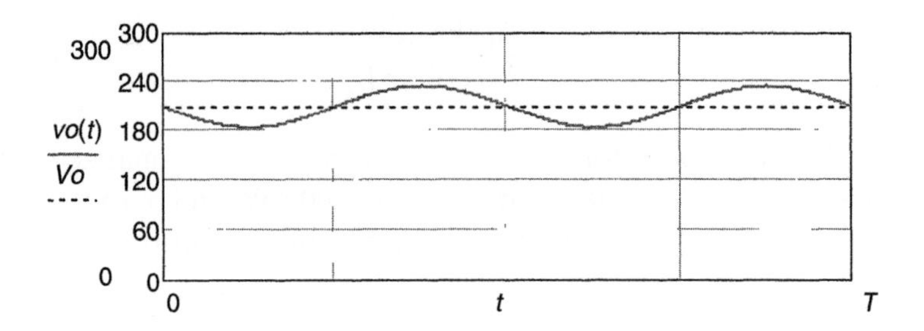

Figure 10.9: Output voltage

10.4 DCM Boost Inductor Selection

Boost PFCs' inductor may operate in two modes—DCM and CCM. In DCM, the pulsating inductor current looks like Figure 10.10, in which the triangular current peaks follow the sinusoid contour and the rectified input voltage.

We first give a detailed examination for one typical current pulse, Figure 10.11, under the assumption of constant switch-on time, t_{on}.

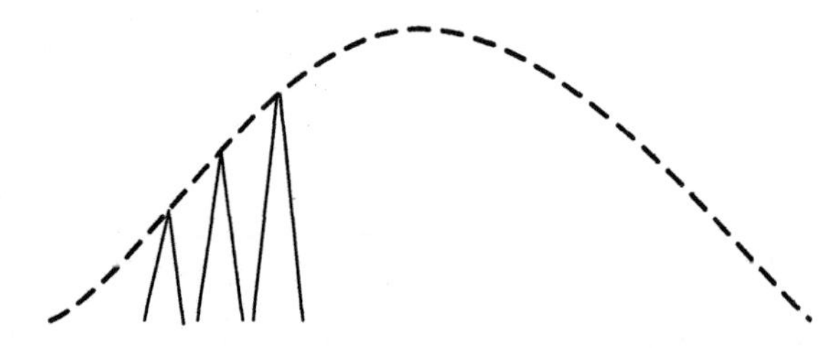

Figure 10.10: Inductor current (solid line) in DCM

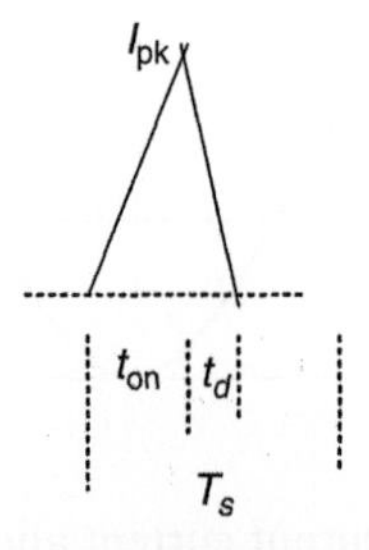

Figure 10.11: Inductor current pulse

If that is the operation mode, the inductor current peak is expressed as

$$I_{pk} = \frac{\sqrt{2}V_{rms}|\sin\omega t|}{L} \bullet t_{on} = \frac{V_{in}|\sin\omega t|}{L} \bullet t_{on} \tag{10.12}$$

At first glance, (10.12) seems to indicate a perfect sinusoidal line current. However, the equation is deceiving. We recall from (10.3) that it is the average that counts. It turns out that the average of (10.12) is distorted and consists of multiple harmonics. We shall try to prove the assertion. First, the volt–second balance demands

$$t_d = \frac{V_{in}|\sin\omega t| \bullet t_{on}}{V_o - V_{in}|\sin\omega t|} = \frac{L \bullet I_{pk}}{V_o - V_{in}|\sin\omega t|} \tag{10.13}$$

The inductor current's average is then given by

$$< i_L(t) > = \frac{I_{pk}(t_{on} + t_d)}{2 \bullet T_s} \tag{10.14}$$

The equation can be easily rewritten in an intermediate step as

$$\begin{aligned} < i_L(t) > &= \frac{V_{in}|\sin\omega t| \bullet t_{on}}{2 \bullet L \bullet T_s}\left(t_{on} + \frac{V_{in}|\sin\omega t| \bullet t_{on}}{V_o - V_{in}|\sin\omega t|}\right) \\ &= \frac{V_{in}|\sin\omega t| \bullet t_{on}^2}{2 \bullet L \bullet T_s}\left(1 + \frac{V_{in}|\sin\omega t|}{V_o - V_{in}|\sin\omega t|}\right) \end{aligned} \tag{10.15}$$

We can gain more insight if (10.15) is placed in the following form:

$$\begin{aligned} < i_L(t) > &= \frac{V_o \bullet t_{on}^2}{2 \bullet L \bullet T_s}\left(1 + \frac{\frac{V_{in}}{V_o}|\sin\omega t|}{1 - \frac{V_{in}}{V_o}|\sin\omega t|}\right)\frac{V_{in}}{V_o}|\sin\omega t| \\ &= \frac{V_o \bullet t_{on}^2}{2 \bullet L \bullet T_s}\left(1 + \frac{\alpha \bullet |\sin\omega t|}{1 - \alpha \bullet |\sin\omega t|}\right)\alpha|\sin\omega t| \\ &= \left[\frac{V_o \bullet t_{on}^2}{2 \bullet L \bullet T_s}\frac{\alpha}{1 - \alpha \bullet |\sin\omega t|}\right]|\sin\omega t| \end{aligned} \tag{10.16}$$

It is very clear, given the magnitude of the term in brackets [ ] in (10.16) that the average line current is distorted. It will follow the input line voltage if the term in the parentheses ( ) is forced to 1, or equivalently the ratio term in the parentheses is forced to 0. This can be done only with $\alpha = V_{in}/V_o = 0$, but then $< iL >$ is also forced to 0. A compromise is a small value for α, or equivalently a high-boost output voltage V_o compared with the input line voltage magnitude V_{in}.

We can also look at the issue at hand by rewriting (10.14) as

$$< i_L(t) >= \frac{I_{pk}(D + D_2)}{2} \tag{10.17}$$

where duty-cycle symbols, not constants, are used. With a little patience, it can be reformulated as

$$< i_L(t) >= \frac{D^2 \bullet V_o \bullet V_{in}|\sin\omega t|}{2 \bullet L \bullet f_s \bullet (V_o - V_{in}|\sin\omega t|)} \tag{10.18}$$

Note that a difference term, $(V_o - V_{in}|\sin\omega t|)$, appears in the denominator. Simple mathematics tells us that the presence of a near zero quantity, if improperly selected, in the denominator of a quotient can blow up hyperbolically and nonlinearly introducing distortion. Given $120\,V_rms$, $60\,Hz$, $L = 1\,mH$, $T_s = T/200$, $T =$ line cycle, $t_{on} = 0.3T_s$, and plotting (10.16), Figure 10.12 gives a clear view as to what happens to the line current (one rectified cycle) when the boost ratio is high and low.

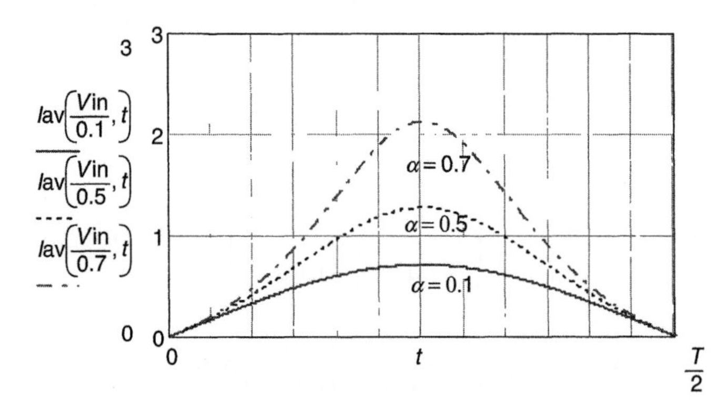

Figure 10.12: Boost ratio ($\alpha = V_{in}/V_o$) effects on the average line current

Evidently, a constant on time, t_{on}, or constant duty cycle, D, at the selected switching frequency, f_s, is not the best strategy. Well, then, what can be done to improve the situation? We know what we want: the average line current proportional to and in phase with the line voltage; that is, making

$$< i_L(t) >= \frac{D^2 \bullet V_o \bullet V_{\text{in}}|\sin\omega t|}{2 \bullet L \bullet f_s \bullet (V_o - V_{\text{in}}|\sin\omega t|)} = k \bullet V_{\text{in}}|\sin\omega t| \tag{10.19}$$

where k is a constant. It is equivalent to saying

$$k = \frac{D^2 \bullet V_o}{2 \bullet L \bullet f_s \bullet (V_o - V_{\text{in}}|\sin\omega t|)} \tag{10.20}$$

In other words, instead of being a constant, the duty cycle is

$$D = \sqrt{\frac{2 \bullet k \bullet L \bullet f_s \bullet (V_o - V_{\text{in}}|\sin\omega t|)}{V_o}} = \sqrt{\frac{2 \bullet k \bullet L \bullet f}{V_o}} \sqrt{V_o - V_{\text{in}}|\sin\omega t|} \tag{10.21}$$

That is, the duty cycle tracks the square root of the instantaneous output and input difference. The correction factor, the second square root, has the form of Figure 10.13.

Up to this point in this section, we have not dealt with the determination of inductor size. Fortunately, (10.13) and (10.21) together point out

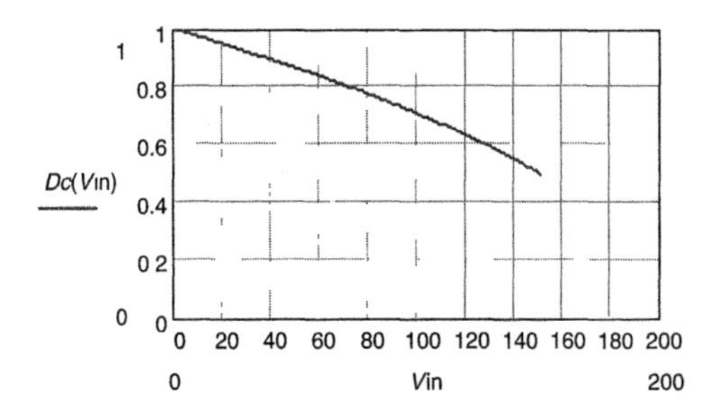

Figure 10.13: Line-voltage-corrected duty cycle

where we should be heading. We recognize that the sum of t_{on} and t_d in Figure 10.11 must be less than T_s:

$$t_{\text{on}} + t_d = t_{\text{on}}\left(1 + \frac{V_{\text{in}}|\sin\omega t|}{V_o - V_{\text{in}}|\sin\omega t|}\right) \leq T_s \tag{10.22}$$

or

$$D\left(\frac{V_o}{V_o - V_{\text{in}}|\sin\omega t|}\right) \leq 1 \tag{10.23}$$

Replacing D with (10.21), we get

$$\sqrt{\frac{2 \cdot k \cdot L \cdot f_s \cdot (V_o - V_{\text{in}}|\sin\omega t|)}{V_o}}\left(\frac{V_o}{V_o - V_{\text{in}}|\sin\omega t|}\right) \leq 1 \tag{10.24}$$

or

$$L \leq \frac{1 - \alpha|\sin\omega t|}{2 \cdot k \cdot f_s} \tag{10.25}$$

Clearly, given a selected boost ratio α, the upper bound of the boost inductor is given as

$$L_{\text{max}} \leq \frac{1 - \alpha}{2 \cdot k \cdot f_s} \tag{10.26}$$

Equation (10.26) ensures that the DCM is maintained throughout every rectified cycle.

10.5 CCM Boost Inductor Selection

In CCM operation, the boost inductor current looks like Figure 10.14. It is bounded by two sinusoids.

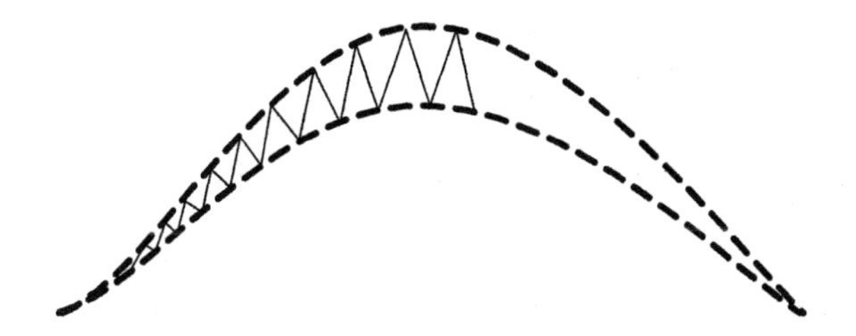

Figure 10.14: Boost inductor current in CCM

Unlike the DCM case, the selection process for the CCM boost inductor is less clear-cut. The most uncertain task in the process is the step selecting the ripple current magnitude. Before we proceed, we look at an interesting property of boost topology in CCM. The time-varying ripple current magnitude can be expressed as

$$\delta i(t) = \frac{V_{\text{in}}|\sin\omega t|D}{L \bullet f_s} = \frac{V_{\text{in}}|\sin\omega t|}{L \bullet f_s}\left(1 - \frac{V_{\text{in}}|\sin\omega t|}{V_o}\right) \tag{10.27}$$

A simple calculus operation shows that the ripple magnitude reaches a maximum when the instantaneous input voltage equals one half of the output. At that instant, the ripple current magnitude equals

$$\delta i_{\text{max}} = \frac{V_o}{4 \bullet L \bullet f_s} \tag{10.28}$$

This gives the ripple current magnitude an interesting profile over one rectified cycle, as shown in Figure 10.15. It reaches maximum twice if the conditions are right.

Given a required output power, P_o, and conversion efficiency, η, the input average current is

$$i_{\text{av}}(t) = \frac{\sqrt{2} \bullet P_o}{V_{\text{rms}} \bullet \eta}|\sin\omega t| = I_{\text{pk}}|\sin\omega t| \tag{10.29}$$

if we assume that the ripple current is riding on the average. The instantaneous input current is enveloped by

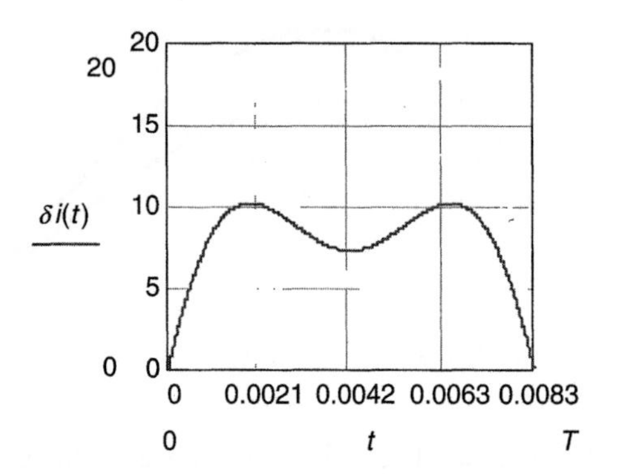

Figure 10.15: Ripple current magnitude in one half cycle

$$
\begin{aligned}
i(t) &= i_{\text{av}}(t) \pm \frac{\delta i(t)}{2} \\
&= \left\{ \frac{\sqrt{2}\cdot P_o}{\eta \cdot V_{\text{rms}}} \pm \frac{V_{\text{in}}}{2L\cdot f_s}\left(1 - \frac{V_{\text{in}}|\sin\omega t|}{V_o}\right)\right\}|\sin\omega t| \qquad (10.30)
\end{aligned}
$$

Ideally, for CCM operation, the instantaneous current during the half cycle never reaches zero. This condition requires

$$
\frac{\sqrt{2}\cdot P_o}{\eta \cdot V_{\text{rms}}} - \frac{V_{\text{in}}}{2L\cdot f_s}\left(1 - \frac{V_{\text{in}}|\sin\omega t|}{V_o}\right) \geq 0 \qquad (10.31)
$$

That is,

$$
L_{\text{min}} \geq \frac{\eta \cdot V_{\text{rms}}^2}{2\cdot P_o \cdot f_s}\left(1 - \frac{\sqrt{2}\cdot V_{\text{rms}}|\sin\omega t|}{V_o}\right) \qquad (10.32)
$$

For practical considerations,

$$
L_{\text{min}} \geq \frac{\eta \cdot V_{\text{rms}}^2}{2\cdot P_o \cdot f_s} \qquad (10.33)
$$

10.6 High-Power PFC and Load Sharing

As of now, most single-phase boost PFC power supplies available as commercial, off-the-shelf modules offer no more than 1 KW. To obtain high power, for instance, 10 KW, multiple modules are placed in parallel with active forward current control to assure load current sharing (Figure 10.16).

But the forward current control alone does not guarantee return current balance among modules. In addition, balance resistors, $R_1, R_2, \ldots, R_n$, are also placed on the load return line to further ensure load sharing. Yet, as shown, the scheme still has a major detriment. The switched currents of individual modules do not return properly. They cross flow as shown (solid dark line, for example) in the figure. The defect disrupts both current shaping sensing and load return current balance and eventually leads to failure.

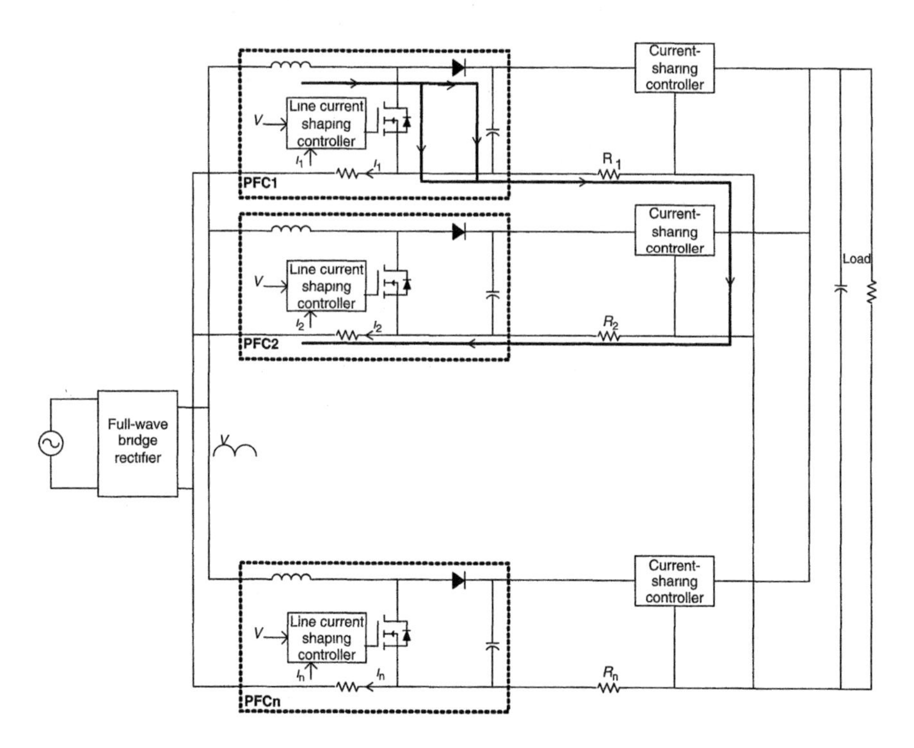

Figure 10.16: Multiple PFCs in parallel with current sharing

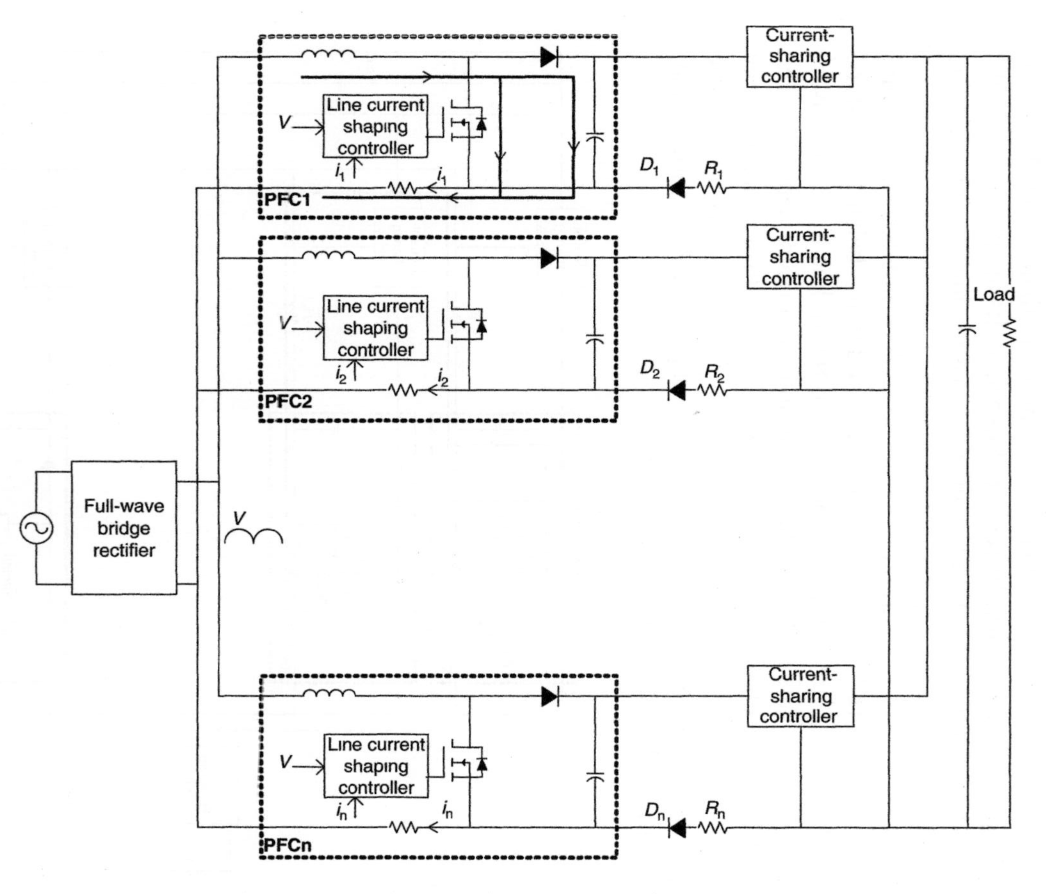

Figure 10.17: Improved load return current balance

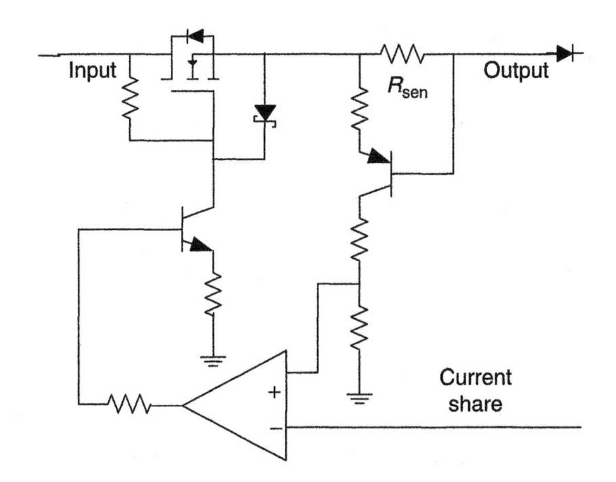

Figure 10.18: Current-sharing controller

The destructive crosstalk can be remedied by incorporating a unidirectional device (US Patent 6,703,946)—a diode—in series with each balance resistor, shown in Figure 10.17. By so doing, the switch current of each module is isolated and load current sharing is assured, including return current equalization and unidirectionality.

Figure 10.18 gives a simple current-sharing controller in which the "current-share" signal, equal to (total load current)/(parallel module number), is common to all controllers while the output current is summed via the diode ORing. A total load current sensor follows the summing diodes and provides the "current share."

The current-sharing controller, as given, suffers a minor deficiency. The current sensor is the combination of an in-line resistor, R_{sen}, and the base-emitter junction of a bipolar transistor. Aside from power dissipation across R_{sen}, a minimum output current in the form of (current × resistance) is required to forward bias the bipolar transistor's base-emitter junction. Prior to this condition, the current-sharing control loop is essentially open, because the bipolar transistor is not conducting. The emitter resistor may be removed or a base resistor may be added. Either way, some minimum voltage must be present to overcome the base-emitter junction. However, at low loads, precise current sharing may not be so imperative. In that case, the total circuit can afford to have some current controllers not operating. Under heavy loads, it would not be an issue.

10.7 Surge Protection

The circuit as given in Figure 10.16 has one more problem which is quite serious during initial turn-on if a capacitive load also is present. The mechanism of concern can be described as follows. Note that all front-end PFCs have an output capacitor. And, in general practice, PFCs are turned on and ready prior to turning on the current-sharing controllers. However, the load capacitors are generally discharged with zero starting voltage. At the moment the current-sharing controllers commence conduction, surge currents result in an instantaneous redistribution of the existing charge in all PFC output capacitors to the load capacitors. The resulting surge may easily overstress the main current-sharing controller element, the MOSFET.

There are numerous ways to protect against such a potentially destructive condition. Figure 10.19 shows a current-sharing controller with a synchronous soft start from a master command. Prior to turning on, the soft-start command places a 5-volt bias voltage at the noninverting input of the error amplifier. By so doing, the bipolar npn-transistor is fully on and the main MOSFET is shut off. At the commencement of the soft start, the soft-start capacitor slowly releases the bias voltage and the main MOSFET builds up conduction gradually.

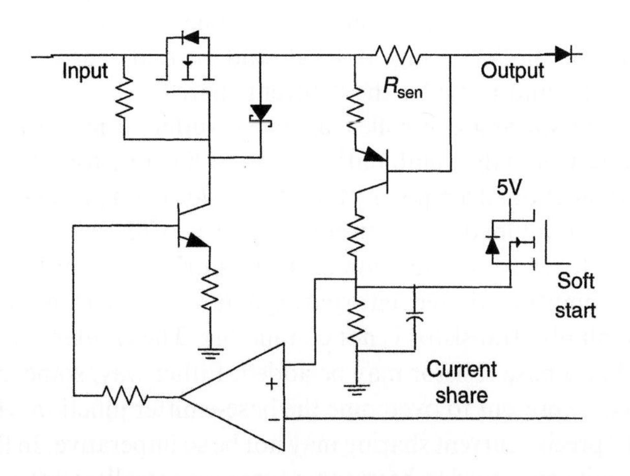

Figure 10.19: Current-sharing controller with soft start

This approach requires several components for each current controller. It is not considered a good solution. In theory, a better approach that has the ability to achieve the same surge suppression is shown in Figure 10.20, in which a saturable reactor (reset mechanism not shown) is employed (patent pending) to slow down the inrush current during turn-on transient. It is understood that the reactor behaves as a variable current limiter in the course of high *di/dt*. It behaves as a short circuit once the through current stabilizes. In essence, it is a nonlinear switch.

Further improvement can be made by precharging the load capacitor bank prior to commencing all current controllers. This is done by tapping the full-wave rectifier output and connecting it through a diode (patent pending) to the load capacitors (Figure 10.21). With this option, the saturable reactor is resized.

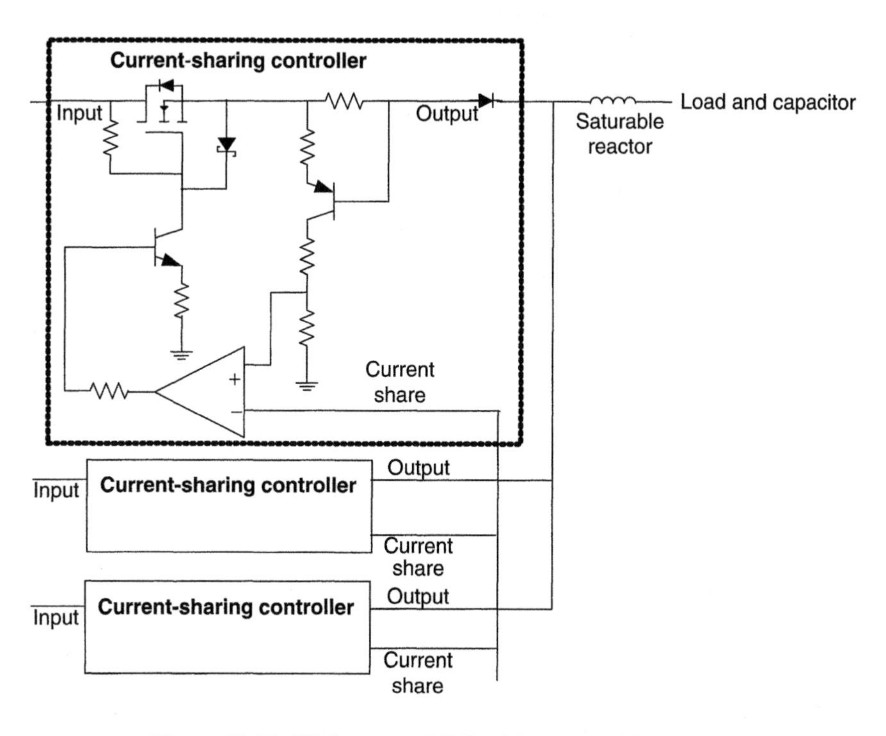

Figure 10.20: High-power PFC with a saturable reactor

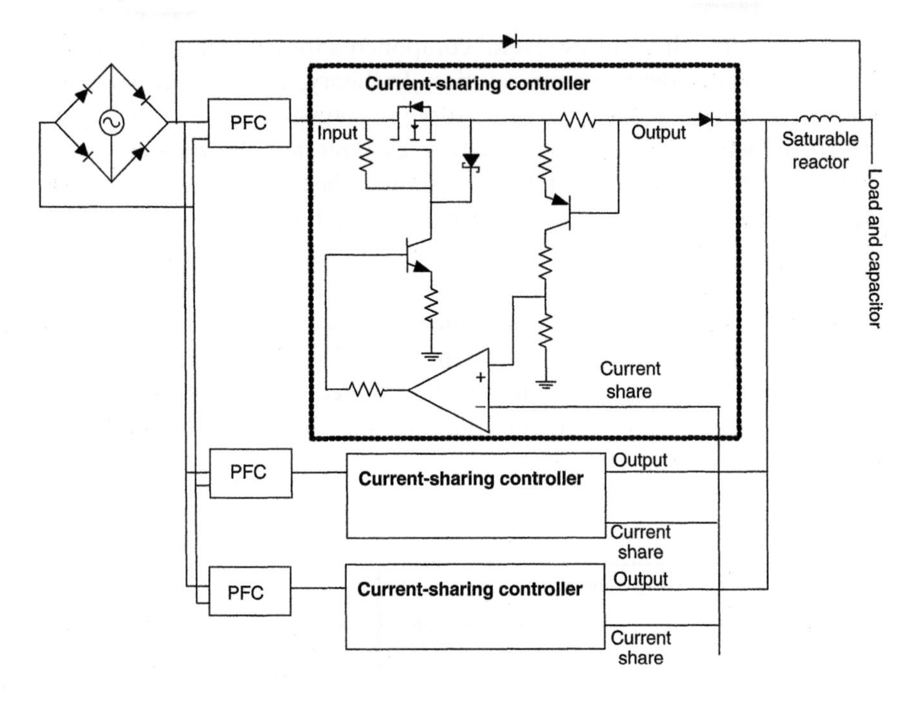

Figure 10.21: High-power PFCs with precharge and surge suppression

10.8 Load Short-Circuit Protection

Like any power converter, the capacitive load is prone to failure. In the case of PFCs, this is even more of a concern because of higher output ripple voltage and, consequently, higher AC current through the load capacitor. Implementing overload and short-circuit protection becomes a necessity. Active overload protection by sensing and limiting the output current is the first line of defense, if the sensor responds fast enough. A gate-to-source clamping for the main MOSFET, as shown in Figure 10.21, provides the second line of defense. This second option also is subject to response speed. A more drastic measure is shown in Figure 10.22, in which a current transformer (patent pending), as part of the precharging circuit, activates the AC line magnetic circuit breaker when a short occurs. It is understood that the precharging diode conducts only briefly during startup. In a steady state, the diode is back biased because of the boost nature of the PFC. However, when the load shorts, a current

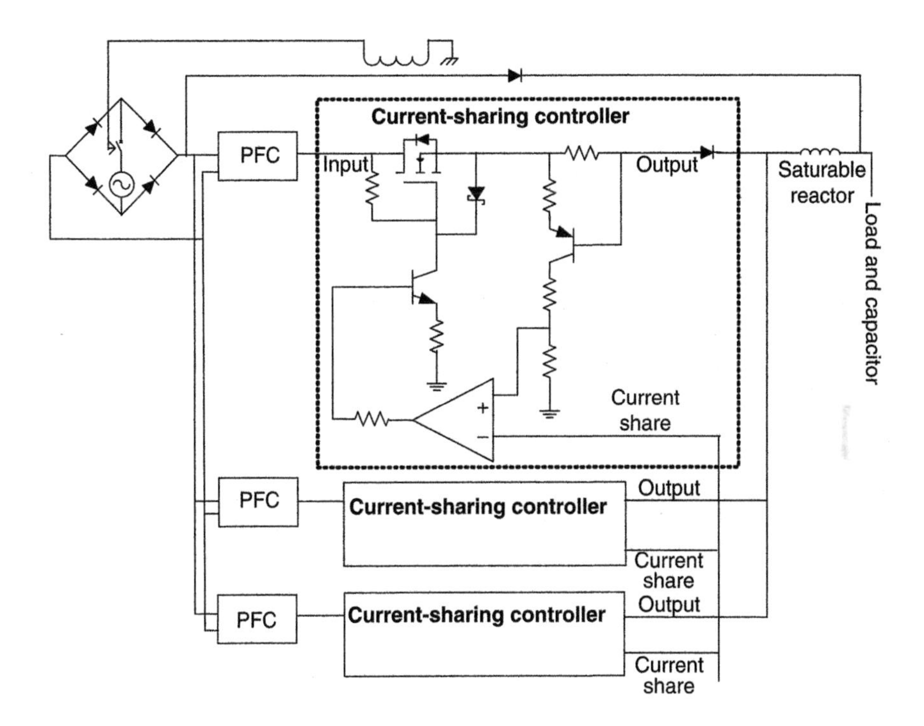

Figure 10.22: High-power PFCs with short-circuit protection

spike develops rapidly through the diode. A portion of that current may trigger the AC circuit breaker and cut off the AC supply.

10.9 Three-Phase PFC

As shown in Figure 10.7, the input power for a single-phase PFC pulsates at twice the line frequency. The situation can be improved with three-phase configuration that offers constant input power. Figure 10.23 gives three individual phases and the sum.

However, implementing the three-phase PFC is not all rosy. The main difficulty arises in routing the neutral terminal for a Y-connected sourcing transformer, which happens to be the most popular isolation transformer available. Here is how the problem sneaks in.

Figure 10.23 and the common desire of keeping everything simple lead one to believe that, by feeding three single-phase PFCs in parallel (Figure 10.24) via a Y-connected transformer, a three-phase PFC comes alive. We wish it were that simple.

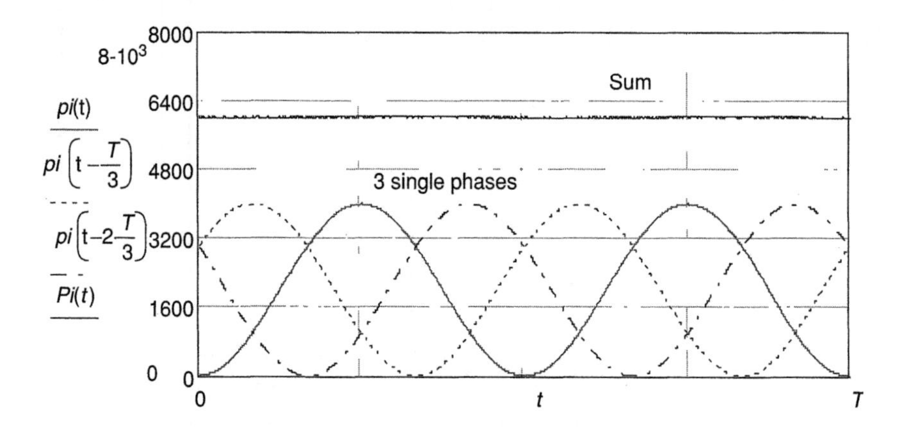

Figure 10.23: Three-phase input power

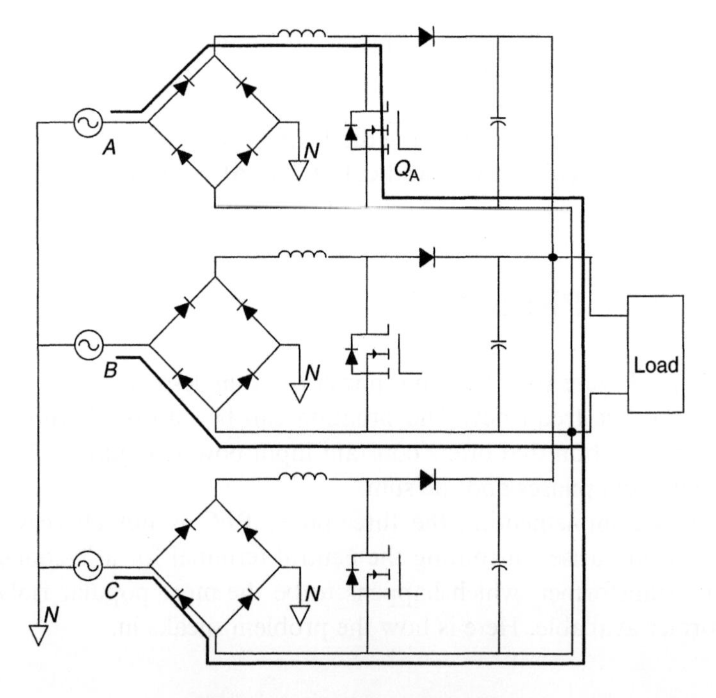

Figure 10.24: Three-phase PFC

There are basically two problems, one when Q_A turns on and the other when Q_A turns off. In the former case, as indicated with the dark solid line, the phase *A* current does not return through neutral, *N*. Instead, it returns through other phases and, consequently, defeats current shaping, the very goal of the PFC. We may incorporate blocking diodes, D_A, D_B, and D_C, as shown in Figure 10.25. Well, the diodes do not solve the problem either; they merely shift the problem to other time segments. That is, when Q_A turns off, the off current does not return through neutral. Again, it returns through the other two phases, as shown by the dark solid line in the figure. Someone suggested a solution by splitting the boost inductor and inserting one half of the inductance along each phase's return leg as shown in Figure 10.26. It is not clear, however, that the solution works, given that inductor cores are notoriously nonlinear and have high tolerances.

Given that, does a Δ-connected transformer (Figure 10.27) fare better on the task? It may. But it is still subjected to the same concern of imbalance of impedance.

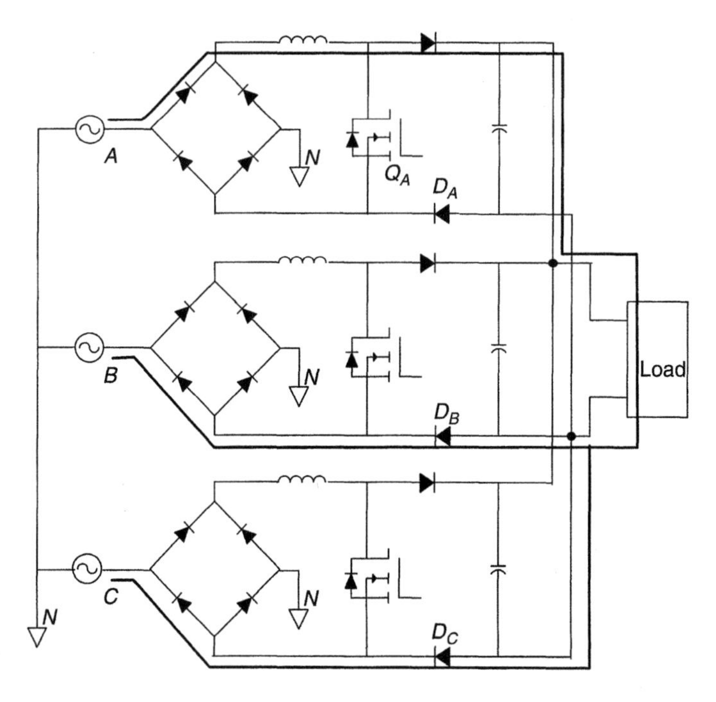

Figure 10.25: Three-phase PFC with blocking diodes

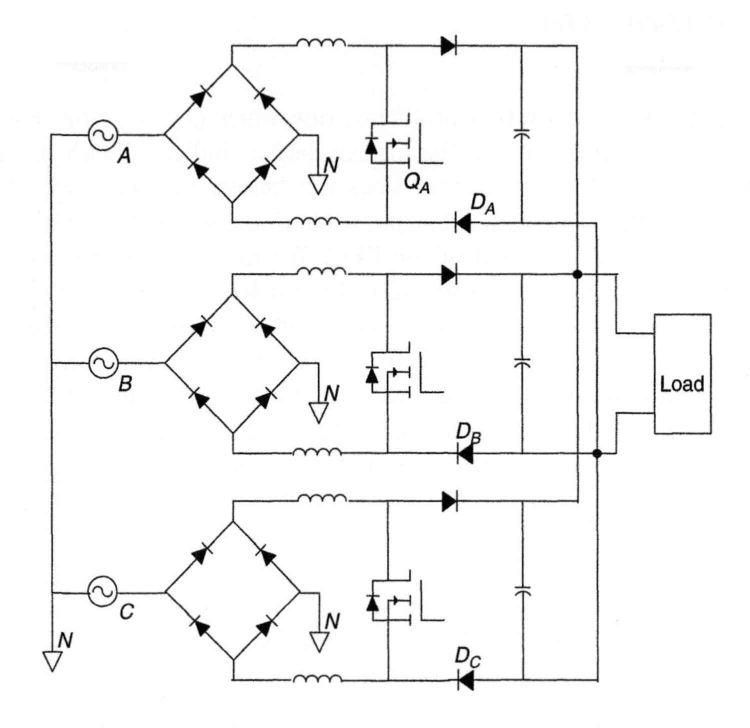

Figure 10.26: Three-phase PFC with splitting inductors

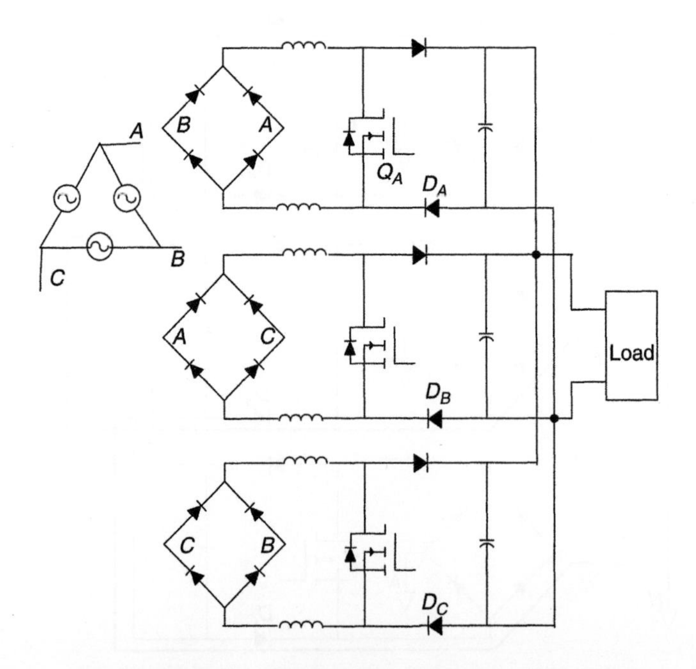

Figure 10.27: Three-phase PFC with a Δ transformer

So far, all three-phase configurations presented require numerous duplicities of circuits and components. The situation is undesirable. Something better should be available. This is what Figure 10.28, from Tu and Chen [11], offers. It possesses simplicity and fewer parts. For the rest of the chapter, we give this interesting scheme thorough coverage.

This highly innovative scheme effectively uses a very important aspect of three-phase inputs, E_a, E_b, and E_c. We can appreciate the observation better if we examine the three inputs (Figure 10.29) in which the input time frame is partitioned into six segments.

Figure 10.28 also uses a unique, nonstandard component symbol, such as S_{ab}. It represents a bidirectional switch as shown in Figure 10.30.

Now, we examine the switch function in detail during T_1, when the relevant switches turn on or off. From Figure 10.29, it is clear that switch S_{ab} is not needed. In other words, during T_1, the power plant looks like Figure 10.31.

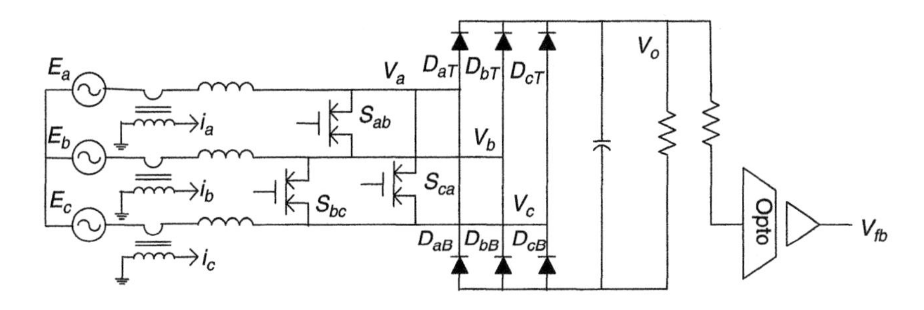

Figure 10.28: Power stage of a novel three-phase PFC

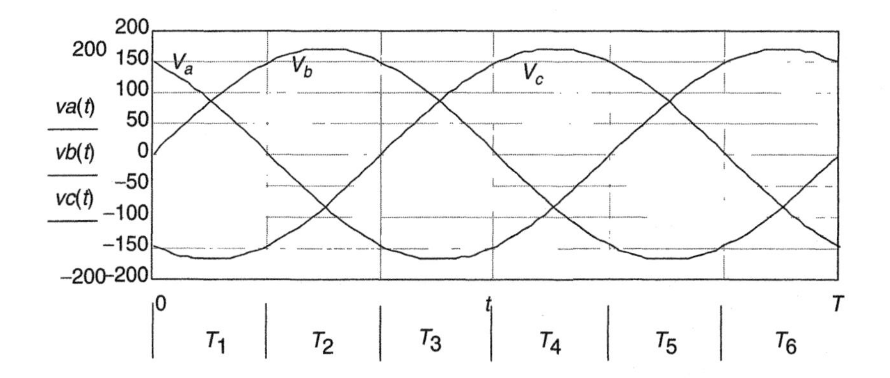

Figure 10.29: Three-phase timing and partition

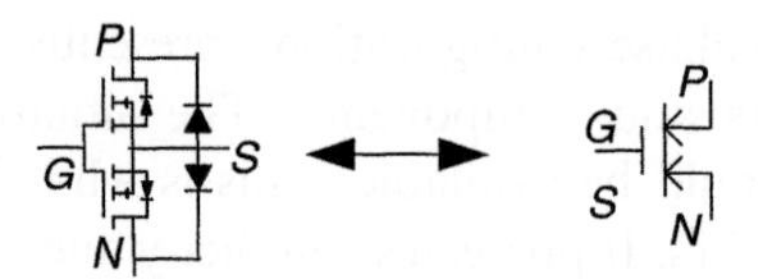

Figure 10.30: Bidirectional switch

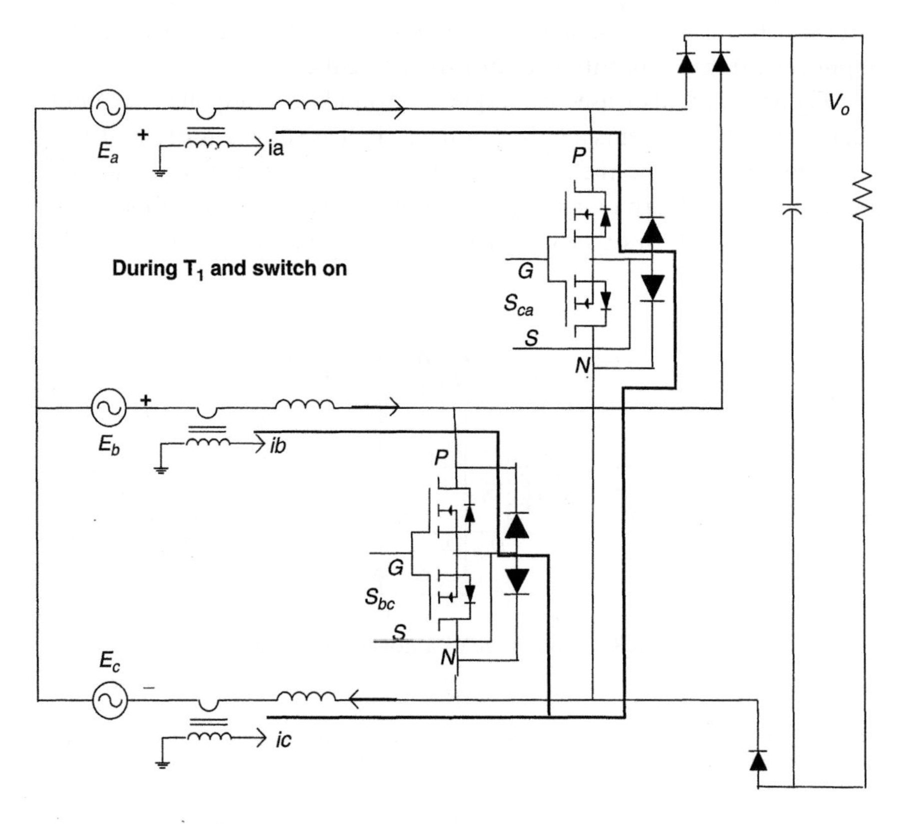

Figure 10.31: Power plant in T_1 and switch on

The drawing also shows the two currents, one from phase *a* and the other from phase *b*—the dark solid line—that charge the in-line boost inductors. When the switches are turned off, the circuit topology changes to Figure 10.32 and the inductor discharges.

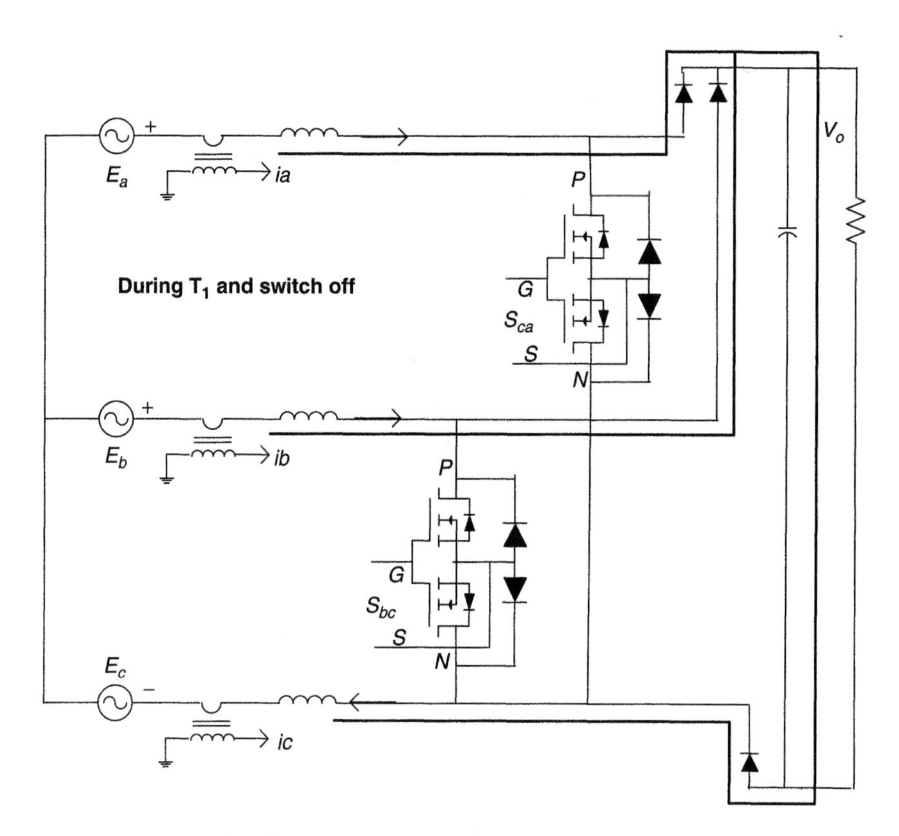

Figure 10.32: Power plant in T_1 and switch off

That is to say, during T_1, the power plant alternates between two configurations, Figures 10.31 and 10.32. While in Figure 10.31, only the upper switch and the lower diode of the bidirectional switches are involved.

In T_2, a significant change takes place. In this second time segment, phases a and c are negative and only phase b is positive. Switch S_{ca} does not conduct. The power plant during this time interval alternates between Figures 10.33 and 10.34. In this time segment, a single-stream positive current comes out of phase b and returns to phases a and c when charging in-line boost inductors. When boost inductors discharge, the discharge current comes out of phase b, goes through the load, and returns to the phase a and c terminals.

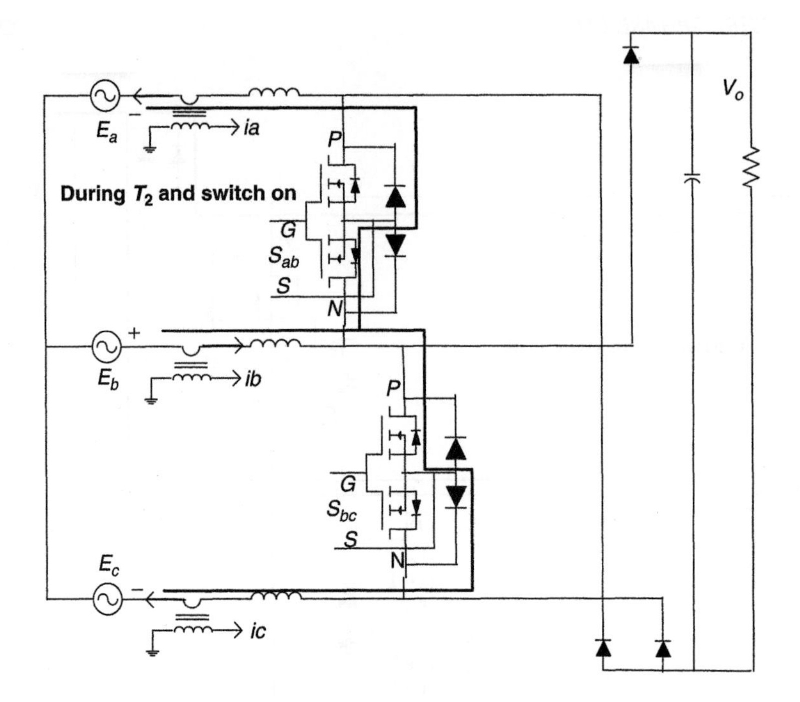

Figure 10.33: Power plant in T_2, switch on and charging

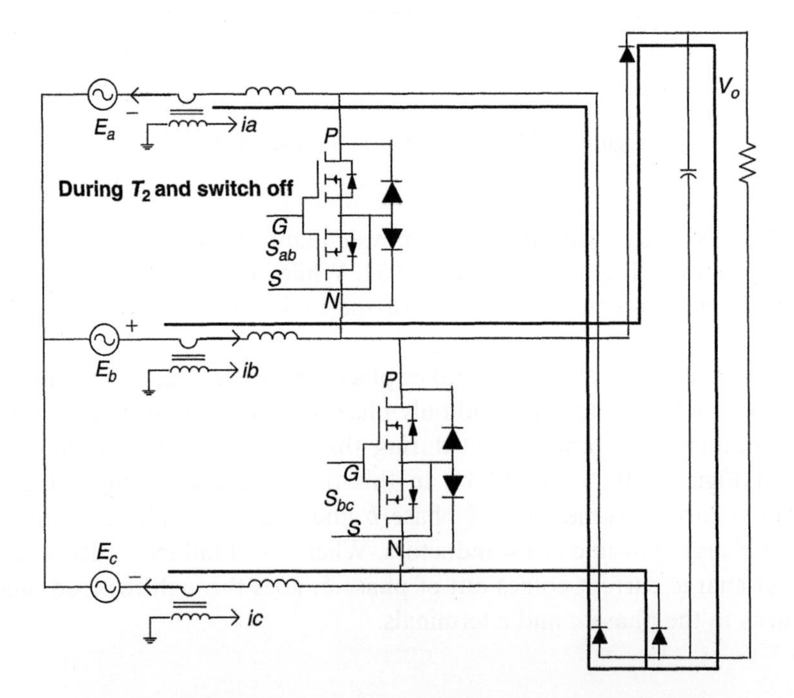

Figure 10.34: Power plant in T_2, switch off and discharging

Readers should have no problem going through the rest of the time segments using Figure 10.29 as a guideline.

Well, one may ask how the time partition function, T_1 through T_6, is generated. This can be accomplished in two steps. Step one detects the polarity of AC input lines. This is done by zero-crossing detectors (Figure 10.35) fed by reduced AC sources through small isolation step-down transformers.

The time partition, Figure 10.36, is done using the polarity information developed in Figure 10.35. The gating logic is quite simple and is not explained here.

Next, we treat the input phase (line) current sensing, one of the key operations for the complete task. As shown in Figure 10.28, three magnetic, isolated current sensors are inserted in series with all three phase lines. The current-sensor outputs feed three hysteretic current comparators (Figure 10.37). The current commands I_a^*, I_b^* and I_c^* used in Figure 10.37 are generated by the output voltage feedback and multiplier

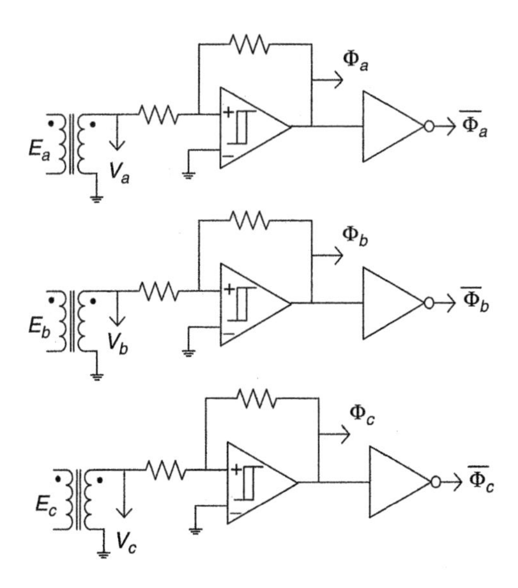

Figure 10.35: AC line voltage zero-crossing/polarity detector

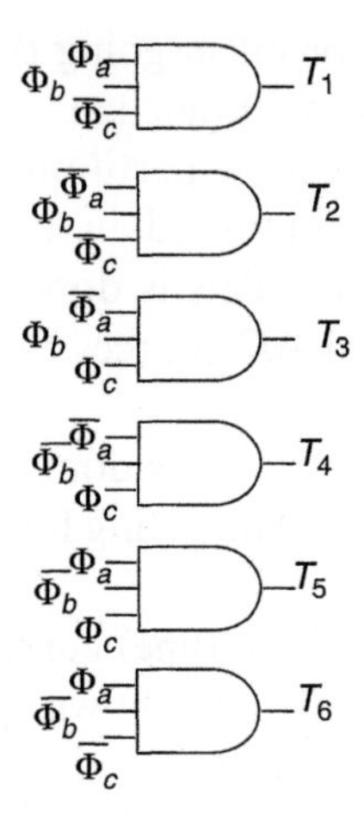

Figure 10.36: Time partition

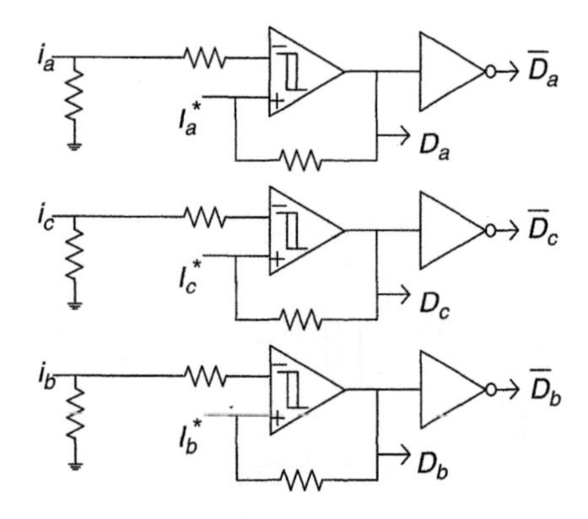

Figure 10.37: Hysteretic current comparators

(Figure 10.38). As shown, the voltage loop error, V_e, rides on small-scale AC phase voltages. The multiplicative combinations shape the line current.

Tables 10.1 through 10.3 give the command states (1 = on) for all three switches.

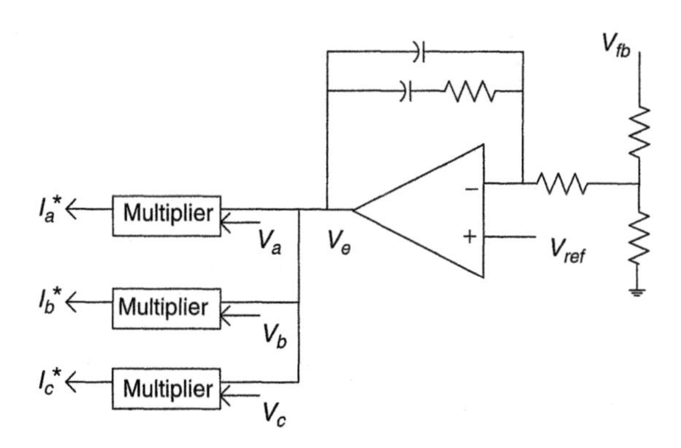

Figure 10.38: Current command generator

Table 10.1 Truth table for switch S_{bc}

	111	110	101	100	011	010	001	000
T_1	1	1			1	1		
T_2		1		1		1		1
T_3								
T_4			1	1			1	1
T_5	1		1		1		1	
T_6								

Note: Line-current sensing and hysteretic-current steering output: D_a, D_b, and D_c.

Table 10.2 Truth table for switch S_{ca}

	111	110	101	100	011	010	001	000
T_1	1	1	1	1				
T_2								
T_3	1		1		1		1	
T_4					1	1	1	1
T_5								
T_6		1		1		1		1

Note: Line-current sensing and hysteretic-current steering output: D_a, D_b, and D_c.

Table 10.3 Truth table for switch S_{ab}

	111	110	101	100	011	010	001	000
T_1								
T_2					1	1	1	1
T_3	1	1			1	1		
T_4								
T_5	1	1	1	1				
T_6			1	1			1	1

Note: Line-current sensing and hysteretic-current steering output: D_a, D_b, and D_c.

From Table 10.1 and employing Boolean algebra, for instance, in the first row corresponding to T_1, the decision logic for turning on switch S_{bc} can be expressed as

$$T_1(D_aD_bD_c + D_aD_b\overline{D}_c + \overline{D}_aD_bD_c + \overline{D}_aD_b\overline{D}_c) = T_1(D_aD_b + \overline{D}_aD_b) = T_1D_b \quad (10.34)$$

Following the same procedure for the rest, the overall switch logic is, for S_{bc},

$$S_{bc} = T_1D_b + T_2\overline{D}_c + T_4\overline{D}_b + T_5D_c \quad (10.35)$$

for S_{ca},

$$S_{ca} = T_1D_a + T_3D_c + T_4\overline{D}_a + T_6\overline{D}_c \quad (10.36)$$

and for S_{ab},

$$S_{ab} = T_2\overline{D}_a + T_3D_b + T_5D_a + T_6\overline{D}_b \quad (10.37)$$

The preceding logic is implemented with switch steering logic (Figure 10.39) including isolation.

At this point, we can certainly assemble all the preceding in a single figure, as in Figure 10.40, that shows the complete circuit in a block diagram form with key variables passed along from one block to another.

Further improvement is possible by considering current advancement in digital signal processing. It looks like at least Figures 10.38 and 10.39

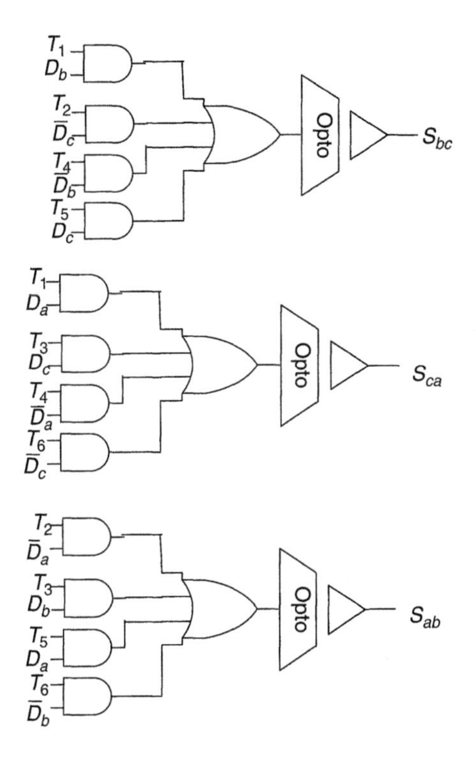

Figure 10.39: Switch steering logic with isolation.

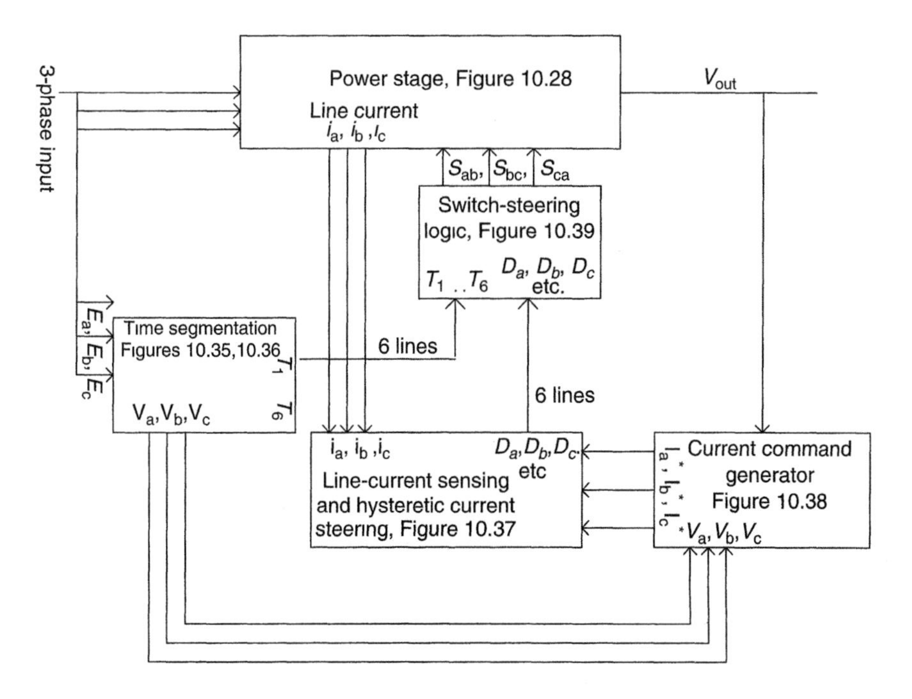

Figure 10.40: Complete three-phase PFC

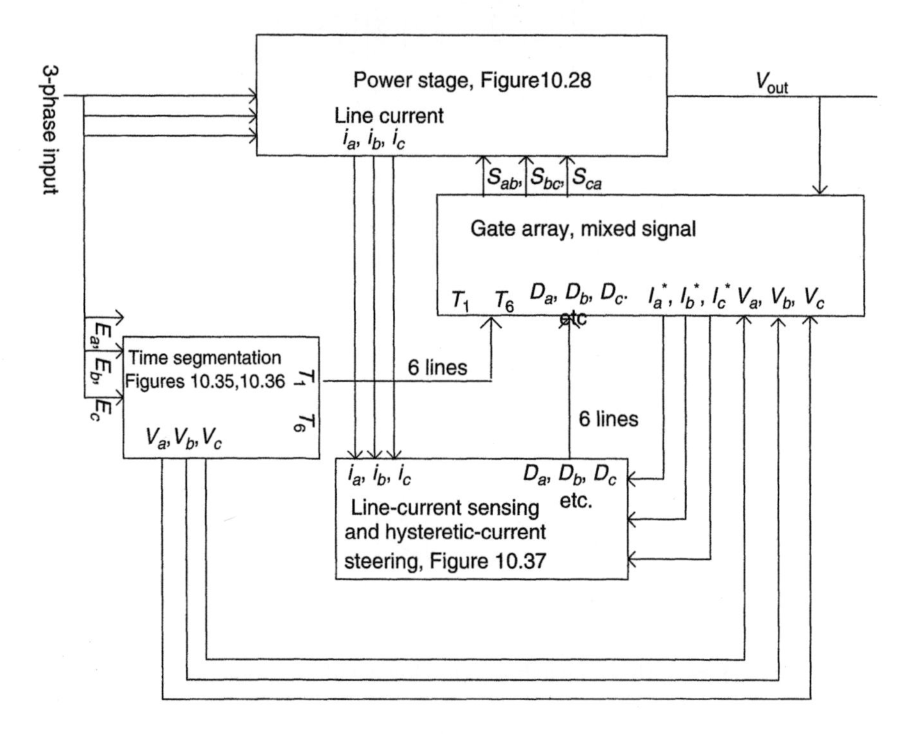

Figure 10.41: Complete three-phase PFC using gate array

can be replaced by a field programmable gate array. This action may simplify Figure 10.40 to Figure 10.41. Of course, note that many digital-to-analog or analog-to-digital interfaces are not shown but implied in the last figure.

Chapter 11

Error Amplifiers

As shown repeatedly in the previous chapters, modern switch-mode power converters are closed-loop control systems in every respect. As such, they all can be represented at the most fundamental level by a block diagram (Figure 11.1) consisting of two interconnected key blocks: power plant (power stage) and feedback controller containing a high gain amplifier. So far we have focused almost exclusively on the closed-loop formulation without actually dealing with the amplifier design, except to represent the block with a symbol $A(s)$ or $EA(s)$. This by no means suggests that the block is unimportant or the task is easy. As a matter of fact, the past experience of many designers paints a grim picture on the subject. To many, the design process for the compensation amplifier—component selection in particular—is so intractable as to border on the trial and error that tends to give engineering a bad name.

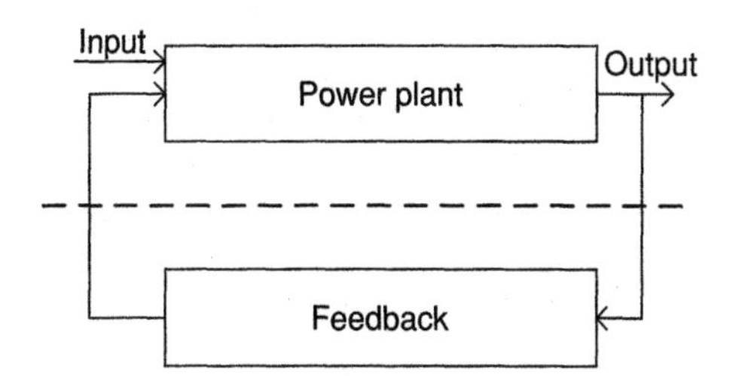

Figure 11.1: Typical converter loop partition

The question is this: Is it really so hard to do? In this chapter, we try to make it easier.

11.1 Amplifier Category

Compensation amplifiers come in many forms, the three most popular ones for general control-loop applications are given in Figure 11.2. Type I amplifier's transfer function is easily identified as

$$\frac{V_o}{V_{\text{in}}} = -\frac{1}{R_1 C_1 s} \tag{11.1}$$

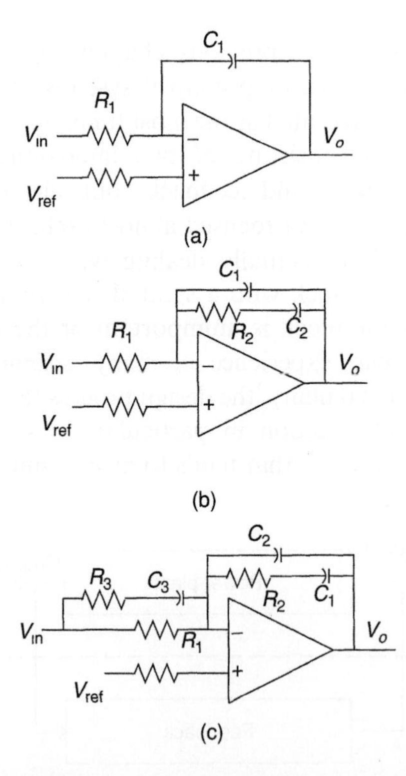

Figure 11.2: Amplifiers: (a) type I, (b) type II, and (c) type III

The amplifier gives a −20dB/decade roll-off starting at zero frequency and a constant −270°, or 90°, phase. Type II amplifier's transfer function is given as

$$\frac{V_o}{V_{\text{in}}} = -\frac{R_2 \bullet C_2 \bullet s + 1}{R_1(C_1 + C_2)s\left(R_2 \bullet \dfrac{C_1 \bullet C_2}{C_1 + C_2} \bullet s + 1\right)} \tag{11.2}$$

The type II amplifier gives three frequency points of importance. If standing alone, the pole at zero frequency crosses the zero dB axis at f_{p0}, while the pole–zero pair offers corner frequencies, f_{p1} and f_{z1}, respectively. The three frequencies are given as

$$\begin{aligned} f_{p0} &= \frac{1}{2 \bullet \pi \bullet R_1(C_1 + C_2)} \\ f_{p1} &= \frac{1}{2 \bullet \pi \bullet R_2 \bullet \dfrac{C_1 \bullet C_2}{C_1 + C_2}} \\ f_{z1} &= \frac{1}{2 \bullet \pi \bullet R_2 \bullet C_2} \end{aligned} \tag{11.3}$$

Important behaviors begin to show up for the type II amplifier (Figure 11.3). Instead of being a constant, the amplifier's phase changes, in our favor, between the pole–zero pair's frequencies. The phase still starts from −270° at low frequency, but it moves toward −180° between f_{p1} and f_{z1}. In other words, a phase boost takes place (see Wu [10]). The amount of boost is a function of the separation of pole–zero pair frequencies. If a frequency, f_c, is picked such that $f_{z1} < f_c < f_{p1}, f_c/f_{z1} = k$, and $f_c/f_{p1} = 1/k$, the amount of phase boost at the selected frequency is

$$\alpha_b = 2 \bullet \tan^{-1} k - \frac{\pi}{2} \tag{11.4}$$

Or, the pole–zero separation, $f_{p1}/f_{z1} = k^2$, satisfies

$$\begin{aligned} k^2 &= \left[\tan\left(\frac{\alpha_b}{2} + \frac{\pi}{4}\right)\right]^2, \quad \alpha_b \text{ in radians} \\ &= \left\{\tan\left[\left(\frac{\alpha_b}{2} + 45\right)\frac{\pi}{180}\right]\right\}^2, \quad \alpha_b \text{ in degrees} \end{aligned} \tag{11.5}$$

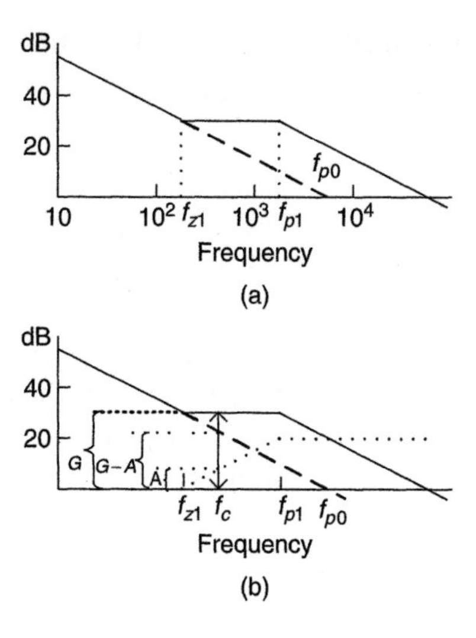

Figure 11.3: Type II amplifier gain

This equation shows that, to recover about 80°, k^2 equals about 130. That is, the pole–zero pair is separated by slightly more than two decades in frequency.

For the type III amplifier, we can follow a similar procedure. The transfer function is given as

$$\frac{V_o}{V_{\text{in}}} = -\frac{(R_2 \bullet C_1 \bullet s + 1)[(R_1 + R_3)C_3 \bullet s + 1]}{R_1(C_1 + C_2)s\left(R_2 \bullet \dfrac{C_1 \bullet C_2}{C_1 + C_2} \bullet s + 1\right)(R_3 \bullet C_3 \bullet s + 1)} \tag{11.6}$$

Consequently, five frequencies are involved—f_{p0}, f_{p1}, f_{z1}, f_{p2}, and f_{z2}:

$$\begin{aligned}
f_{p0} &= \frac{1}{2 \bullet \pi \bullet R_1(C_1 + C_2)} \\
f_{p1} &= \frac{1}{2 \bullet \pi \bullet R_2 \bullet \dfrac{C_1 \bullet C_2}{C_1 + C_2}} \\
f_{p2} &= \frac{1}{2 \bullet \pi \bullet R_3 \bullet C_3} \\
f_{z1} &= \frac{1}{2 \bullet \pi \bullet R_2 C_1} \\
f_{z2} &= \frac{1}{2 \bullet \pi \bullet (R_1 + R_3) \bullet C_3}
\end{aligned} \tag{11.7}$$

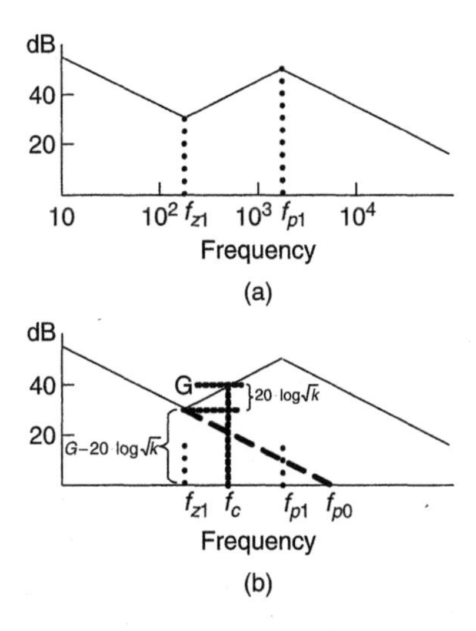

Figure 11.4: Type III amplifier gain

In practical application, it is advantageous to make $f_{p1} = f_{p2}$ and $f_{z1} = f_{z2}$. In effect, a double pole and a double zero are created (Figure 11.4). The phase still starts from −270° at low frequency, but it moves toward −90° between f_{p1} and f_{z1}, see Wu [10]. In other words, a greater phase boost takes place with type III. The amount of boost is again a function of the separation of pole–zero pair frequencies. If a frequency, f_c, is picked such that $f_{z1} < f_c < f_{p1}$, $f_c/f_{z1} = \sqrt{k}$, and $f_c/f_{p1} = 1/\sqrt{k}$, the amount of phase boost at the selected frequency is

$$\alpha_b = 2 \cdot \tan^{-1} \sqrt{k} - \frac{\pi}{2} \tag{11.8}$$

Or, the pole–zero separation, $f_{p1}/f_{z1} = k$, is

$$\begin{aligned} k &= \left[\tan\left(\frac{\alpha_b}{2} + \frac{\pi}{4}\right)\right]^2, \quad \alpha_b \text{ in radians} \\ &= \left\{\tan\left[\left(\frac{\alpha_b}{2} + 45\right)\frac{\pi}{180}\right]\right\}^2, \quad \alpha_b \text{ in degrees} \end{aligned} \tag{11.9}$$

This equation shows that, to recover about 160°, k equals about 110. That is, the double pole–zero is separated by slightly more than two decades in frequency.

11.2 Innate Phase of the Control Loop

Prior to actually cutting in an error amplifier with the suitable gain and phase boost, we must decide how much gain compensation and phase correction are needed. The decision can be made by examining the most essential parts for a typical control loop. We redraw Figure 11.1 and pay attention to its phase property around the loop as shown Figure 11.5. Let us assume that the consolidated block (modulator) consisting of the power plant, the PWM, and the output filter, if any, has a phase α_M. The remaining block is the compensation amplifier. From the last section, it was noted that the error amplifier operates in an inverting mode and has a pole at zero frequency. The inverting mode of operation immediately yields a -180° phase while the pole at zero frequency adds an additional -90°. If a desired phase margin, α_m, is selected, the resultant phase relationship around the loop can be described as the sum of all contributing factors, including the phase boost, α_b, required:

$$\alpha_m = \alpha_M - 180 - 90 + \alpha_b, \quad \alpha_b = \alpha_m - 90 - \alpha_M \tag{11.10}$$

Next, we need to know the gain and phase behavior for the consolidated block (modulator, Figure 11.5) at the selected frequency, f_c, as the designer wishes the final loop gain to cross over. This information can be obtained by two approaches, theoretical and measurement. The theoretical approach employs all the analytical techniques given in the previous chapters. The measurement approach is straightforward if the closed loop is already stable and measurement can take place. If the closed loop is unstable, measurement cannot be taken. In this case, the loop can be

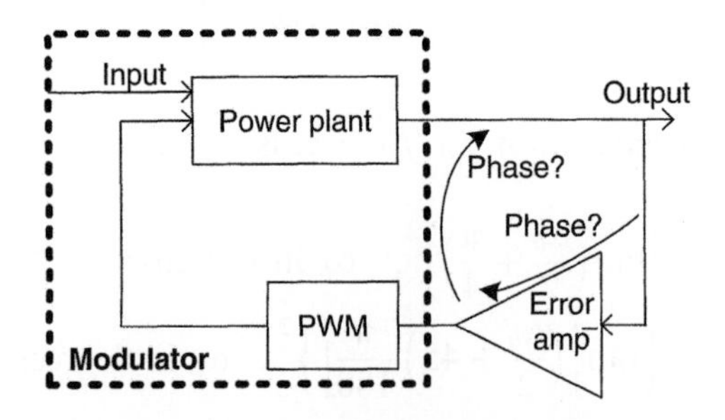

Figure 11.5: Closed loop, repartitioned

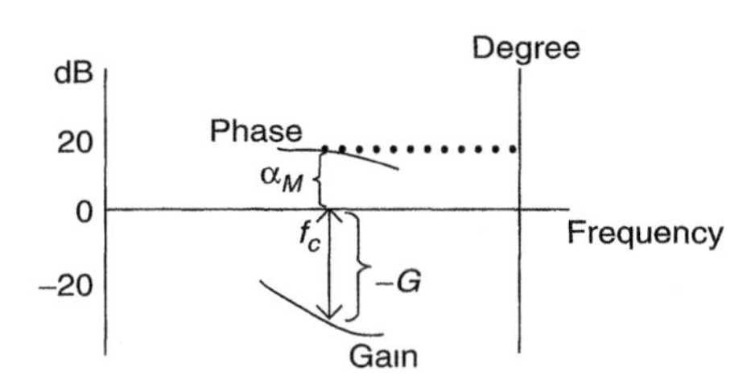

Figure 11.6: Modulator gain/phase

temporarily made stable by forcing the error amplifier as an integrator with a single pole at very low, near zero frequency. This is done with a large capacitor in the local feedback surrounding the error amplifier under study. Once this temporary, low-bandwidth mode is forced, the measurement can proceed. Either way, a gain/phase plot resembling Figure 11.6 is available. Now, we are ready to implement type II and type III amplifiers.

11.3 Type II Amplifier Implementation

Based on (11.10) and given a desired phase margin, if the phase boost requirement can be met with a type II amplifier, the following steps lead to the component selection with high confidence.

Step 1. Given Figure 11.6, select a desired frequency, f_c, at which the final closed loop crosses 0 decibles. Identify the modulator gain deficiency, $-G$ dB, and the modulator phase, α_M, at f_c.

Step 2. Given a desired phase margin, α_m, use (11.10) and compute the phase boost, α_b, needed.

Step 3. Given α_b from step 2, use (11.5) and compute the pole–zero separation factor, k. From Figure 11.3(b), we recognize that both the integrator pole (pole at zero) and the zero are capable of providing gain. Together they make up the gain deficiency at the selected crossover frequency, f_c. In other words, at the crossover frequency, the type II amplifier pulls up the modulator gain and makes it cross 0 dB. Therefore, the modulator gain deficiency is divided into A and $(G - A)$, in decibels: A from the zero and $(G - A)$ from the integrator pole.

Step 4. Since Figure 11.3 is in log-scale, the following can be established for f_{z1}:

$$20 \bullet \log\left(\frac{f_c}{f_c/k}\right) = 20 \bullet \log(k) = A \tag{11.11}$$

Step 5. Similarly, the following holds for f_{po}, the integrator:

$$\begin{gathered} 20 \bullet \log\left(\frac{f_{p0}}{f_c}\right) = G - A = G - 20 \bullet \log(k) \\ f_{p0} = 10^{\left[\frac{G}{20} - \log(k)\right]} \bullet f_c \end{gathered} \tag{11.12}$$

Step 6. Rewrite (11.3) as

$$f_{z1} = \frac{1}{2 \bullet \pi \bullet R_2 \bullet C_2} = \frac{f_c}{k} \tag{11.13}$$

$$f_{p1} = \frac{1}{2 \bullet \pi \bullet R_2 \bullet \dfrac{C_1 \bullet C_2}{C_1 + C_2}} = k \bullet f_c \tag{11.14}$$

$$f_{p0} = \frac{1}{2 \bullet \pi \bullet R_1(C_1 + C_2)} = 10^{\left[\frac{G}{20} - \log(k)\right]} \bullet f_c \tag{11.15}$$

Step 7. Take the ratio of (11.13) to (11.14). C_2 is expressed in terms of C_1. Consequently, $(C_1 + C_2)$ is also expressed in terms of C_1:

$$C_2 = (k^2 - 1)C_1, \quad C_1 + C_2 = k^2 C_1 \tag{11.16}$$

In general, R_1 is preselected for bias current and offset voltage considerations.

Step 8. Plug in the second part of (11.16) in (11.15) and we have the following:

$$C_1 = \frac{1}{2 \bullet \pi \bullet R_1 \bullet k^2 \bullet f_c \bullet 10^{\left[\frac{G}{20} - \log(k)\right]}} \tag{11.17}$$

Step 9. Then, C_2 follows from the first part of (11.16).
Step 10. Use (11.13) and calculate the last part R_2:

$$R_2 = \frac{1}{2 \cdot \pi \cdot \frac{f_c}{k} \cdot C_2} \tag{11.18}$$

11.4 Type III Amplifier Implementation

Based on (11.10) and given a desired phase margin, if the phase boost requirement can be met with a type III amplifier, the following steps lead to the component selection with high confidence.

Step 1. Given Figure 11.6, select a desired frequency, f_c, at which the final closed loop crosses 0 dB. Identify the modulator gain deficiency, –G dB, and the modulator phase, α_M, at f_c.

Step 2. Given a desired phase margin, α_m, use (11.10) and compute the phase boost, α_b, needed.

Step 3. Given α_b from step 2, use (11.9) and compute the pole–zero separation factor, k. From Figure 11.4(b), we recognize that both the integrator pole (pole at zero) and the double zero are capable of gain contribution. Together they make up the gain deficiency at the selected crossover frequency, f_c.

Step 4. Since Figure 11.4 is in log scale, the following can be established for f_{z1}:

$$20 \cdot \log\left(\frac{f_c}{f_c/\sqrt{k}}\right) = 20 \cdot \log\sqrt{k} \tag{11.19}$$

Step 5. Similarly, the following holds for f_{po}, the integrator:

$$\begin{gathered} 20 \cdot \log\left(\frac{f_{p0}}{f_c/\sqrt{k}}\right) = G - 20 \cdot \log\sqrt{k} = G - 10 \cdot \log k \\ f_{p0} = 10^{\frac{G-10 \cdot \log k}{20}} \cdot \frac{f_c}{\sqrt{k}} \end{gathered} \tag{11.20}$$

Step 6. Rewrite (11.7) as

$$f_{p0} = \frac{1}{2 \cdot \pi \cdot R_1(C_1 + C_2)} = 10^{\frac{G - 10 \cdot \log k}{20}} \cdot \frac{f_c}{\sqrt{k}}$$

$$f_{p1} = \frac{1}{2 \cdot \pi \cdot R_2 \cdot \dfrac{C_1 \cdot C_2}{C_1 + C_2}} = \sqrt{k} \cdot f_c$$

$$f_{p2} = \frac{1}{2 \cdot \pi \cdot R_3 \cdot C_3} = f_{p1} \tag{11.21}$$

$$f_{z1} = \frac{1}{2 \cdot \pi \cdot R_2 C_1} = \frac{f_c}{\sqrt{k}}$$

$$f_{z2} = \frac{1}{2 \cdot \pi \cdot (R_1 + R_3) \cdot C_3} = f_{z1}$$

Step 7. Take the ratio of f_{p1} to f_{z1}. C_1 is expressed in terms of C_2. Consequently $(C_1 + C_2)$ is also expressed in terms of C_2:

$$C_1 = (k - 1)C_2, \quad C_1 + C_2 = k \cdot C_2 \tag{11.22}$$

In general, R_1 is preselected for bias current and offset voltage considerations.

Step 8. Plug in the second part of (11.22) in f_{p0} of (11.21) and we have the following:

$$C_2 = \frac{1}{2 \cdot \pi \cdot R_1 \cdot \sqrt{k} \cdot 10^{\frac{G - 10 \cdot \log k}{20}}} \tag{11.23}$$

Step 9. Then, C_1 follows from the first part of (11.22).

Step 10. With C_1 from step 9, use the f_{z1} equation in (11.21) to calculate R_2:

$$R_2 = \frac{\sqrt{k}}{2 \cdot \pi \cdot f_c \cdot C_1} \tag{11.24}$$

Step 11. Take the ratio of f_{p2} to f_{z2}; R_3 is expressed in terms of R_1:

$$R_3 = \frac{R_1}{k - 1} \tag{11.25}$$

Step 12. The last part is

$$C_3 = \frac{1}{2 \cdot \pi \cdot R_3 \cdot \sqrt{k} \cdot f_c} \tag{11.26}$$

11.5 Example for Type II Amplifier Implementation

In Chapter 13, a Cuk converter (Figure 13.1) gives a closed-loop bandwidth of only about 3 KHz (Figures 13.7 and 13.8). Following the steps given in Section 11.3, the type II error amplifier is improved with new values, $R_{323} = 23\,\mathrm{K}$, $C_{311} = 1.9\,\mathrm{nF}$, $C_{313} = 290\,\mathrm{pF}$. Figure 11.7 shows that the bandwidth is extended to 5 KHz without phase margin or gain margin degradation.

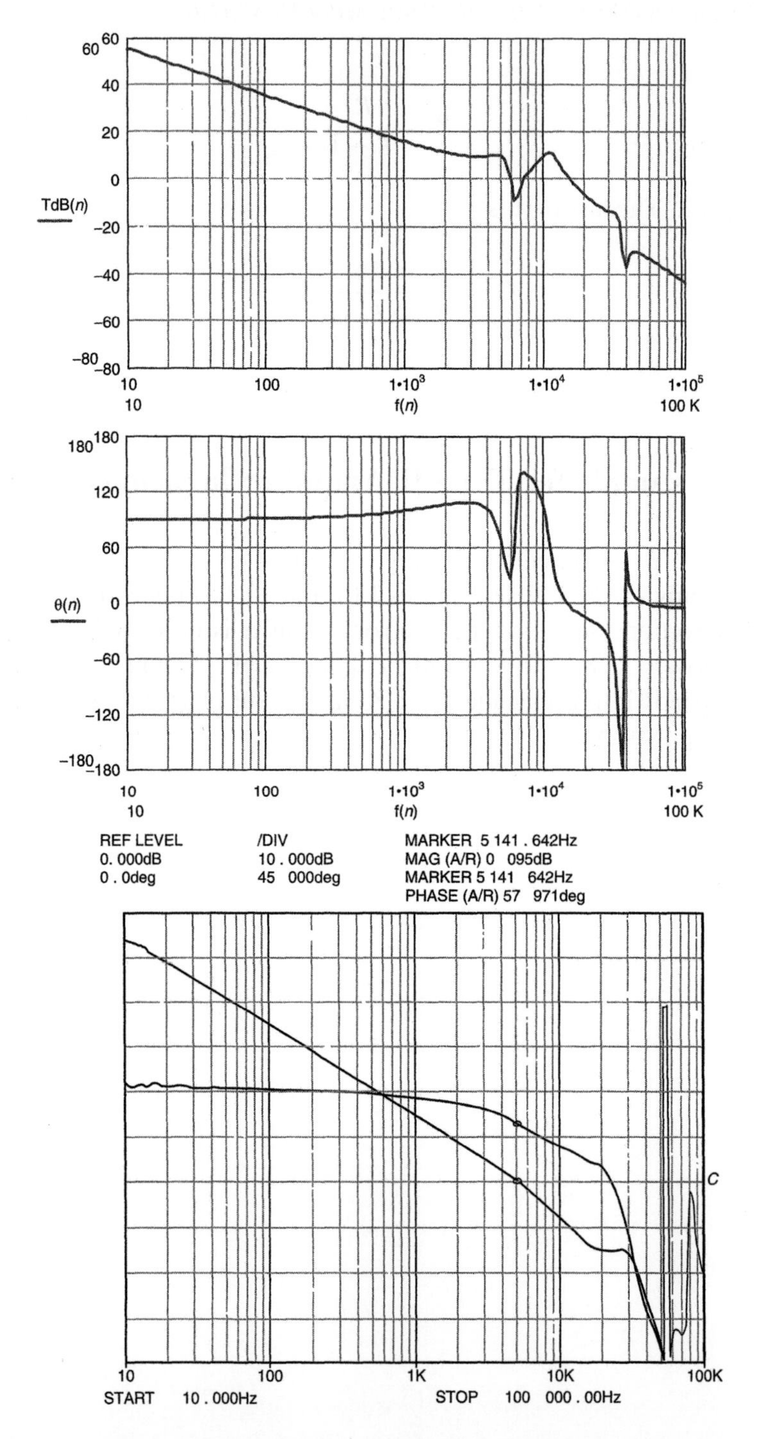

Figure 11.7: Improved loop gain for a Cuk converter

Chapter 12

Supporting Circuits

So far, we have focused our attention mainly on the power plant, the control modes, and the error amplifiers. No doubt these are important topics. But, without the supporting circuits, the power plant may not operate under optimal conditions For instance, the power switch may turn on or off in an undesirable manner, incur excessive losses (power dissipation), complicate thermal management, or reduce reliability.

With those in mind, in this chapter, we cover the bipolar switch drivers, MOSFET drivers, dissipative snubbers, lossless snubbers, feedback isolators, and soft start.

12.1 Bipolar Switch Drivers

In the early 1970s, switch-mode power converters employed bipolar junction transistors as the main power switch exclusively, since MOSFET were not yet mature for power application. Bipolar transistor switches at that time, in general, also operated at switching frequencies of lower tens of kilohertz. The latter is attributed to the slower response time as a result of distributed junction capacitances in the bipolar transistor's junctions. More important, the bipolar transistors are current-controlled devices. Effective control of the transistor requires timely injection, or removal, of charge carriers (current) into, or out of, the base terminal. Figure 12.1 gives such a circuit, which was often named the transformer-coupled (-isolated) Darlington driver.

As shown, the Q_2 base terminal is driven by a rectangular waveform with two alternating states. Therefore, the circuit also switches between two alternating structures corresponding to Q_2 on and Q_2 off (Figure 12.2). During the on state, the emitter capacitor provides an AC path at the turn-on edge. This mechanism gives a high-base surge current and Q_2 turns on rapidly. However, the drive current spike subsides once the emitter capacitor is charged and the emitter resistor R_1 takes over the DC path and maintains the on state.

During the off state and with a null base drive, Q_2 turns off. The turning-off mechanism is sped up by turning on Q_1, which derives its

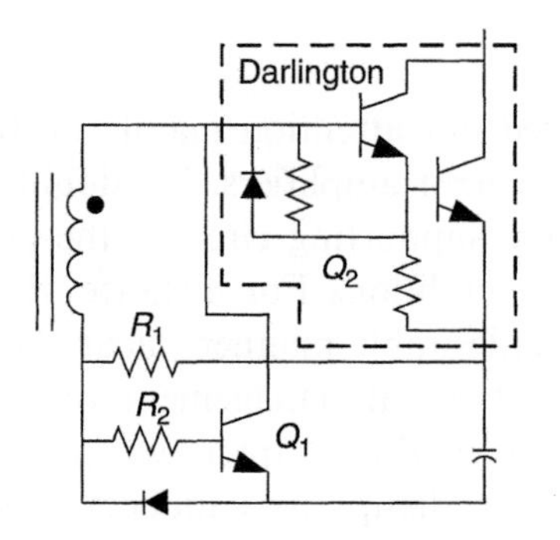

Figure 12.1: Transformer-coupled Darlington driver

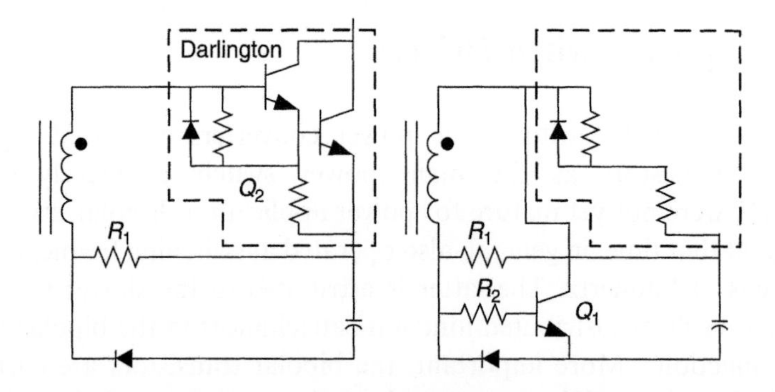

Figure 12.2: On- and off-state circuits for Figure 12.1

base drive via R_2 and the charge retained in the emitter capacitor. In effect, the capacitor discharges via the emitter–base junction of transistor Q_1 and in the process creates a reverse base current for Q_2. The latter reduces the storage turn-off delay for Q_2. For a better unidirectional control, a diode may be added in series with R_1 (Figure 12.3). More improvement in shaping turn-off time is possible if an additional resistor-diode combination is added across the Q_2 base–emitter junction (Figure 12.4).

This circuit, however, is limited to less than 50% duty cycle because the base drive transformer core is operated in the first quadrant of the B–H curve. The deficiency can be remedied with another modification, Figure 12.5, in which the base drive transformer core is operated in four quadrants.

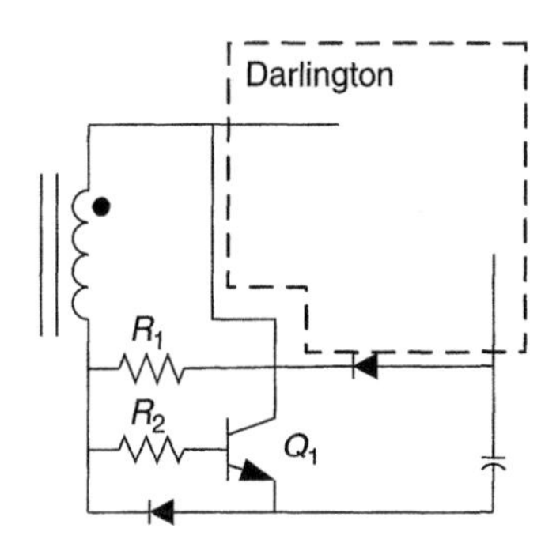

Figure 12.3: Modification with unidirectional control

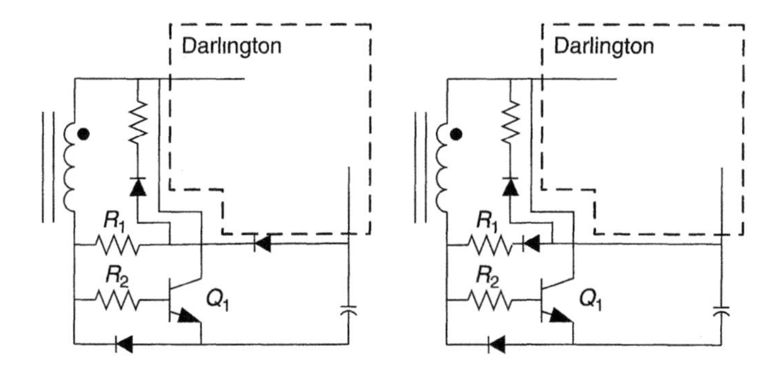

Figure 12.4: Adding turn-off shaping

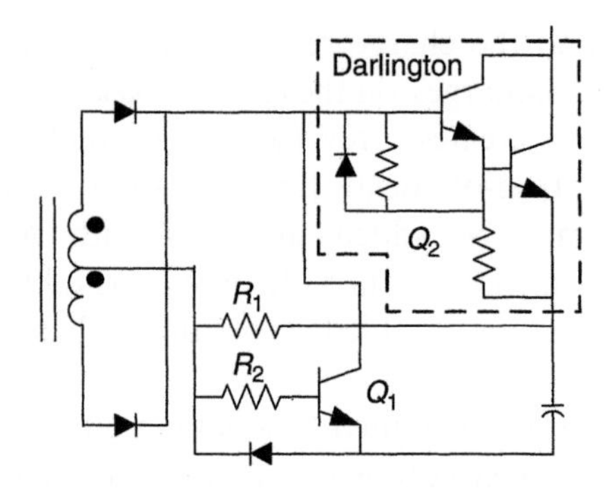

Figure 12.5: Extend duty cycle beyond 50%

Also note that these circuits are suitable for either half-bridge or full-bridge switches. Precisely because of this, the circuits are presented as floating drives with no explicit return connections shown.

In case grounded-emitter switches are desirable, Figure 12.6 gives such a drive without isolation. Again, the circuit alternates between two configurations, on and off.

During the on configuration, Figure 12.7, the base capacitor presents a momentary AC short circuit and a high base drive (current) into Q_4. Once the voltage across the capacitor reaches the zener breakdown, the zener diode takes over and keeps Q_4 in an on state.

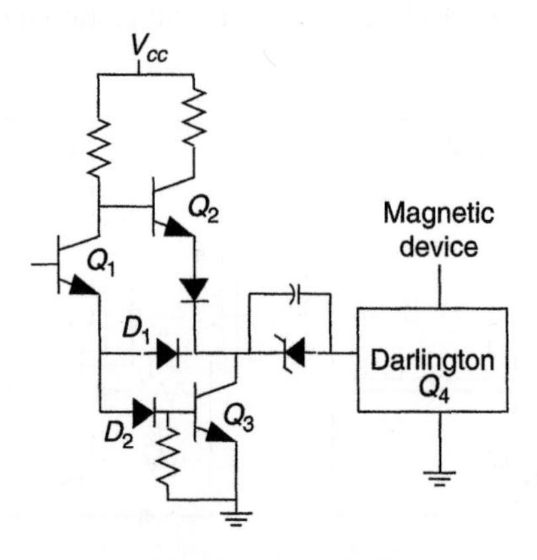

Figure 12.6: Switch with grounded emitter

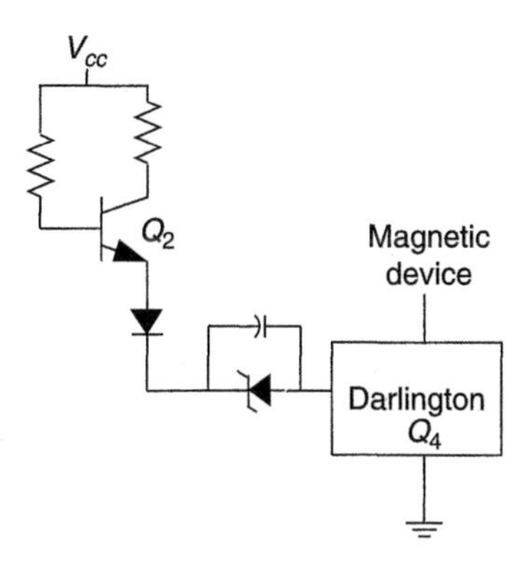

Figure 12.7: Grounded emitter switch when turning on

During the off configuration (Figure 12.8), Q_1 and Q_3 turn on and discharge the base capacitor. Again, a reverse current is generated and turns off Q_4 rapidly. Q_3 of this configuration uses a very clever circuit, a Baker clamp consisting of diodes at the base terminal and across the base–collector junction. These diodes ensure that Q_3 turns on without saturation; Q_3's collector–emitter voltage equals at least one diode drop. By so doing, Q_3 can turn off quickly without storage delay and prepare Q_4 for the next on cycle.

An important class of drivers should be mentioned before we close this section: the proportional current driver (Figure 12.9(a)) using

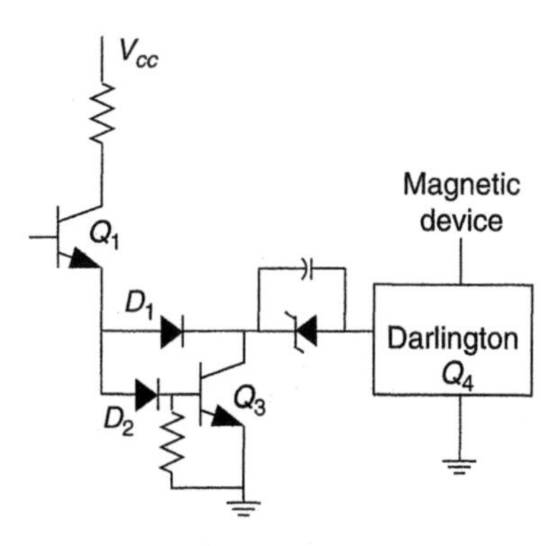

Figure 12.8: Grounded emitter switch when turning off

a current transformer. Like the drivers given previously, this driver operates in two distinctive configurations (Figure 12.9(b) and (c)). In the (Q_2) on state, the Darlington transistor base drive current is proportional to the collector current via the turn ratio, $I_B = (N_{78}/N_{56})I_C$, and an important constraint is also imposed: the on-time is limited to

$$\left(\frac{N_{78}}{N_{56}} + 1\right)(V_{BE} + V_z)t_{\text{on}} \leq 2 \cdot \lambda_s \tag{12.1}$$

If N_{78} is chosen, one turn and given a selected transformer core with known saturation flux, λ_s, and a desired on time, t_{on}, (12.1) gives N_{56}. The lower bound of the off-time, t_{off}, can be approximated with λ_s/V_s. The values of t_{on} and t_{off} together give some indication of switching frequency. However, if the switching frequency is fixed and selected, either the on-time or the off-time becomes a variable subjected to closed-loop control.

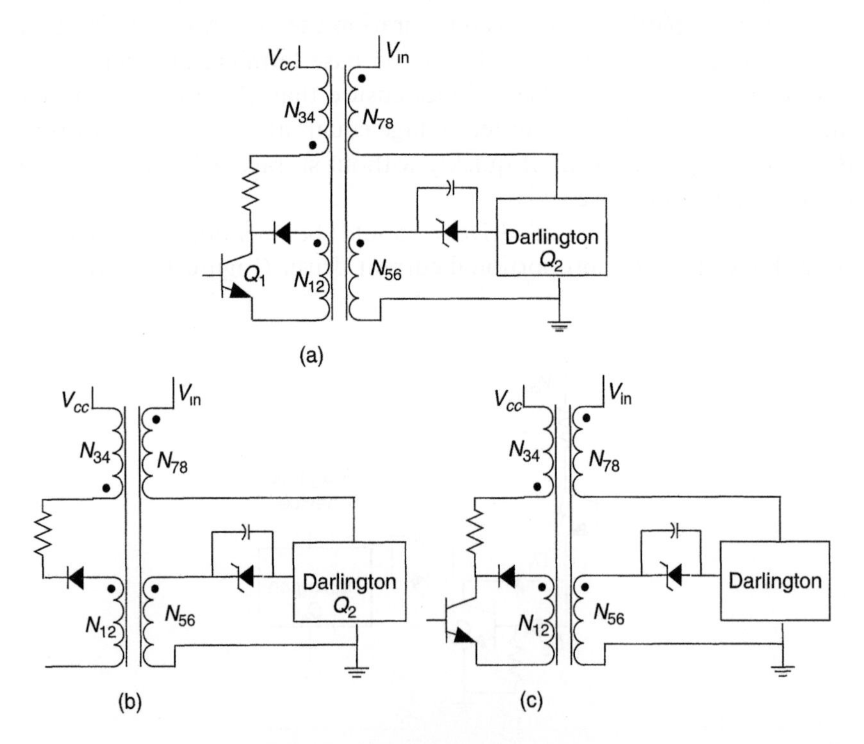

Figure 12.9: Proportional drives

One unique feature of the transformer is that the core is intentionally driven to saturation during the off time when Q_1 is on. However, it is not saturated when driven for the on state. In other words, the core is operated in four quadrants but skewed to one end of magnetic core saturation. Also keep in mind that the Darlington driver operates in an on/off switch mode instead of a linear mode. When in the on-state, the concept of current gain is murky. This makes the selection of ratio N_{78}/N_{56} and the circuit design a whole lot less straightforward.

12.2 MOSFET Switch Drivers

In contrast to the bipolar devices, the MOSFETs are voltage-controlled devices. Insofar as controlling the switch on/off state, the gate voltage control for MOSFET resembles to some extent the base current control for bipolar transistors. As a matter of fact, to achieve switch on/off for MOSFET at high speed, drivers with high forward and reverse current capability are often employed. In this section, we present several circuits for that purpose without any in-depth analysis. Readers can refer to Wu [10] for the relevant discussions.

For an N-channel MOSFET, Figure 12.10 is the simplest driver. The upper bipolar transistor is responsible for turning on while the lower is for turning off. By introducing a capacitor, Figure 12.11, a capacitor-coupled driver similar to Figure 12.6 is created. The improved version offers a higher, positive on-surge current charging the built-in gate-to-source capacitance when turning on. When turning off, the coupling capacitor and the lower driver provide a transitory negative voltage at the MOSFET gate. It is also possible to split the gate resistor, Figure 12.12. The circuit gives designer more degrees of freedom.

More shaping for the gate drive can be done by introducing a diode, Figure 12.13, which gives two different on/off drive-time constants. With a transformer incorporated, a floating drive is also feasible, Figure 12.14. With transformer coupling, isolation is also achieved. Without doubt, those series of modifications show the power and the beauty of analog design.

For the P-channel MOSFET used as power switches, Figure 12.15 gives the simplest version. Switching on and off the bipolar driver turns the P-channel on and off, too.

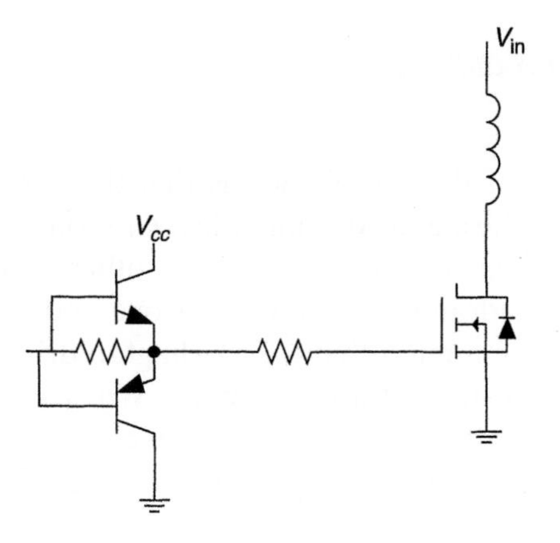

Figure 12.10: Nonisolated *N*-channel MOSFET driver

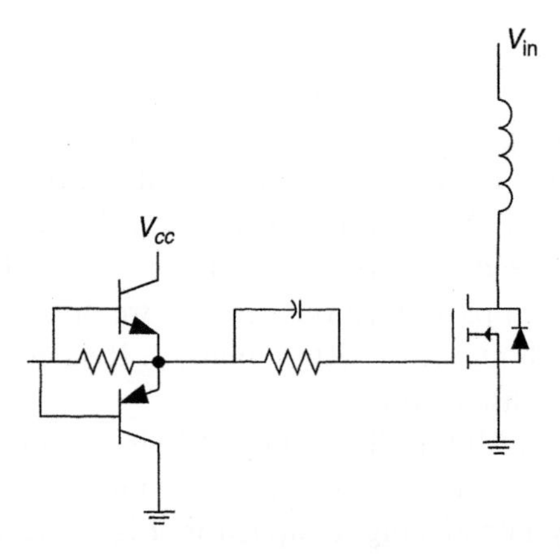

Figure 12.11: Capacitor coupled *N*-channel MOSFET driver

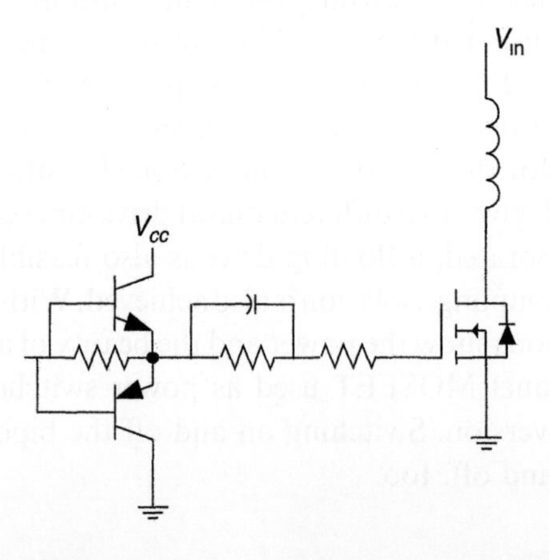

Figure 12.12: Modified capacitor coupled *N*-channel MOSFET driver

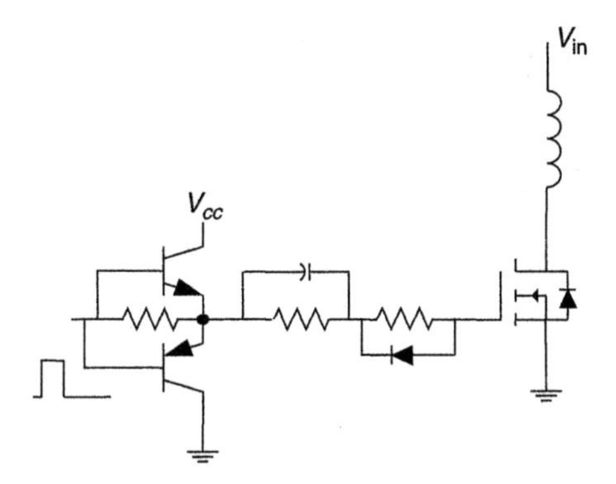

Figure 12.13: More modifications for the *N*-channel MOSFET driver

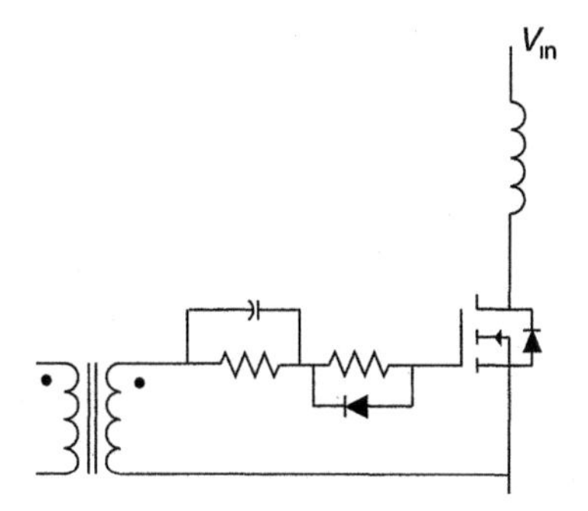

Figure 12.14: Transformer-coupled *N*-channel MOSFET driver

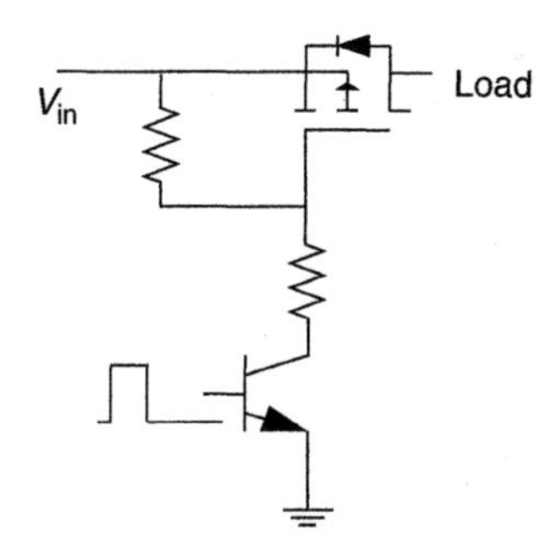

Figure 12.15: *P*-channel MOSFET driver

Surge suppression may be included if a capacitor is added across the gate and source terminals (Figure 12.16). The transient property of capacitor voltage makes the *P*-channel device behave like a variable resistor during the turn-on phase. In case the input voltage varies over a wide range, the driver network, as shown, may not work for low input. Then Figure 12.17(a) provides a way out.

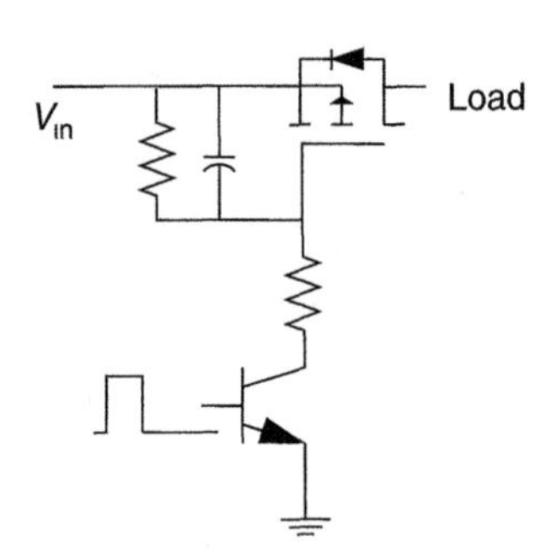

Figure 12.16: *P*-channel MOSFET switch with a soft-starting driver

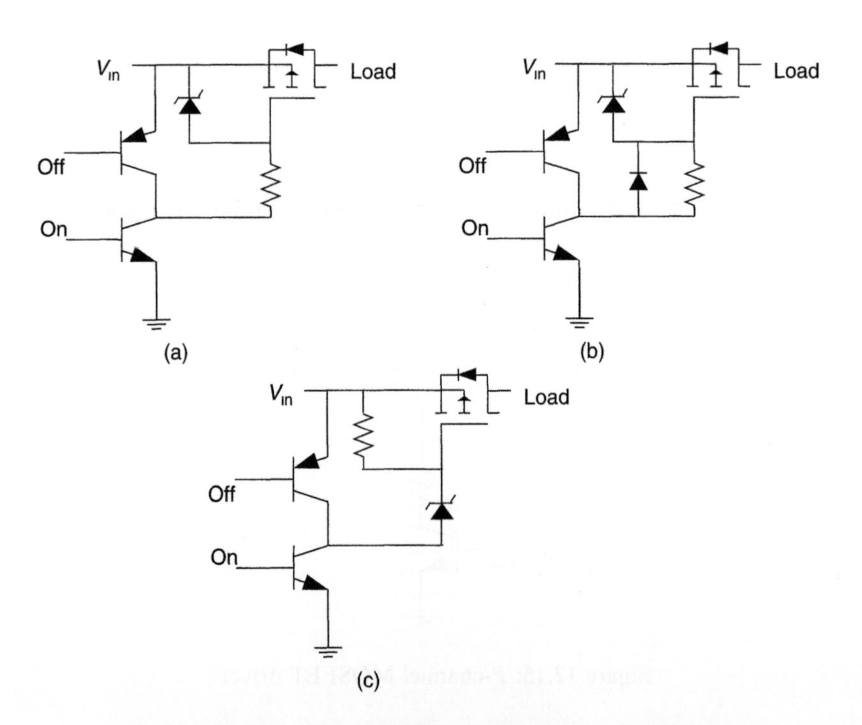

Figure 12.17: *P*-channel MOSFET switch for variable input

The combination of a zener diode and a resistor take the input variation out of the picture and the driver is now capable of riding through a low-input condition. Furthermore, a diode placed strategically as shown in Figure 12.17(b) gives the circuit enhanced speed in switching.

12.3 Dissipative Snubber

The load lines encountered in a switch-mode power converter, in general, are inductive. On the *i–v* coordinate plane, the power switch current and voltage traverse a nonlinear trajectory. This load behavior exerts significant stress on the switch in the form of local power dissipation that elevates the device's operating temperature and eventually reduces reliability. The switch power dissipation can be roughly partitioned into two parts: saturation losses and switching losses. The saturation losses are attributed to the switch-on voltage and current and considered more of a static nature, whereas the switching losses are associated more with the transitory nature of rising and falling of switch voltage and current at the very moment of on/off state changes. The latter, although dynamic in nature, is more accessible to the designer and can be managed to some extent. The voltage rising due to an inductive *di/dt* kickback is more pronounced when the switch turns off. We, therefore, focus on the turn-off snubbers, which are intended to manage the rate of falling current. However, be cautioned that the process of snubber design is not easily subjected to exact analysis. On the one hand, it treats circuits under highly dynamic conditions in which stray elements involved in the process are not totally accounted for. On the other hand, the parasitic elements participating in the process are also nonlinear and the circuit structure changes are not clearly tractable.

Conventional snubbers are made of passive elements. To direct current flow in a preferred, guided way, diodes are also used. Figure 12.18(a) gives a turn-off RC-diode snubber.

The operating sequence of the circuit can be briefly described as follows. When the power switch terminates conduction, the inductive load voltage flips polarity (Figure 12.18(b)). The clamping diode conducts and keeps the magnetizing current flowing without interruption; albeit at a decreasing rate. The action moves the energy previously

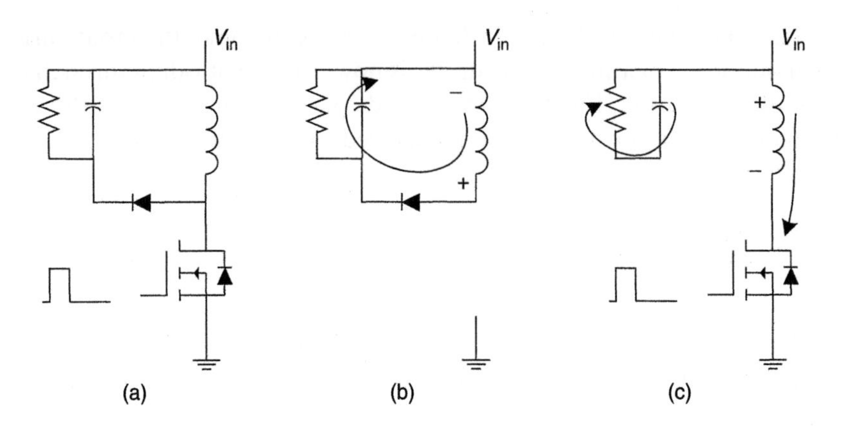

Figure 12.18: RC-diode turn-off snubber

stored in the magnetic core to the capacitor. When the next on-cycle commences, the clamping diode disengages the RC network and the capacitor discharges its contents (Figure 12.18(c)). Readers are referred to Wu [10] for a detailed mathematical analysis.

12.4 Lossless Snubber

It is clearly indicated in the previous section that the stored magnetic energy is eventually dumped and dissipated in the resistive element. That is the basis for calling the mechanism *dissipative*. A slight modification alters the mechanism and makes it possible to recover some, if not all, of the energy burned in the RC-diode clamping. Figure 12.19 shows the general configuration of such a lossless snubber consisting of an LC-diode.

The key to the circuit operation is twofold. First, the stored (magnetizing) energy is transferred unidirectionally to the clamping capacitor while the main switch is in the off-state (Figure 12.20(a)). When the main switch turns on, the capacitor releases its contents to the clamping inductor (Figure 12.20(b)).

The mathematics involved in the design of this type of snubber is not as precise as we would like it to be. We do not attempt to cover it. Readers interested in the topic may refer to Smith and Smedley [12] for further studies.

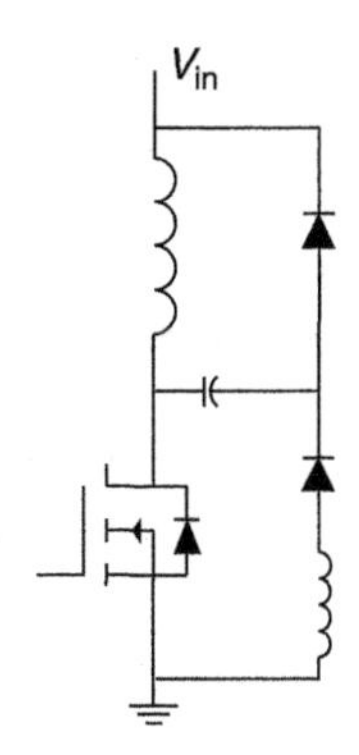

Figure 12.19: LC-diode energy-recovering snubber

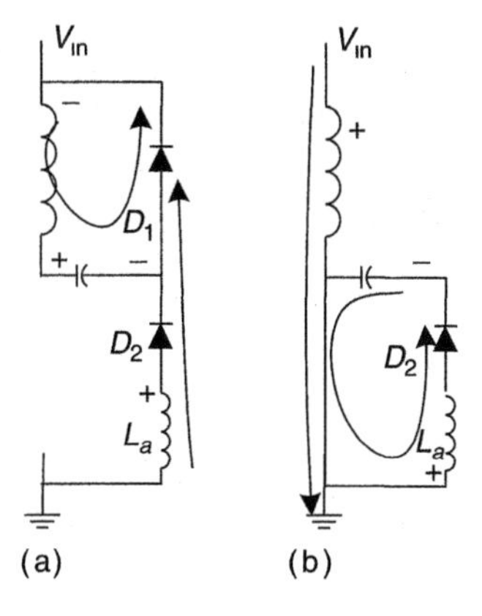

Figure 12.20: Lossless snubber in (a) off and (b) on states

12.5 Isolated Feedback

Both voltage and current feedbacks are used effectively in modern power converters. Early on, voltage feedback had the upper hand, since it was easy to implement. Nonisolated voltage feedback is even easier. It takes just a node voltage and two resistors as a voltage divider. Of course, its simplicity hides its shortcoming—no isolation. To implement isolated voltage feedback, either magnetic or optical means must be employed. The magnetic approach requires that the voltage being fed back must first be converted into AC form. The optical approach requires that the feedback voltage be in current form to drive an optical element. You see, the requirement of isolation wipes out the advantage voltage feedback has over current feedback. The latter stands out as a better choice for that reason and others. In addition to the ability of offering isolation, current feedback can accommodate many auxiliary tasks: soft start, current limiting, and current sharing, to name a few. Current feedback is also less prone to noise. A voltage node with certain impedance can be easily influenced by radiated emissions with no direct connection. In contrast, it is not so easy to inject unwanted current into a current-sensing branch without direct physical contact. It is with these understandings that Figure 12.21 is shown.

In the figure, the voltage feedback is converted into current. An isolated magnetic current transformer with a ratio of 1:n takes the main switch's pulsating current and scales it by $1/n$ across the isolation barrier. Both are summed, diodes D_1 and D_2, and injected across the current-to-voltage conversion resistor, R_{sen}. Since R_{sen} has the lowest impedance,

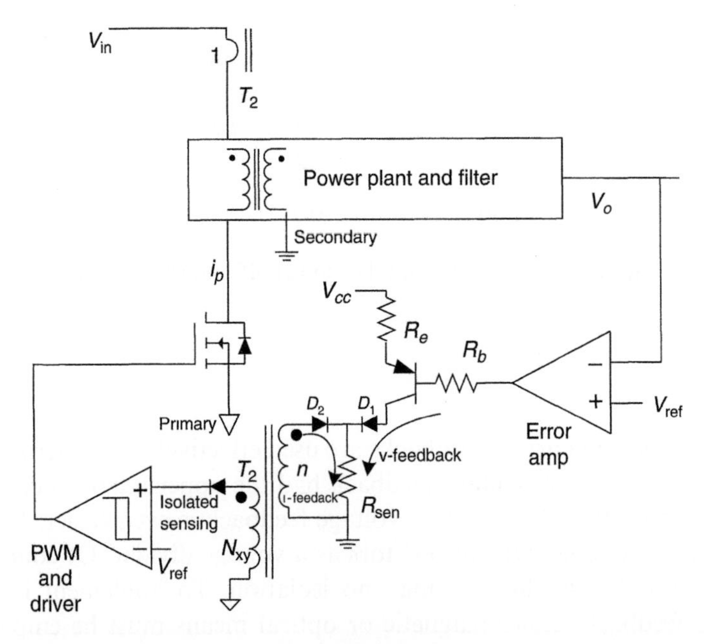

Figure 12.21: Isolated current feedback (U.S. patent 6,285,234)

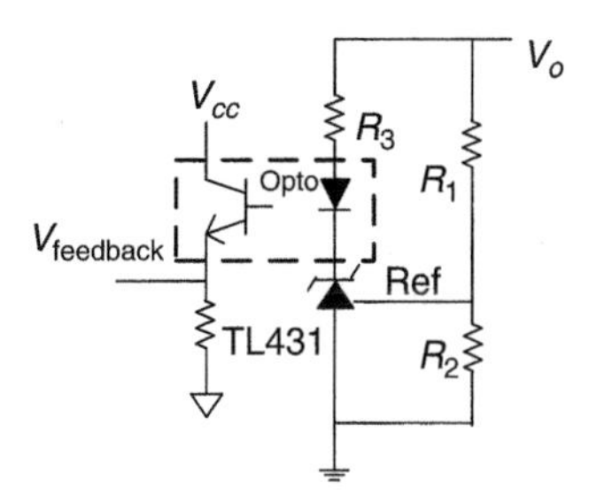

Figure 12.22: Optical isolator

the voltage of winding N_{xy}, where the current feedback and filter reside, is slaved to that of winding n. In effect, the voltage of winding N_{xy} is given by

$$N_{xy} \bullet \frac{R_{\text{sen}}}{n} \bullet \left[h_{FE} \frac{V_{CC} - V_{BE} - A(V_{\text{ref}} - V_o)}{(1 + h_{FE})R_e + R_b} + \frac{i_p}{n} \right] \tag{12.2}$$

Readers can refer to the patent for an alternative approach in which current subtraction is implemented.

The opto-coupler is yet another device providing isolation. As shown in Figure 12.22, the output to be regulated feeds a voltage divider—R_1 and R_2—that samples the output. The sampled output is compared with a reference voltage, $V_{\text{ref}=2.5\,\text{V}}$, residing in TL431, a precision shunt regulator. The error voltage is converted into drive current for the opto-coupler, an LED and phototransistor combination. The current transfer ratio (CTR) of the opto-coupler transfers the drive current. The opto-coupler output current is converted into a feedback voltage via a resistor.

12.6 Soft Start

Based on Figure 12.21, a soft-start mechanism can be easily implemented by injecting a control current pulse at the current summing point of D_1 and D_2. The pulse current injector is just a current source with an RC

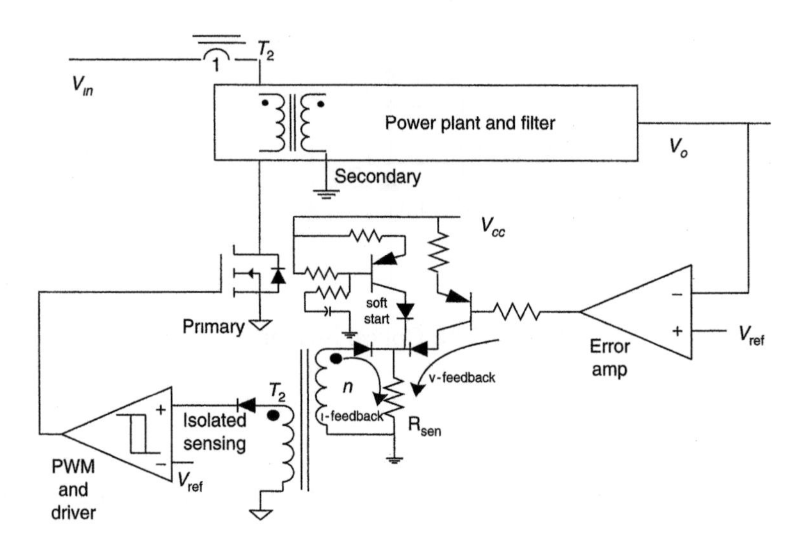

Figure 12.23: Isolated current feedback with soft start

timer (Figure 12.23). At the initial turn-on, the soft-start current forces a high current across R_{sen} and develops a high voltage. The high voltage forces the same across winding N_{xy} and in turn forces a minimum duty cycle. The soft-start current diminishes gradually and normal function takes over.

12.7 Negative-Charge Pump

Quite often, a negative supply at low current is needed for auxiliary circuits. Figure 12.24(a) gives such a simple circuit.

When Q_1 is turned off as in Figure 12.24(b), C_1 is charged through D_1 with the polarity shown. When Q_1 is turned on, the contents of C_1 are reversed and transferred via D_2 to C_2. The off configuration gives two equations:

$$R \bullet C_1 \frac{dv_1}{dt} + v_1 = V_s - V_{\text{cr}}, \quad C_2 \frac{dv_2}{dt} + \frac{v_2}{R_L} = 0 \tag{12.3}$$

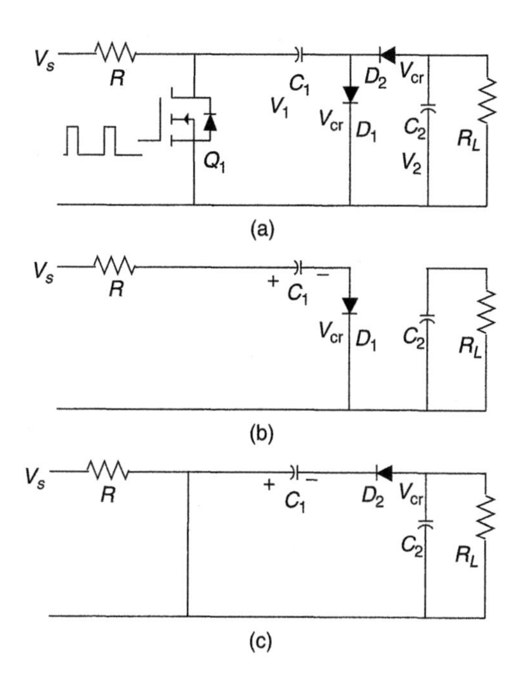

Figure 12.24: (a) Negative charge pump with (b) Q_1 off and (c) Q_1 on

The on configuration gives

$$v_2 = V_{\text{cr}} - v_1, \quad (C_1 + C_2)\frac{dv_2}{dt} + \frac{v_2}{R_L} = 0 \tag{12.4}$$

Equation (12.4) also implies

$$(C_1 + C_2)\frac{dv_1}{dt} = \frac{V_{\text{cr}} - v_1}{R_L} \tag{12.5}$$

Again, by using the technique of continuity of states, we arrive at the steady-state solution. The voltage across C_1 is given as

$$\begin{aligned} v_1(t) = {} & \left[V_{1a}e^{-t/R\cdot C} + (V_s - V_{\text{cr}})\left(1 - e^{-t/R\cdot C}\right)\right][u(t) - u(t_{\text{off}})] \\ & + \left[V_{1b}e^{\frac{t-t_{\text{off}}}{R_L(C_1+C_2)}} + V_{\text{cr}}\left(1 - e^{\frac{t-t_{\text{off}}}{R_L(C_1+C_2)}}\right)\right][u(t - t_{\text{off}}) - u(t - T)] \end{aligned} \tag{12.6}$$

where

$$V_{1a} = \frac{\begin{vmatrix} -(V_s - V_{cr})\left(1 - e^{\frac{t_{off}}{R \cdot C_1}}\right) & -1 \\ -V_{cr}(1 - e^{\frac{T - t_{off}}{R_L(C_1+C_2)}}) & e^{\frac{T - t_{off}}{R_L(C_1+C_2)}} \end{vmatrix}}{\begin{vmatrix} e^{\frac{t_{off}}{R \cdot C_1}} & -1 \\ -1 & e^{\frac{T - t_{off}}{R_L(C_1+C_2)}} \end{vmatrix}}$$

$$V_{1b} = \frac{\begin{vmatrix} e^{\frac{t_{off}}{R \cdot C_1}} & -(V_s - V_{cr})\left(1 - e^{\frac{t_{off}}{R \cdot C_1}}\right) \\ -1 & -V_{cr}(1 - e^{\frac{T - t_{off}}{R_L(C_1+C_2)}}) \end{vmatrix}}{\begin{vmatrix} e^{\frac{t_{off}}{R \cdot C_1}} & -1 \\ -1 & e^{\frac{T - t_{off}}{R_L(C_1+C_2)}} \end{vmatrix}} \quad (12.7)$$

The output voltage is given as

$$\begin{aligned} v_2(t) = {} & V_{2a} e^{\frac{t}{R_L \cdot C_2}} [u(t) - u(t_{off})] \\ & + \left\{ V_{cr} - \left[V_{1b} e^{\frac{t - t_{off}}{R_L(C_1+C_2)}} + V_{cr}\left(1 - e^{\frac{t - t_{off}}{R_L(C_1+C_2)}}\right)\right]\right\} \\ & [u(t - t_{off}) - u(t - T)] \end{aligned} \quad (12.8)$$

where

$$V_{2a} = V_{cr} - \left[V_{1b} e^{\frac{T - t_{off}}{R_L(C_1+C_2)}} + V_{cr}\left(1 - e^{\frac{T - t_{off}}{R_L(C_1+C_2)}}\right)\right] \quad (12.9)$$

We give an example to illustrate and confirm the working of such a circuit. Given $R = 30$, $C_1 = 10\,\mu\text{F}$, $C_2 = 10\,\mu\text{F}$, $R_L = 20$, $V_s = 10$ volts, $f_s = 100\,\text{KHz}$, and $t_{off} = 6\,\mu\text{s}$, the output voltage looks like Figure 12.25.

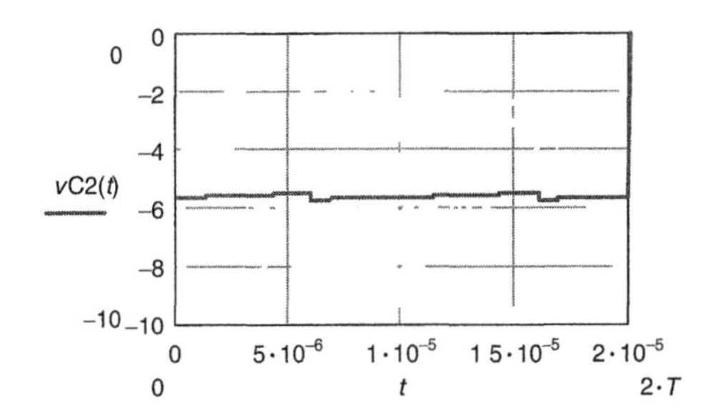

Figure 12.25: Example output of a negative charge pump

12.8 Single-Phase Full-Wave Rectifier with RC Filter

Both half-wave and full-wave rectifiers with RC filters have been in use for years. The circuit given in Figure 12.26 has been known to exhibit waveforms, as shown in Figure 12.27. However, as of this date (July 2004), no analytical solution in symbolic, closed-form expression has been worked out for the rectifier conduction starting time, t_1; the conduction end time, t_2; the rectified input line current, i_s; the capacitor current, i_c; and the output voltage, v. Wu [10] gave the first brief symbolic solution with many intermediate steps skipped. This section gives a complete, detailed analysis.

Under the steady state, every rectified cycle (half cycle at AC line frequency) is divided into two time segments: conducting $t_2 - t_1$ and nonconducting $T/2 - (t_2 - t_1)$. During rectifier conducting time, the capacitor is charged while the load R is also fed. During the nonconducting

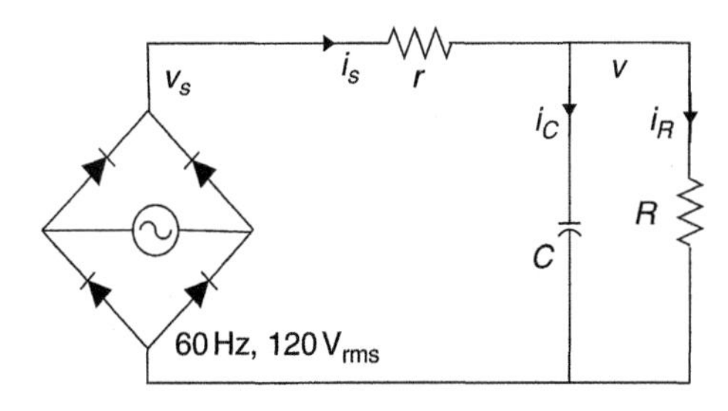

Figure 12.26: Single-phase full-wave rectifier with RC

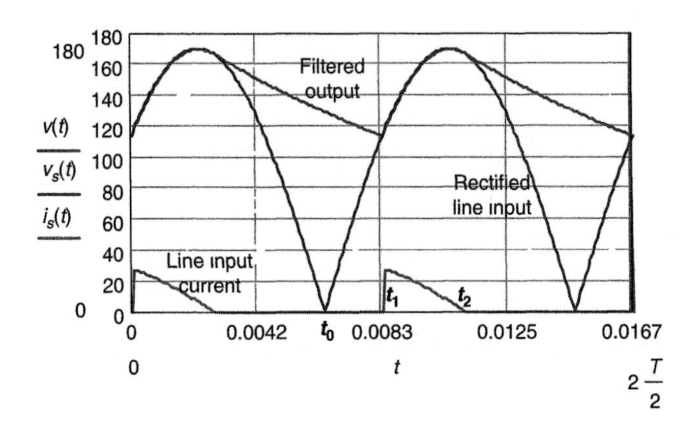

Figure 12.27: Waveforms for a single-phase full-wave rectifier with RC

phase, the capacitor discharges to support the R load while the line current ceases.

Two first-order differential equations are given for both time segments:

$$
\begin{aligned}
&\frac{dv}{dt} + \frac{1}{\tau_p} v = \frac{1}{\tau_1} v_s \quad \text{conducting} \quad \tau_p = \frac{R \bullet r}{R + r} C, \quad \tau_1 = r \bullet C \\
&\frac{dv}{dt} + \frac{1}{\tau_2} v = 0 \quad \text{nonconducting} \quad \tau_2 = R \bullet C
\end{aligned}
\tag{12.10}
$$

To solve the conducting equation, the driving source must be time-shifted to the output voltage, v, in the time scale in which t_1 is treated as the origin time. In other words, the driving source, v_s, is $A|\sin\omega(t+t_1)|$, instead of i$|\sin\omega t|$.

Next, we take the Laplace transform of the conducting equation with the assumption of an unknown starting state, V_{0a}. This step leads to

$$
\begin{aligned}
V(s) = &\frac{V_{0a}}{s + 1/\tau_p} + \frac{A \sin\omega t_1}{\tau_1} \frac{s}{(s + 1/\tau_p)(s^2 + \omega^2)} \\
&+ \frac{A \cos\omega t_1}{\tau_1} \frac{\omega}{(s + 1/\tau_p)(s^2 + \omega^2)}
\end{aligned}
\tag{12.11}
$$

By setting $t_2 - t_1 = t_{\text{on}}$, the inverse Laplace transform yields the output voltage during conducting time:

$$v_a(t) = \left\{ V_{0a}\mathrm{e}^{-t/\tau_p} + \frac{A \bullet \tau_p \bullet \sin \omega t_1}{\tau_1 \bullet Z}[-\mathrm{e}^{-t/\tau_p} \sin\alpha + \sin(\omega t + \alpha)] \right.$$

$$\left. + \frac{A \bullet \tau_p \bullet \cos \omega t_1}{\tau_1 \bullet Z}[\mathrm{e}^{-t/\tau_p} \sin\beta + \sin(\omega t - \beta)] \right\} [u(t) - u(t - t_{\text{on}})],$$

$$\alpha = \tan^{-1}\left(\frac{1}{\omega\tau_p}\right), \quad \beta = \tan^{-1}(\omega\tau_p), \quad Z = \sqrt{1 + (\omega\tau_p)^2} \qquad (12.12)$$

By also assuming an unknown starting state, V_{0b}, for the nonconducting time, the nonconducting differential equation yields a solution:

$$v_b(t) = V_{0b}\mathrm{e}^{-\frac{t - t_{\text{on}}}{\tau_2}} \left[u(t - t_{\text{on}}) - u\left(t - \frac{T}{2}\right) \right] \qquad (12.13)$$

Then, the continuity of state demands

$$v_a(t_{\text{on}}) = V_{0a}\mathrm{e}^{-\frac{t_{\text{on}}}{\tau_p}} + \frac{A \bullet \tau_p \bullet \sin \omega t_1}{\tau_1 \bullet Z} \left[-\mathrm{e}^{-\frac{t_{\text{on}}}{\tau_p}} \sin\alpha + \sin(\omega t_{\text{on}} + \alpha) \right]$$

$$+ \frac{A \bullet \tau_p \bullet \cos \omega t_1}{\tau_1 \bullet Z} \left[\mathrm{e}^{-\frac{t_{\text{on}}}{\tau_p}} \sin\beta + \sin(\omega t_{\text{on}} - \beta) \right] = V_{\text{ob}}$$

$$v_b\left(\frac{T}{2}\right) = V_{0b}\mathrm{e}^{-\frac{\frac{T}{2} - t_{\text{on}}}{\tau_2}} = V_{0a} \text{ or } V_{0b} = V_{0a}\mathrm{e}^{\frac{\frac{T}{2} - t_{\text{on}}}{\tau_2}} \qquad (12.14)$$

Equation (12.14), of course, can be solved for V_{0a} and V_{0b}; that is,

$$V_{0a} = \frac{\left\{ \begin{array}{l} \frac{A \bullet \tau_p \bullet \sin \omega t_1}{\tau_1 \bullet Z}[-\mathrm{e}^{-\frac{t_{\text{on}}}{\tau_p}} \sin\alpha + \sin(\omega t_{\text{on}} + \alpha)] \\ + \frac{A \bullet \tau_p \bullet \cos \omega t_1}{\tau_1 \bullet Z}[\mathrm{e}^{-\frac{t_{\text{on}}}{\tau_p}} \sin\beta + \sin(\omega t_{\text{on}} - \beta)] \end{array} \right\}}{\mathrm{e}^{\frac{\frac{T}{2} - t_{\text{on}}}{\tau_2}} - \mathrm{e}^{-\frac{t_{\text{on}}}{\tau_p}}} \qquad (12.15)$$

Then, V_{0b} follows.

Furthermore, there are two boundary conditions:

$$A \sin \omega t_1 = V_{0a}, \quad A \sin \omega (t_1 + t_{\text{on}}) = V_{0b} \tag{12.16}$$

the first boundary condition of (12.16) and (12.15) can be further combined. By regrouping and separating terms, the cut-in time, t_1, is expressed as

$$\tan \omega t_1 = \frac{\frac{\tau_p}{\tau_1 \cdot Z}\left[e^{-t_{\text{on}}/\tau_p} \sin \beta + \sin(\omega t_{\text{on}} - \beta)\right]}{e^{\frac{\frac{T}{2} - t_{\text{on}}}{\tau_2}} - e^{-t_{\text{on}}/\tau_p} - \frac{\tau_p}{\tau_1 \cdot Z}\left[- e^{-t_{\text{on}}/\tau_p} \sin \alpha + \sin(\omega t_{\text{on}} + \alpha)\right]} \tag{12.17}$$

or

$$t_1 = \frac{\tan^{-1} \dfrac{\frac{\tau_p}{\tau_1 \cdot Z}\left[e^{-t_{\text{on}}/\tau_p} \sin + \sin(\omega t_{\text{on}} - \beta)\right]}{e^{\frac{\frac{T}{2} - t_{\text{on}}}{\tau_2}} - e^{-t_{\text{on}}/\tau_p} - \frac{\tau_p}{\tau_1 \cdot Z}\left[-e^{-t_{\text{on}}/\tau_p} \sin \alpha + \sin(\omega t_{\text{on}} + \alpha)\right]}}{\omega} \tag{12.18}$$

By taking the ratio of the two boundary conditions in (12.16) and grouping the terms, we get

$$\frac{\sin \omega (t_1 + t_{\text{on}})}{\sin \omega t_1} = e^{\frac{\frac{T}{2} - t_{\text{on}}}{\tau_2}}, \quad \tan \omega t_1 = \frac{\sin \omega t_{\text{on1}}}{e^{\frac{\frac{T}{2} - t_{\text{on}}}{\tau_2}} - \cos \omega t_{\text{on}}} \tag{12.19}$$

Evidently, (12.17) and (12.19) can be consolidated to give a nonlinear equation for t_{on} alone:

$$\frac{\frac{\tau_p}{\tau_1 \cdot Z}\left[e^{-t_{\text{on}}/\tau_p} \sin \beta + \sin(\omega t_{\text{on}} - \beta)\right]}{e^{\frac{\frac{T}{2} - t_{\text{on}}}{\tau_2}} - e^{-t_{\text{on}}/\tau_p} - \frac{\tau_p}{\tau_1 \cdot Z}\left[- e^{-t_{\text{on}}/\tau_p} \sin \alpha + \sin(\omega t_{\text{on}} + \alpha)\right]} - \frac{\sin \omega t_{\text{on1}}}{e^{\frac{\frac{T}{2} - t_{\text{on}}}{\tau_2}} - \cos \omega t_{\text{on}}} = 0 \tag{12.20}$$

The transcendental equation (12.20) cannot be solved explicitly for t_{on}. Rather it can be solved numerically with very high accuracy. However, the point is that the conduction duration, and consequently the conduction cut-in time, can be expressed in symbolic, closed form. Once t_{on} and t_1 are known, V_{0a} and V_{0b} follow. Given V_{0a} and V_{0b}, the steady-state output voltage is given by (12.12) and (12.13), $v(t) = v_a(t) + v_b(t)$. The rectified line current is then $i_s(t) = [v_s(t) - v(t)]/r$ if $v_s(t) > v(t)$ [$i_s(t) = 0$ if $v_s(t) \leq v(t)$]; the load current by $i_L(t) = v(t)/R$; and the capacitor current $i_c(t) = i_s(t) - i_L(t)$.

Moreover, with the pulsating, rectified line current expressed in symbolic, closed form, other in-depth studies, for instance, harmonic content and power factor, can be conducted easily.

The procedure just given can be extended to include the equivalent series resistance that is omnipresent with electrolytic capacitors frequently used in high-power AC rectifiers. By so doing, Figure 12.26 changes to Figure 12.28.

With capacitor esr (equivalent series resistance), the governing equations also change. They are

$$\begin{gathered}
\frac{dv}{dt} + \frac{1}{\tau_3} v = \frac{1}{k} v_s \quad \textit{conducting} \quad \tau_3 = \left(\frac{R \bullet r}{R + r} + r_c\right) C \\
\frac{dv}{dt} + \frac{1}{\tau_4} v = 0 \quad \textit{nonconducting} \quad \tau_4 = (R + r_c) \bullet C \\
k_1 = \frac{\dfrac{R \bullet r_c}{R + r_c}}{\dfrac{R \bullet r_c}{R + r_c} + r} \qquad k_2 = \frac{\dfrac{R \bullet r}{R + r}}{\dfrac{R \bullet r}{R + r} + r_c} \qquad k = \frac{k_1}{r_c C}
\end{gathered} \tag{12.21}$$

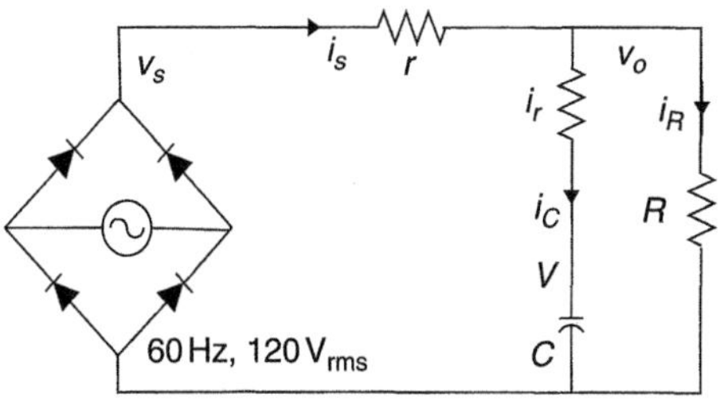

Figure 12.28: Single-phase full-wave rectifier with RC and capacitor esr

The Laplace transform of (12.21) generates four transfer functions with the driving source treated as before:

$$F_1(s) = \frac{1}{s+\frac{1}{\tau_3}}, \quad F_2(s) = \frac{s}{\left(s+\frac{1}{\tau_3}\right)(s^2+\omega^2)},$$
$$F_3(s) = \frac{\omega}{\left(s+\frac{1}{\tau_3}\right)(s^2+\omega^2)}, \quad F_4(s) = \frac{1}{s+\frac{1}{\tau_4}} \tag{12.22}$$

Again, we assume two unknown starting states, V_{0a} and V_{0b}. The capacitor voltage in the transformed domain during the conducting phase is

$$V_a(s) = V_{0a}F_1(s) + k \bullet A \bullet \sin \omega t_1 \bullet F_2(s) + k \bullet A \bullet \cos \omega t_1 \bullet F_3(s) \tag{12.23}$$

and during the nonconducting phase is

$$V_a(s) = V_{0b}F_4(s)\mathrm{e}^{-t_{\mathrm{on}} \bullet s} \tag{12.24}$$

Next we assume that the inverse transform of (12.22) leads to $f_1(t), f_2(t), f_3(t)$, and $f_4(t)$. The continuity of state requires the following

$$V_{0a}f_1(t_{\mathrm{on}}) - V_{0b} = -k \bullet A[\sin \omega t_1 \bullet f_2(t_{\mathrm{on}}) + \cos \omega t_1 \bullet f_3(t_{\mathrm{on}})]$$
$$V_a(s) - V_{0b}f_4\left(\frac{T}{2} - t_{\mathrm{on}}\right) = 0 \tag{12.25}$$

The two cyclic starting states, V_{0a} and V_{0b}, as functions of t_1 and t_{on} are

$$V_{0a}(t_1, t_{\mathrm{on}}) = \frac{-k \bullet A[\sin \omega t_1 \bullet f_2(t_{\mathrm{on}}) + \cos \omega t_1 \bullet f_3(t_{\mathrm{on}})] \bullet f_4\left(\frac{T}{2} - t_{\mathrm{on}}\right)}{f_1(t_{\mathrm{on}})f_4\left(\frac{T}{2} - t_{\mathrm{on}}\right) + 1}$$
$$V_{0b}(t_1, t_{\mathrm{on}}) = \frac{k \bullet A[\sin \omega t_1 \bullet f_2(t_{\mathrm{on}}) + \cos \omega t_1 \bullet f_3(t_{\mathrm{on}})]}{f_1(t_{\mathrm{on}})f_4\left(\frac{T}{2} - t_{\mathrm{on}}\right) + 1} \tag{12.26}$$

Therefore, the (ideal) capacitor voltage is

$$
\begin{aligned}
v(t, t_1, t_{on}) &= v_a(t, t_1, t_{on})[u(t) - u(t - t_{on})] \\
&\quad + v_b(t, t_1, t_{on})\left[u(t - t_{on}) - u\left(\frac{T}{2} - t_{on}\right)\right] \\
v_a(t, t_1, \ldots, t_{on}) &= [V_{0a}(t_1, t_{on})f_1(t) + k \bullet A \bullet \sin \omega t_1 \bullet f_2(t) \\
&\quad + k \bullet A \bullet \cos \omega t_1 \bullet f_3(t)] \\
v_b(t, t_1, t_{on}) &= V_{0b}(t_1, t_{on})f_4(t - t_{on})
\end{aligned}
\tag{12.27}
$$

Now, we are ready to find the cut-in time, t_1, and the conduction duration, t_{on}. However, the presence of capacitor esr slightly complicates the matter; that is, the output voltage is no longer equal to the ideal capacitor voltage (12.27). Instead, the output is given by the superposition of both the capacitor voltage and the rectified line source:

$$
\begin{aligned}
v_o(t, t_1, t_{on}) &= [k_1 \bullet A \bullet \sin \omega(t + t_1) + k_2 \bullet v_a(t, t_1, t_{on})][u(t) - u(t - t_{on})] \\
&\quad + k_3 v_b(t, t_1, t_{on})\left[u(t - t_{on}) - u\left(t - \frac{T}{2}\right)\right]
\end{aligned}
\tag{12.28}
$$

Again, two boundary conditions, at $t = 0$ and $t = t_{on}$, are given:

$$
\begin{aligned}
k_1 \bullet A \bullet \sin \omega(t_1) + k_2 \bullet v_a(0, t_1, t_{on}) &= A \bullet \sin \omega(t_1) \\
k_3 v_b(t_{on}, t_1, t_{on}) &= A \bullet \sin \omega(t_1 + t_{on})
\end{aligned}
\tag{12.29}
$$

In theory, one can substitute (12.26) in (12.27) and, in turn, plug (12.27) in (12.29). However, in doing so, the enormous complexity quickly arises and denies us the chance to further simplify the equation as we did in (12.16) to (12.20). We just use the equation chains and solve the two unknowns, t_1 and t_{on}, numerically.

12.9 Duty-Cycle Clamping

In conventional pulse-width-modulated switching DC–DC converters, various conditions exist in which the modulating pulse tends to exceed a certain undesirable maximum. In such a situation, critical circuit components are quite often overstressed, such as the magnetic core is saturated and circuit performance is impaired, leading to poor voltage regulation, for example. Therefore, it is highly desirable to have duty ratio clamping such that component overstress is prevented.

As shown in Figure 12.29, methods 1 and 2 are the most popular choice of duty-cycle clamping. Both techniques take advantage of semiconductor active junction properties. Subtle differences exist between two methods. Method 1 uses the reverse junction breakdown, whereas method 2 employs the forward junction saturation. In either approach, the error signal at node A is clamped statically, however, at V_z or $(V_{be} + V_m)$ as shown in, Figure 12.30, if conditions such as low line input, output short, or V_{ref} failure arises.

In applying these approaches, since the active junction of semiconductor interacts directly with the error signal node, they all suffer a drawback associated with the semiconductor junction; that is, a prolonged device recovery time caused by storage charge and consequently longer circuit recovery, as shown in Figure 12.30, when conditions causing the long duty cycle are removed. Meanwhile, elaborate schemes of soft start, which ensures a gradual increase of operating duty cycle, are needed for this kind of duty-cycle clamping. In the absence of such a soft-start mechanism, the transient operating duty cycle, clamped at the maximum width, may extend over many tens of switching cycles and unnecessarily overstress the parts.

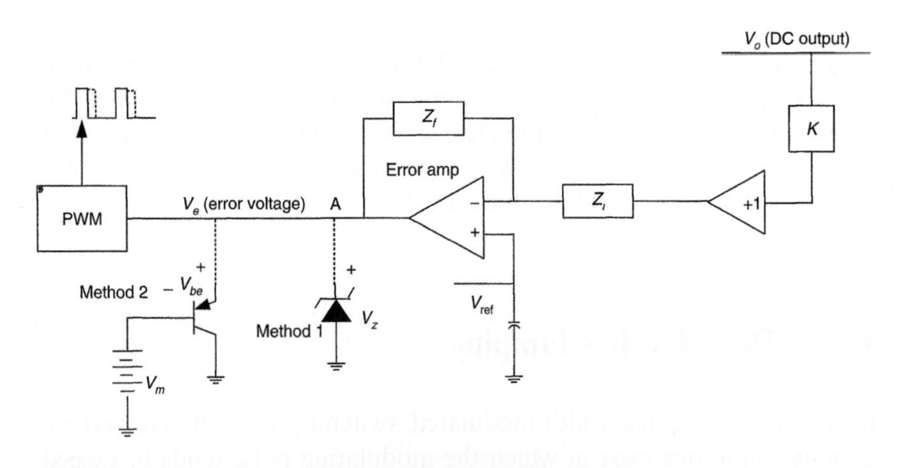

Figure 12.29: Static clamping

A better approach (US Patent 6,894,910) was conceived in 2002. Referring to Figure 12.31, the network enclosed in the dotted line constitutes the main body of the invention. Standing alone, it is a simple and rather straightforward voltage amplifier/comparator. It is how the network is connected to the rest of the circuit that is considered unique.

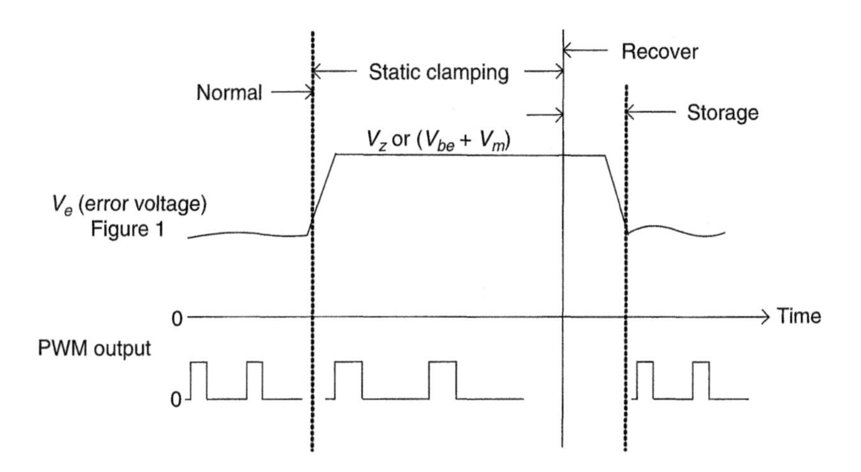

Figure 12.30: Waveforms corresponding to Figure 12.29

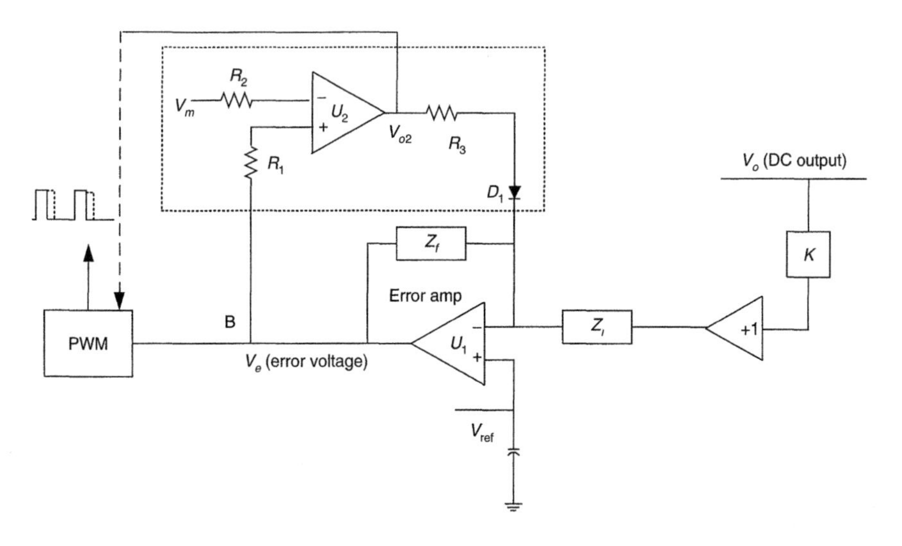

Figure 12.31: Dynamic clamping (patent pending)

As can be seen from Figure 12.31, Z_f serves originally as the negative feedback network of the error amplifier, U_1. With the incorporation of amplifier/comparator U_2 and its associated components, Z_f serves a double purpose as the positive hysteresis feedback network of U_2. This latter role Z_f speeds up the dynamic action of amplifier/comparator U_2.

A. Normal Operation

During normal operation, the amplifier/comparator output is quiescent at low voltage (Figure 12.32), since the error signal is less than V_m. The existence of diode D_1 ensures that the amplifier/comparator output does not interfere with the networks, Z_i and Z_f, surrounding the error amplifier. In addition to this isolation effect in the normal state, diode D_1 also makes it unnecessary to have a strictly zero potential at the amplifier/comparator output. Instead, a voltage higher than zero but less than the forward voltage of D_1 is acceptable, which also enables a single supply operation for the amplifier/comparator U_2.

B. Dynamic Clamping by Averaging (U_2 Is a Linear Amplifier)

As shown in Figure 12.32, the error signal exceeding the maximum set voltage V_m is amplified and fed back negatively through R_3 and D_1. This mechanism creates an oscillatory action circling the local loop consisting

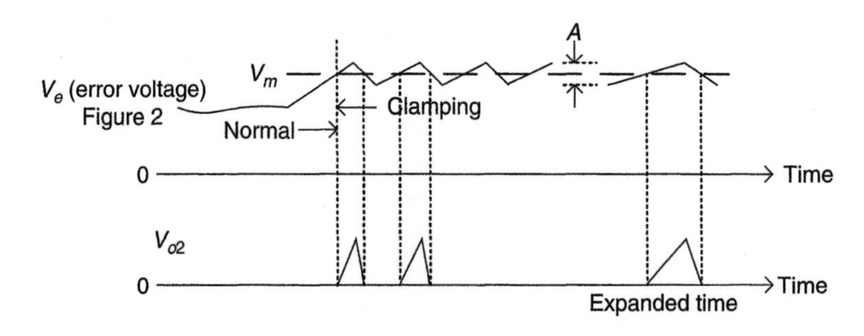

Figure 12.32: Waveforms corresponding to Figure 12.27, U_2 linear

of U_1 and U_2. The oscillating motion persists as long as the agent causing the action remains. Depending on individual requirements, the depth of excursion (A in Figure 12.32) at U_1 output can be either a shadow one centering around V_m or a large swing from zero volts to V_m. The former case yields a cyclic reduction of effective average duty cycle, while the latter suppresses a group of PWM pulses periodically in addition to a periodic reduction of the average duty cycle. In this mode of operation, the time period of local oscillation is in the millisecond range and depends on both the property of amplifier U_2 and local feedback network, Z_f.

This configuration also offers another useful signal, V_{o2} at i_2 output. It may be gated by a power-on-reset signal such that initial startup is guaranteed. The gated signal may also be used either as a synchronization pulse or as a duty suppression pulse.

C. Dynamic Cycle-by-Cycle Clamping (U_2 Using a High-Speed Voltage Comparator with Discrete Output)

This configuration employs a high-speed voltage comparator with open collector output that requires a pull-up resistor. It also produces an oscillation when the error voltage exceeds the maximum set point V_m. However, rather than generating a linear output, the comparator output is a discrete large signal (Figure 12.33). Since the comparator can operate at extremely high speed, the oscillation period is very short, in the

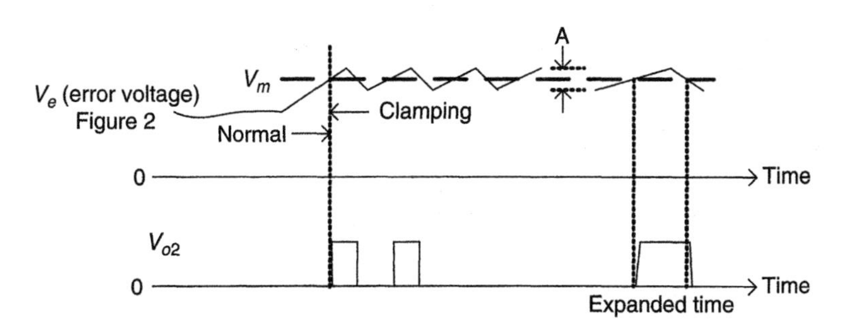

Figure 12.33: Waveforms corresponding to Figure 12.31, U_2 nonlinear

microsecond range, and the error amplifier output excursion is much smaller than that of Figure 12.32. If the oscillation period is in the vicinity of the PWM clock period, a cycle-by-cycle duty limit is made possible. Again, the comparator output may be used for the other purposes described earlier in addition to the local loop.

Chapter 13

State-Space Averaging and the Cuk Converter

The theory of state-space averaging applied in the switch-mode power converter was conceived in the early 1970s and well developed in the early 1980s. Dr. Robert Middlebrook and his then graduate student Dr. Slobodan Cuk were credited with the concept and techniques associated with it. Therefore, the switch-mode converter models generated on the basis of the theory have oftentimes been named *Middlebrook models*.

Over the years, the theory and the numerous models it created have been employed extensively by both academic researchers and industrial practitioners. Over this time, misuse of the theory also has arisen. In this chapter, we try to clarify this misapplication and present a complete example exhibiting the flow for using this powerful technique.

13.1 State-Space Averaging

In theory, a dynamic system can be described by a set of first-order differential equations. For example, the circuit of Figure 1.29 is analytically described by equations (1.60)–(1.62). The equation set can be further placed in a compact form using matrix symbolic representation, the state equation

$$\frac{dX}{dt} = A \cdot X + B \cdot E \tag{13.1}$$

For instance, see (1.115), where X is the state vector representing all state variables in the system. The implication is that, if the system structure switches cyclically between two alternating topologies within a preset period of time T, both topologies are governed by two distinctive state equations:

$$\frac{dX}{dt} = A_1 \cdot X + B_1 \cdot E, \quad \frac{dX}{dt} = A_2 \cdot X + B_2 \cdot E \tag{13.2}$$

and the output by

$$Y = C_1 \cdot X + H_1 \cdot E, \quad Y = C_2 \cdot X + H_2 \cdot E \tag{13.3}$$

The state-space averaging technique claims that, if a low (frequency)-pass control element exists in the system, the low-frequency behavior of the switching system under the steady state can be described by a single, weighted sum of both topologies:

$$\frac{dX}{dt} = A \cdot X + B \cdot E, \quad Y = C \cdot X + H \cdot E \tag{13.4}$$

where

$$A = D \cdot A_1 + (1 - D) \cdot A_2, \quad B = D \cdot B_1 + (1 - D) \cdot B_2,$$
$$C = D \cdot C_1 + (1 - D) \cdot C_2, \quad H = D \cdot H_1 + (1 - D) \cdot H_2$$

and DT and $(1 - D)T$ are the dwell times of two topologies. In general, A and B are square matrices while C and H are row matrices.

At the steady state and given a small-signal perturbation, the state variables and the operating point can be partitioned into two parts, DC and AC. Accordingly, the average equation is rewritten as

$$\frac{dx}{dt} = \frac{dX}{dt} + \frac{d\hat{x}}{dt} = [(D + \hat{d})A_1 + (1 - D - \hat{d})A_2(X + \hat{x}) \\ + (D + \hat{d})B_1 + (1 - D - \hat{d})B_2(E + \hat{e})] \tag{13.5}$$

The right-hand side can be expanded and regrouped so that the DC and the AC parts are separated. We also recognize that the nonlinear terms—the product terms consisting of perturbation quantities—can be safely ignored. In addition, the derivative of the DC part yields zero. The process then gives the steady state:

$$X = -A^{-1} \bullet B \bullet E \tag{13.6}$$

the AC duty-to-state and source-to-state transfer functions:

$$\begin{aligned} \hat{x}(s) &= \{(s \bullet I - A)^{-1}[(A_1 - A_2)X + (B_1 - B_2)E]\}\hat{d} \\ &\quad + \{(s \bullet I - A)^{-1}B\}\hat{e} \\ &= G_{xd}(s) \bullet \hat{d} + G_{xg}(s) \bullet \hat{e} \end{aligned} \tag{13.7}$$

the steady-state output:

$$Y = -C \bullet A^{-1} \bullet B \bullet E + H \bullet E \tag{13.8}$$

and the duty-to-output and source-to-output transfer functions:

$$\begin{aligned} \hat{y}(s) &= \left\{ \begin{array}{l} C \bullet (s \bullet I - A)^{-1}[(A_1 - A_2)X + (B_1 - B_2)E] \\ +(C_1 - C_2)X + (H_1 - H_2)E \end{array} \right\} \hat{d} \\ &\quad + \{C(s \bullet I - A)^{-1}B + H\}\hat{e} \\ &= G_{vd}(s) \bullet \hat{d} + G_{vg}(s) \bullet \hat{e} \end{aligned} \tag{13.9}$$

where I is the unity matrix and X is the steady state (13.6).

Readers are strongly urged to refer to Middlebrook and Cok [1] for a more in-depth discussion about this procedure since the original innovator is given the credit.

13.2 General Procedure

To use the technique effectively, the following steps are recommended.

Step 1. Identify the power stage that contains switches and energy storage elements.

Step 2. Identify the state variables.

Step 3. Assign state variable symbols and variable sign/polarity.

Step 4. Identify the two alternating topologies.

Step 5. Write the dynamic equations (first-order differential equations) for topology 1.

Step 6. Identify A_1, B_1, C_1, and H_1 from step 5.

Step 7. Repeat step 5 for topology 2.

Step 8. Identify A_2, B_2, C_2, and H_2 from step 7.

Step 9. Use (13.8) to find the steady-state duty cycle D by solving

$$\begin{aligned} V_o = &- [D \cdot C_1 + (1 - D) \cdot C_2] \cdot [D \cdot A_1 + (1 - D) \cdot A_2]^{-1} \\ &\cdot [D \cdot B_1 + (1 - D) \cdot B_2] \cdot E + [D \cdot H_1 + (1 - D) \cdot H_2] \cdot E \end{aligned} \tag{13.10}$$

Step 10. Compute the steady-state vector X, using (13.6), and the steady-state duty cycle D from step 9.

Step 11. Obtain the output transfer functions $G_{vd}(s)$ and $G_{vg}(s)$, using (13.9).

Step 12. Obtain the PWM gain.

Step 13. Obtain the feedback ratio and error amplifier gain as shown in many previous sections.

Step 14. Compute the loop gain.

13.3 Example: Cuk Converter

The Cuk converter (U.S. Patent 4,184,197; 4,274,133) was conceived in 1977 by Slobodan M. Cuk and Robert D. Middlebrook. It is considered the optimum topology with nonpulsating input current, nonpulsating output current, minimum storage elements, minimum switch number, and high energy-storage density using a capacitor instead of an inductor. A typical converter of this optimum topology is given in Figure 13.1.

We follow the steps outlined in the previous section. Step 1 leads to Figure 13.2, in which the power stage is clearly identified.

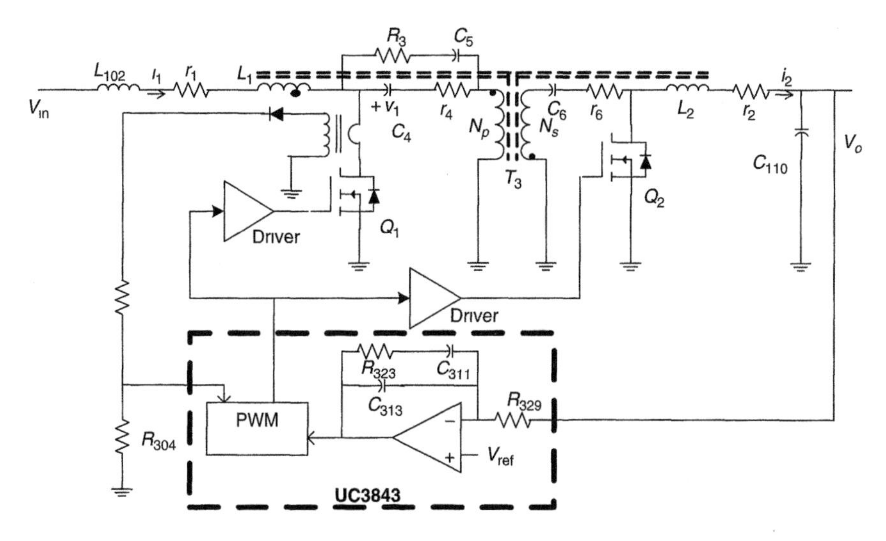

Figure 13.1: Cuk converter

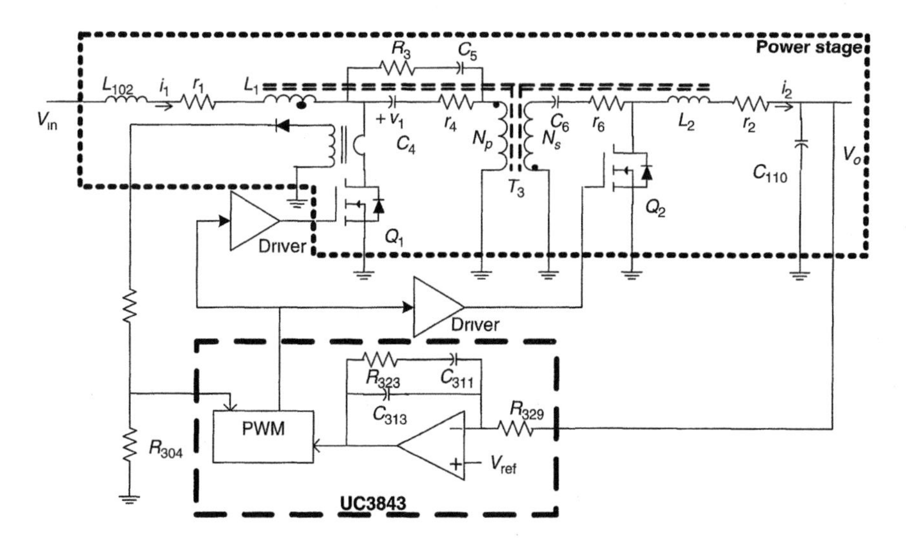

Figure 13.2: Power stage identified

Steps 2 and 3 (Figure 13.3) identify and assign state variables with signs. The state (column) vector is selected $(i_1, i_2, i_3, v_1, v_2, v_3, v_4)^T$

Step 4 (Figures 13.4 and 13.5) gives two alternating topologies.

Steps 5 and 6 yield

$$A_1 = \begin{bmatrix} k_1 r_1 & -k_2 a_1 & -k_2 a_2 & -k_2 a_3 & -k_2 & -k_2 a_4 & -k_2 a_5 \\ -k_4 r_1 & k_3 a_1 & k_3 a_2 & k_3 a_3 & k_3 & k_3 a_4 & k_3 a_5 \\ k_5 r_1 & -k_6 a_1 + k_7 a_2 & -k_6 a_2 + k_7 a_6 & -k_6 a_3 + k_7 a_7 & -k_6 & -k_6 a_4 & -k_6 a_5 + k_7 a_8 \\ 0 & -\frac{\tau_5}{N \bullet k_c} & \frac{\tau_5}{k_c} & -\frac{C_5}{k_c} & 0 & 0 & -\frac{C_5}{k_c} \\ 0 & -\frac{1}{C_6} & 0 & 0 & 0 & 0 & 0 \\ 0 & \frac{r_k}{C_{110}} & 0 & 0 & 0 & \frac{-1}{(r_c + R_L) C_{110}} & 0 \\ 0 & \frac{\tau_4}{N \bullet k_c} & -\frac{\tau_4}{k_c} & -\frac{C_4}{k_c} & 0 & 0 & -\frac{C_4}{k_c} \end{bmatrix} \tag{13.11}$$

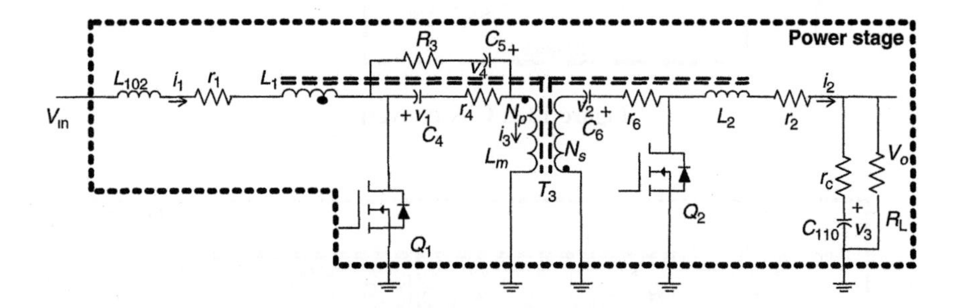

Figure 13.3: Power stage with all the state variables identified

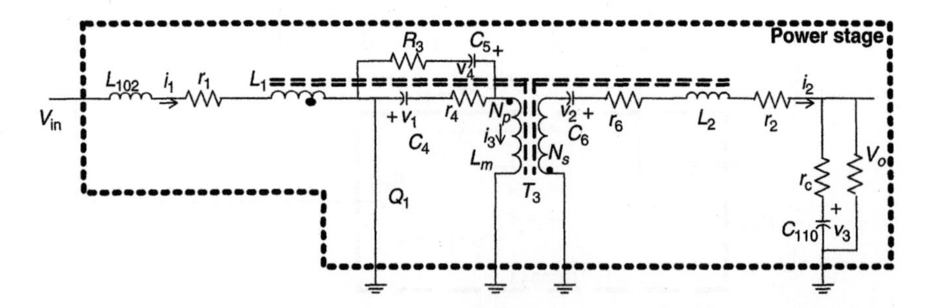

Figure 13.4: Power stage, topology 1, Q_1 on, Q_2 off

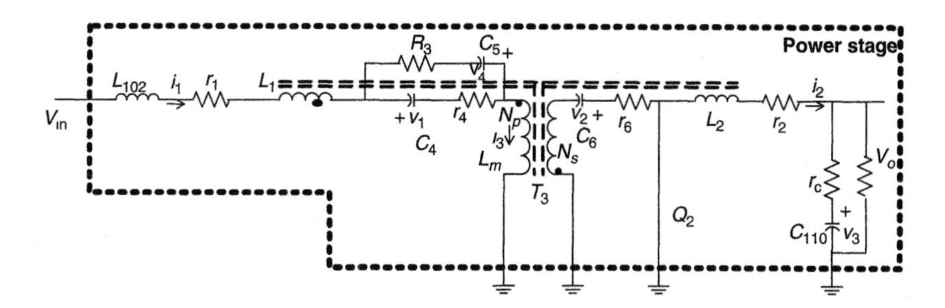

Figure 13.5: Power stage, topology 2, Q_1 off, Q_2 on

$$B_1 = \begin{bmatrix} k_1 & 0 & 0 & 0 & 0 & 0 & 0 \\ -k_4 & 0 & 0 & 0 & 0 & 0 & 0 \\ k_5 & 0 & 0 & 0 & 0 & 0 & 0 \\ 0 & 0 & 0 & 0 & 0 & 0 & 0 \\ 0 & 0 & 0 & 0 & 0 & 0 & 0 \\ 0 & 0 & 0 & 0 & 0 & 0 & 0 \\ 0 & 0 & 0 & 0 & 0 & 0 & 0 \end{bmatrix}, \quad E = \begin{bmatrix} V_{\text{in}} \\ 0 \\ 0 \\ 0 \\ 0 \\ 0 \\ 0 \end{bmatrix},$$

$$C_1 = [0 \quad r_p \quad 0 \quad 0 \quad 0 \quad r_k \quad 0] \tag{13.12}$$

where

$$r_p = \frac{r_c \cdot R_L}{r_c + R_L},\ r_k = \frac{R_L}{r_c + R_L},\ \tau_4 = r_4 \cdot C_4,\ \tau_5 = R_3 \cdot C_5, \quad k_c = C_4\tau_5 + C_5\tau_4,$$

$$a_1 = -\left(r_2 + r_6 + r_p + \frac{\tau_4\tau_5}{N^2 k_c}\right), \quad a_2 = \frac{\tau_4\tau_5}{Nk_c},\ a_3 = \frac{1}{N} - \frac{\tau_4 C_5}{Nk_c},\ a_4 = -r_k,$$

$$a_5 = \frac{\tau_4 C_5}{Nk_c}, \quad a_6 = -\frac{\tau_4\tau_5}{k_c},\ a_7 = \frac{\tau_4 C_5}{k_c} - 1, \quad a_8 = \frac{\tau_4 C_5}{k_c},$$

$$k_1 = \frac{L_m}{D_1}\left(L_2 + \frac{M_{23}}{N}\right), \quad k_2 = \frac{L_m}{D_1}\left(M_{12} + \frac{M_{13}}{N}\right), \quad k_3 = \frac{L_m}{D_1}(L_{102} + L_1),$$

$$k_4 = \frac{L_m}{D_1} M_{21},\ k_5 = \frac{M_{31}\left(L_2 + \frac{M_{23}}{N}\right) - M_{21}M_{32}}{D_1}$$

$$k_6 = \frac{M_{31}\left(M_{12} + \dfrac{M_{13}}{N}\right) - M_{32}(L_{102} + L_1)}{D_1}$$

$$k_7 = \frac{(L_{102} + L_1)\left(L_2 + \dfrac{M_{23}}{N}\right) - M_{21}\left(M_{12} + \dfrac{M_{13}}{N}\right)}{D_1}$$

$$D_1 = \begin{vmatrix} L_{102} + L_1 & M_{12} + \dfrac{M_{13}}{N} & 0 \\ M_{21} & L_2 + \dfrac{M_{23}}{N} & 0 \\ -M_{31} & -M_{32} & L_m \end{vmatrix}$$

Capital M with subscripts stands for mutual inductance and the N turn ratio.

Steps 7 and 8 yield

$$A_2 = \begin{bmatrix} p_1b_1 + p_3b_7 & -p_2b_5 & p_1b_2 + p_3b_8 & p_1b_3 & (-p_1 + p_3)N & -p_2b_6 & p_1b_4 \\ p_4b_1 - p_6b_7 & p_5b_5 & p_4b_2 - p_6b_8 & p_4b_3 & -(p_4 + p_6)N & p_5b_6 & p_4b_4 \\ p_7b_7 & 0 & p_7b_8 & 0 & p_7N & 0 & 0 \\ \dfrac{\tau_5}{k_c} & 0 & 0 & \dfrac{C_5}{k_c} & 0 & 0 & \dfrac{C_5}{k_c} \\ \dfrac{N}{C_6} & 0 & \dfrac{N}{C_6} & 0 & 0 & 0 & 0 \\ 0 & \dfrac{r_k}{C_{101}} & 0 & 0 & 0 & \dfrac{-1}{(r_c + R_L)C_{101}} & 0 \\ \dfrac{\tau_4}{k_c} & 0 & 0 & \dfrac{C_4}{k_c} & 0 & 0 & \dfrac{C_4}{k_c} \end{bmatrix} \tag{13.13}$$

$$B_2 = \begin{bmatrix} p_1 & 0 & 0 & 0 & 0 & 0 & 0 \\ p_4 & 0 & 0 & 0 & 0 & 0 & 0 \\ 0 & 0 & 0 & 0 & 0 & 0 & 0 \\ 0 & 0 & 0 & 0 & 0 & 0 & 0 \\ 0 & 0 & 0 & 0 & 0 & 0 & 0 \\ 0 & 0 & 0 & 0 & 0 & 0 & 0 \\ 0 & 0 & 0 & 0 & 0 & 0 & 0 \end{bmatrix}, \quad C_2 = C_1 \tag{13.14}$$

where

$$b_1 = -\left(r_1 + N^2 r_6 + \frac{\tau_4 \tau_5}{k_c}\right),\ b_2 = N^2 r_6,\ b_3 = \frac{\tau_4 C_5}{k_c} - 1,\ b_4 = \frac{\tau_4 C_5}{k_c},$$

$$b_5 = -(r_2 + r_p),\ b_6 = -\frac{R_L}{r_c + R_L},\ b_7 = N^2 r_6,\ b_8 = -b_7,$$

$$p_1 = \frac{L_2 L_m}{D_2},\ p_2 = \frac{M_{12} L_m}{D_2},\ p_3 = \frac{M_{12} M_{23} - L_2 M_{13}}{D_2},$$

$$p_4 = \frac{-(M_{21} - M_{23}) L_m}{D_2},\ p_5 = \frac{(L_{102} + L_1 - M_{13}) L_m}{D_2},$$

$$p_6 = \frac{(L_{102} + L_1 - M_{13}) M_{23} - (M_{21} - M_{23}) M_{13}}{D_2},$$

$$p_7 = \frac{(L_{102} + L_1 - M_{13}) L_2 - (M_{21} - M_{23}) M_{12}}{D_2},$$

$$D_2 = \begin{vmatrix} L_{102} + L_1 - M_{13} & M_{12} & M_{13} \\ M_{21} - M_{23} & L_2 & M_{23} \\ 0 & 0 & L_m \end{vmatrix}$$

Given $V_{\text{in}} = 70$, $r_1 = 0.005$, $r_4 = 0.005$, $r_c = 0.005$, $r_2 = 0.005$, $r_6 = 0.005$, $R_3 = 3.4$, $C_5 = 4.5\ \mu\text{F}$, $L_{102} = 7\,\mu\text{H}$, $L_1 = 6.37\,\mu\text{H}$, $L_2 = 2.05\,\mu\text{H}$, $L_m = 6.31\,\mu\text{H}$, $C_4 = 1.5\,\mu\text{F} \times 12$, $C_6 = 6.8\,\mu\text{F} \times 12 + 4.7\ \mu\text{F} \times 2$, $M_{12} = 2.85\ \mu\text{H}$, $M_{13} = 6.16\,\mu\text{H}$, $M_{23} = 2.98\ \mu\text{H}$, $M_{21} = M_{12}$, $M_{31} = M_{13}$, and $M_{32} = M_{23}$, step 9 finds the steady-state duty cycle $D = 0.08$.

Step 10 computes the steady-state vector $X = [0.998, 23, 0, 59.023, 3.445, 2.5, -59.023]^T$.

Step 11 proceeds with the computation of duty-to-output and source-to-output transfer functions.

Step 12 derives the PWM gain. This step requires the establishment of the duty-cycle determination algorithm that employs the peak current control:

$$f(V_{\text{er}}, V_{\text{in}}, V_o) = \frac{R_{304}}{n_s}\left[\frac{V_{\text{in}}}{L_m} D T_s + \frac{1}{N} \bullet \frac{V_o}{R_L}\left(1 - \frac{r_2 + r_6}{2 \bullet L_2} D \bullet T_s\right)\right] - V_{\text{er}} \tag{13.15}$$

Then, a gain factor connecting the error voltage and the PWM is given as

$$F_m = \frac{\partial f(V_{\text{er}}, V_{\text{in}}, V_o)/\partial V_{\text{er}}}{\partial f(V_{\text{er}}, V_{\text{in}}, V_o)/\partial D} \tag{13.16}$$

Since current-mode control is used, the output is also involved in the PWM function and that must be considered; that is,

$$F_v = \frac{\partial f(V_{\text{er}}, V_{\text{in}}, V_o)/\partial V_o}{\partial f(V_{\text{er}}, V_{\text{in}}, V_o)/\partial D} \tag{13.17}$$

Given $R_{329} = 10\,\text{K}$, $R_{323} = 15\,\text{K}$, $C_{311} = 4.7\,\text{nF}$, $C_{313} = 1\,\text{nF}$, step 13 obtains the feedback ratio, $K_f = 1$, and the error amplifier, $E_A(s)$. We already dealt with these types of activities many times and so do not duplicate those efforts. However, it should be mentioned that the current loop is absorbed to give

$$G_i(s) = \frac{G_{vd}(s)}{1 - G_{vd}(s) \bullet F_v} \tag{13.18}$$

The final step, step 14, computes the loop gain:

$$T(s) = K_f \bullet E_A(s) \bullet \frac{1}{3} \bullet F_m \bullet \frac{G_{vd}(s)}{1 - G_{vd}(s) \bullet F_v} \tag{13.19}$$

where the factor 1/3 comes from the UC3843 internal circuit.

The theoretical prediction (Figure 13.6) compares well against the actual measurement (Figure 13.7) except for some resonant peaking near 5 KHz.

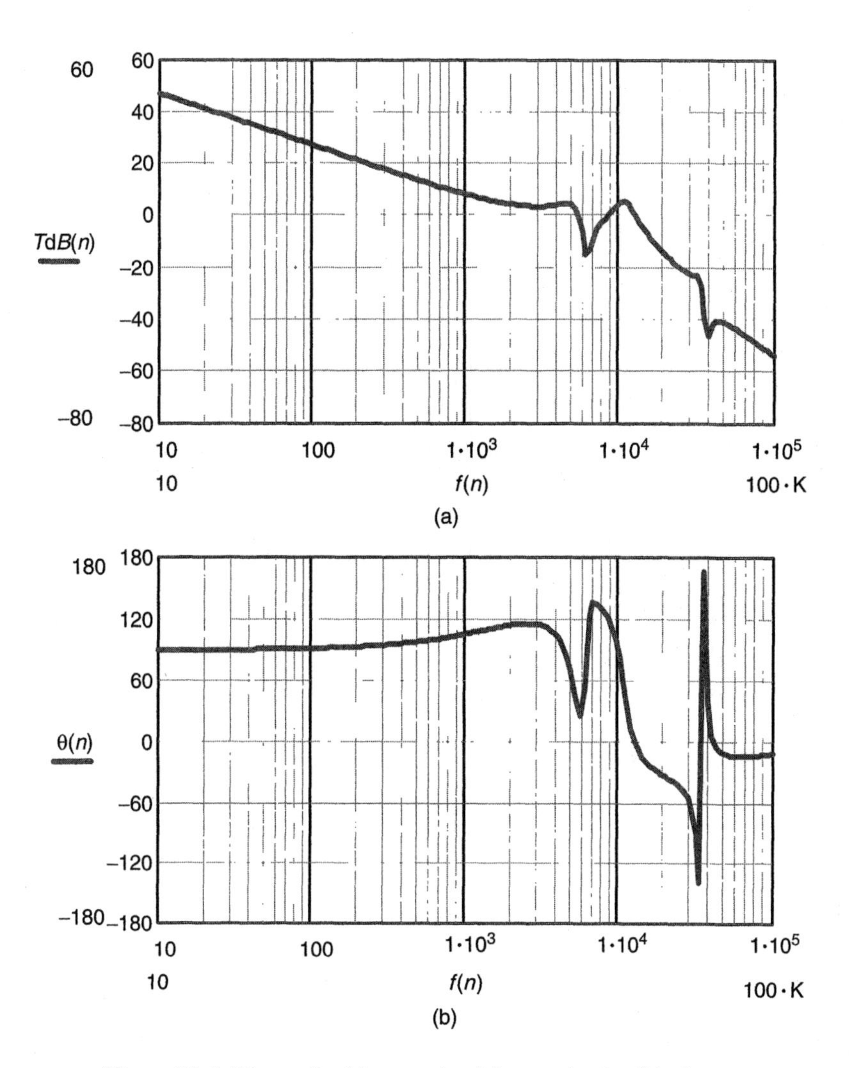

Figure 13.6: Theoretical loop gain: (a) magnitude, (b) phase

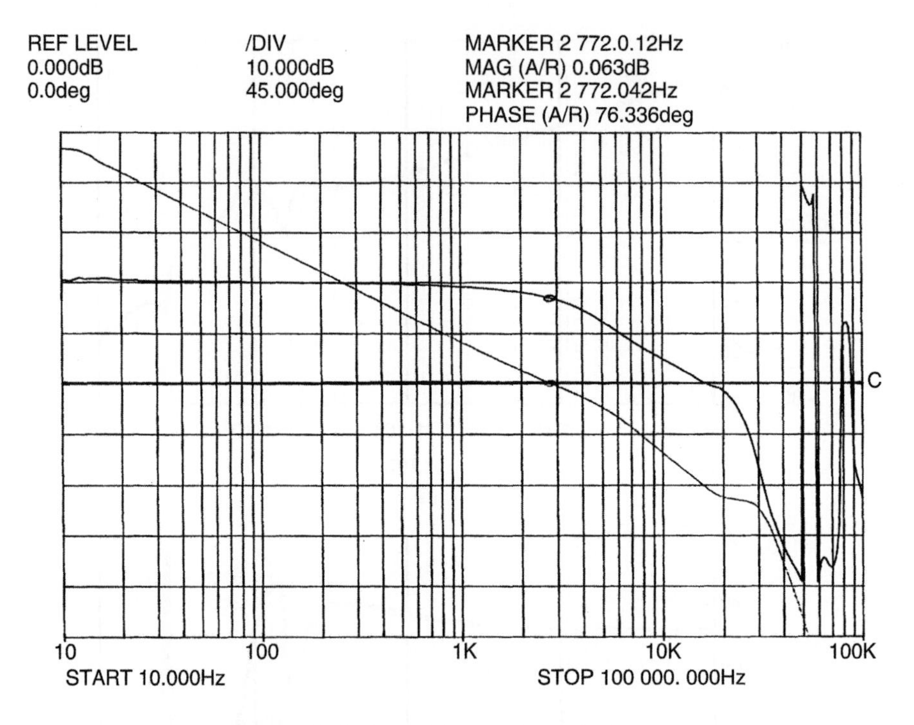

Figure 13.7: Actual loop gain and phase

Chapter 14

Simulation

With the advance of digital computers, simulation has made giant strides in the past 20 or 30 years in all scientific and technical fields. In the late 1970s, for electrical circuit analysis using computer, one dialed the Control Data Corporation (CDC) with a modem-equipped portable terminal/keyboard, edited the circuit file one component (one line) at a time, debugged the file for syntax errors, ran the file for half an hour or so and ran up a phone bill by the minute, read the dot-matrix printout, and repeated the simulation until the desired, appropriately formatted hard copy was obtained. In the end, the whole effort might take hours, if not all night. By 1990, the same study was regularly done on a personal computer and took only a few minutes to produce a colorful printout. This advancement prompted many new generations of electrical engineers to flock to PC-based circuit simulators and to believe that all circuits can be studied using simulation software like Spice or its derivatives. Following the same line of logic, many designers working on switch-mode power converters also quickly converged or on the idea: electronic breadboarding and testing.

However, as rapidly the simulation wildfire spread, numerous difficulties arose. Among the most troublesome was the convergence issue the simulator faced when solving tens of differential equations that described a complex circuit based on Kirchoff's current and voltage laws. Basically, those differential equations in discrete form were solved at finite time points starting from time zero and zero initial conditions, zero voltage across capacitors, and zero current through inductors. At every time step, every circuit loop must give a zero voltage sum while every node must give a zero current sum. If one step out of tens of equations fails to meet the constraint, it is considered a violation and the simulation just stops and reports "fails to converge at time..." Switch-mode converters are particularly prone to such a simulation failure for multiple reasons.

By now, we understand that switch-mode converters are highly nonlinear because of magnetic devices with nonlinear magnetic cores, diodes, and transistors operating in on/off mode. It is also not uncommon for modern converters to operate at hundreds of kilohertz. At that frequency, a switching cycle lasts only a fraction of a microsecond. Then, to observe circuit behavior accurately, the sampling rate is selected to be high. That is to say, a computation time resolution on the order of only several nanoseconds is the rule rather than exception. Even more demanding is that fixed-step technology often does not handle switch transition edges well. Because of this latter concern, a variable time step is often invoked autonomously at transitions. In theory, the variable step should work better. But in reality, the variable step is still constrained by an absolute minimum step that is the nature of digital computers. Yes, variable-step methods may better capture transitional behavior, but it also unwillingly introduces a higher probability of discontinuity in state transitions and precipitates divergence.

In this chapter, we introduce a compromise, an approach that avoids divergence but suffers minor inaccuracy. We dive into the problem without further verbiage.

14.1 Dynamic Equations for a Forward Converter with Voltage-Mode Control

Figure 14.1 gives an isolated forward converter with voltage-mode control using an external oscillator.

A total of 10 state variables need to be accounted for: the input inductor current, i_1; the primary voltage, v_p; the output inductor current,

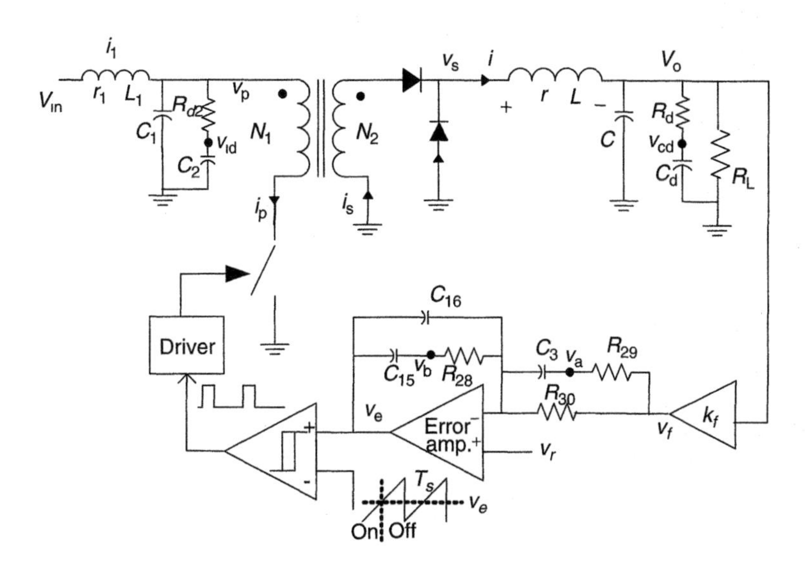

Figure 14.1: Isolated forward converter with voltage-mode control

i; the output voltage, v_o; the output filter damping voltage, v_{cd}; three voltages associated with the error amplifier, v_a, v_b, and v_e; the transformer magnetizing current, i_m; and the input filter damping voltage, v_{id}. The input loop gives

$$L_1 \frac{di_1}{dt} + r_1 \bullet i_1 + v_p = V_{\text{in}} \tag{14.1}$$

This can be easily placed in a discrete form for a small time step δt and step index j:

$$(i_1)_{j+1} = \left[1 - \frac{r_1 \bullet \delta t}{L_1}\right](i_1)_j - \frac{\delta t}{L_1}(v_p)_j + \frac{\delta t}{L_1} V_{\text{in}} \tag{14.2}$$

We can do the same for the primary node:

$$C_1 \frac{dv_p}{dt} + \frac{v_p - v_{id}}{R_{d2}} + i_p = i_1 \tag{14.3}$$

the output loop:

$$L\frac{di}{dt}+r\bullet i+v_o=v_s \tag{14.4}$$

the output node:

$$C\frac{dv_o}{dt}+\frac{v_o-v_{cd}}{R_d}+\frac{v_o}{R_L}=i \tag{14.5}$$

the output damping node:

$$C_d\frac{dv_{cd}}{dt}=\frac{v_o-v_{cd}}{R_d} \tag{14.6}$$

the feedback path (v_r constant):

$$\begin{aligned}
v_f &= k_f\bullet v_o\\
C_3\frac{d(v_a-v_r)}{dt} &= C_3\frac{dv_a}{dt}=\frac{v_f-v_a}{R_{29}}\\
C_{15}\frac{d(v_b-v_e)}{dt} &= C_{15}\frac{dv_b}{dt}-C_{15}\frac{dv_e}{dt}=\frac{v_r-v_b}{R_{28}}\\
C_{16}\frac{d(v_r-v_e)}{dt}+\frac{v_r-v_b}{R_{28}} &= -C_{16}\frac{dv_e}{dt}+\frac{v_r-v_b}{R_{28}}\\
&= \frac{v_f-v_r}{R_{30}}+\frac{v_f-v_a}{R_{29}}
\end{aligned} \tag{14.7}$$

the primary magnetization loop:

$$L_m\frac{di_m}{dt}+R_{\text{switch}}\left(i_m+\frac{N_2}{N_1}i_s\right)=v_p \tag{14.8}$$

and the input damping node:

$$C_2 \frac{dv_{id}}{dt} = \frac{v_p - v_{id}}{R_{d2}} \tag{14.9}$$

It is understood that four variables—i_p, i_s, i_m, and v_s—are switched quantities. Readers are cautioned that the last equation of equation group (14.7) is plugged into the third to make the third a single differential.

Next we generate the external clock with a $v_H - v_L$ swing; a single cycle first, then a repetitive stream:

$$\begin{aligned} v_1(t) &= \left(v_L + \frac{v_H - v_L}{0.98 \cdot T} \cdot t\right)[u(t) - u(t - 0.98 \cdot T)] \\ &\quad + \left[v_H - \frac{v_H - v_L}{0.02 \cdot T}(t - 0.98 \cdot T)\right]\{u(t - 0.98 \cdot T) - u(t - T)\} \\ sw(t) &= \sum_{n=0,1,\ldots} v_1(t - n \cdot T) \end{aligned} \tag{14.10}$$

To simulate a switching operation, we use an "if" statement (MathCAD®, Mathsoft, Inc., Cambridge, MA) with syntax "if (condition, true, false)." By so doing and by comparing the error signal against the clock, the following are established for the four switched quantities:

$$\begin{aligned} v_s &= \text{if}\left(v_e > sw,\ \frac{v_p N_2}{N_1} - V_D,\ -V_D\right) \\ i_s &= \text{if}\ (v_e > sw,\ i,\ 0) \\ i_m &= \text{if}\ (v_e > sw,\ i_m,\ 0) \\ i_p &= i_m + \frac{N_2}{N_1} i_s \end{aligned} \tag{14.11}$$

Some practical facts must also be considered. These include the power supply limiting for the error amplifier and consequently the error voltage <15 V and built-in duty-cycle limiting <3.3 V for 50%. Once these limiting factors are incorporated, (14.11) becomes

$$\begin{aligned}
v_s &= \text{if}\left[\text{if}(v_e > 3.3,\ 3.3,\ v_e) > sw, \frac{v_p N_2}{N_1} - V_D, -V_D\right] \\
i_s &= \text{if}\ [\text{if}(v_e > 3.3,\ 3.3,\ v_e) > sw,\ i,\ 0] \\
i_m &= \text{if}\ [\text{if}(v_e > 3.3,\ 3.3,\ v_e) > sw,\ i_m,\ 0] \\
i_p &= \frac{N_2}{N_1} i_s + i_m
\end{aligned} \tag{14.12}$$

Furthermore, considering the existence of losses in the primary side, the effective voltage available for energy transfer is reduced. In other words, the effective primary voltage accounts for losses and is represented in the following form

$$v_p - R_{\text{switch}}\left\{\frac{N_2}{N_1}\text{if}\ [\text{if}(v_e > 3.3,\ 3.3,\ v_e) > sw,\ i,\ 0] + i_m\right\} \tag{14.13}$$

With all the pertinent equations and switching functions formulated, (14.3) through (14.13) should also be placed in discrete form, just as (14.2) did for (14.1). All discrete equations are then arranged in a seeded iteration form. (Since the iteration formulation involves both the primary and the secondary subscripts that do not use typing space effectively, the following ignores the primary subscript and keeps only the time step index, j, as a subscript.) The form is ready for simulation with the following components and operating parameters: $R_{\text{on}} = R_{\text{switch}} = 0.01$, $L_m = 1\,\text{mH}$, $V_{\text{in}} = 28\,\text{V}$, $L_1 = 3.5\,\text{mH}$, $r_1 = 0.01$, $C_1 = 68\,\mu\text{F}$, $C_2 = 3 \times C1$, $R_{d2} = 4.7$, $V_d(\text{rectifier}) = 0.7$, $N_1 = 13$, $N_2 = 18$, $V_r = 2.5$, $\text{i}_{28} = 63.4\,\text{K}$, $R_{29} = 1.13\,\text{K}$, $R_{30} = 12.7\,\text{K}$, $\text{i}_3 = 4700 \times 3\,\text{pF}$, $L = 5.76\,\text{mH}$, $\text{r} = 0.005$, $R_L = 33$, $C = 22\,\mu\text{F}$, $C_d = 120\,\mu\text{F}$, $R_d = 12.7$, $T = 10.4\,\mu\text{s}$, $\delta t = 0.02\,\text{T}$, $k_f = 0.2515$, $v_H = 5.692$, and $v_L = 1$.

$$\begin{pmatrix} il_{j+1} \\ vp_{j+1} \\ i_{j+1} \\ vo_{j+1} \\ vcd_{j+1} \\ va_{j+1} \\ vb_{j+1} \\ ve_{j+1} \\ im_{j+1} \\ vid_{j+1} \end{pmatrix} = \begin{bmatrix} i1_j + [Vin - (r1 \cdot il_j + vp_j)] \cdot \dfrac{\delta t}{L1} \\ vp_j + \left[il_j - \left[\dfrac{vp_j - vid_j}{Rd2} + \left(\dfrac{N2}{N1} \cdot \text{if}(\text{if}(ve_j \geq 3.3, 3.3, ve_j) > sw_j, i_j, 0) + im_j\right)\right]\right] \cdot \dfrac{\delta t}{C1} \\ i_j + \dfrac{\delta t}{L1} \cdot \left[\text{if}\left[\text{if}(ve_j \geq 3.3, 3.3, ve_j) > sw_j, \dfrac{N2}{N1} \cdot \left[vp_j - R_{\text{on}}\left(\dfrac{N2}{N1} \cdot \text{if}(\text{if}(ve_j \geq 3.3, 3.3, ve_j) > sw_j, i_j, 0) + im_j\right)\right] - Vd, -Vd\right] - (r \cdot i_j + vo_j)\right] \\ vo_j + \dfrac{\delta t}{C} \cdot \left[i_j - \left(\dfrac{vo_j - vcd_j}{Rd} + \dfrac{vo_j}{RL}\right)\right] \\ vcd_j + \dfrac{\delta t}{Cd} \cdot \dfrac{vo_j - vcd_j}{Rd} \\ va_j + \dfrac{\delta t}{C3} \cdot \dfrac{kf \cdot vo_j - va_j}{R29} \\ vb_j + \delta t \cdot \left(\dfrac{Vr - vb_j}{R28C16} - \dfrac{kf \cdot vo_j - Vr}{R30C16} - \dfrac{kf \cdot vo_j - va_j}{R29C16} + \dfrac{Vr - vb_j}{R28C15}\right) \\ \text{if}\left[ve_j > 15, 15, ve_j + \dfrac{\delta t}{C16} \cdot \left(\dfrac{Vr - vb_j}{R28} - \dfrac{kf \cdot vo_j - Vr}{R30} - \dfrac{kf \cdot vo_j - va_j}{R29}\right)\right] \\ \text{if}\left[\text{if}(ve_j \geq 3.3, 3.3, ve_j) > sw_j, im_j + \dfrac{\delta t}{Lm} \cdot \left[vp_j - \left(\dfrac{N2}{N1} \cdot \text{if}(\text{if}(ve_j \geq 3.3, 3.3, ve_j) > sw_j, i_j, 0) + im_j\right) \cdot R_{\text{on}}\right], 0\right] \\ vid_j + \dfrac{\delta t}{C2} \cdot \dfrac{vp_j - vid_j}{Rd2} \end{bmatrix}$$

14.2 Turn-on Forward Converter with Voltage-Mode Control

The forward converter is turned on with an all-zero initial state. The simulation runs for 200 clock cycles and produces an end state. Figure 14.2 shows major node voltages and branch currents identified in Figure 14.1.

14.3 Steady-State Forward Converter with Voltage-Mode Control

The end state produced by the initial turn-on run is passed along to the next 200 cycles as the starting state. The end state of the second 200-cycle run is then used as the starting state for the third 200-cycle run. This repeats until the end state no longer changes. A consistent end state signifies the steady state.

14.4 Steady State, Zoomed In

To observe the waveform details, a 10 cycle run under the steady state is performed. Figure 14.4 shows the end result.

Figure 14.4 shows signs of subharmonic oscillation at about 10 KHz. This is attributed to the low-phase margin at that closed-loop frequency. Certainly, the error amplifier should be improved. However, that is not the goal of this chapter. This chapter shows another way to perform simulation. Before closing the present section, it is very instructive to observe the actions of voltage-mode pulse-width modulation. This is done by superimposing the clock and the error voltage on the same plot (Figure 14.5).

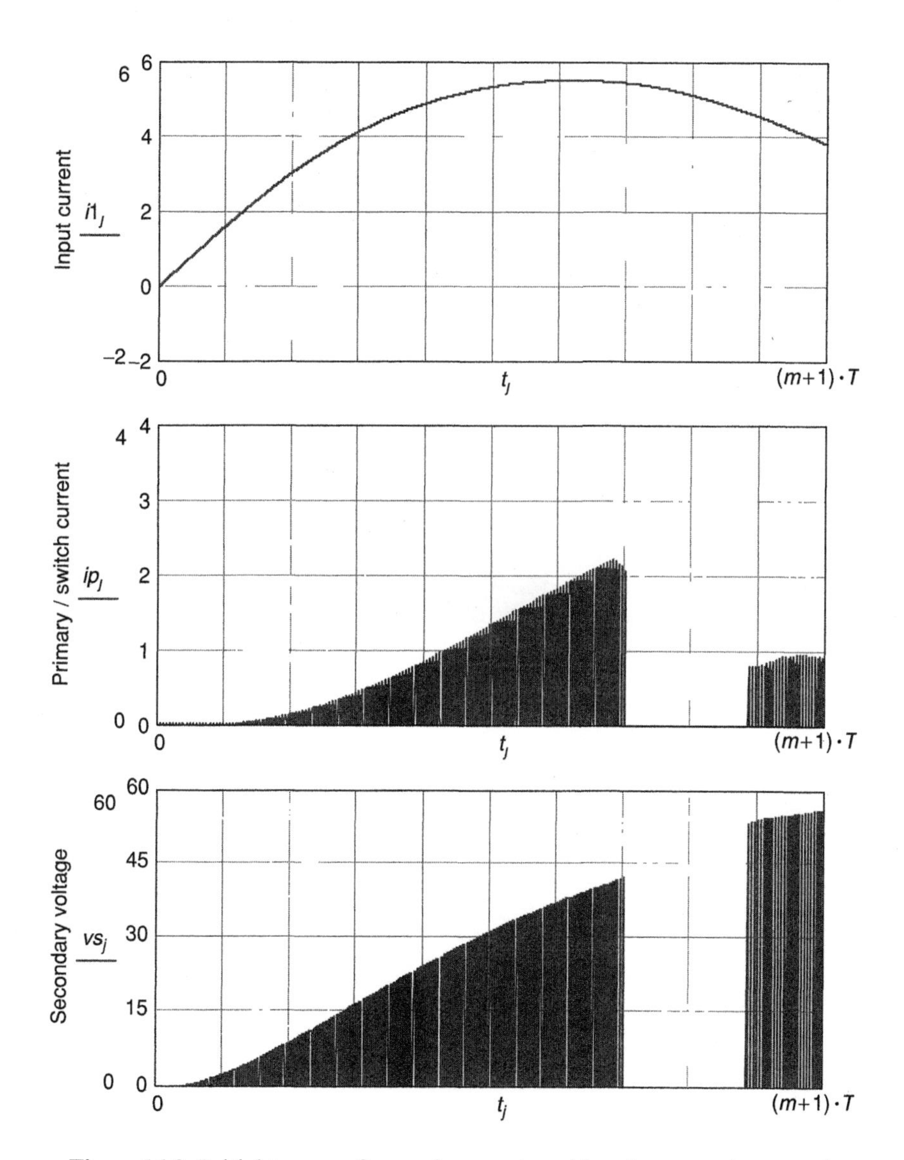

Figure 14.2: Initial turn-on forward converter with voltage-mode control

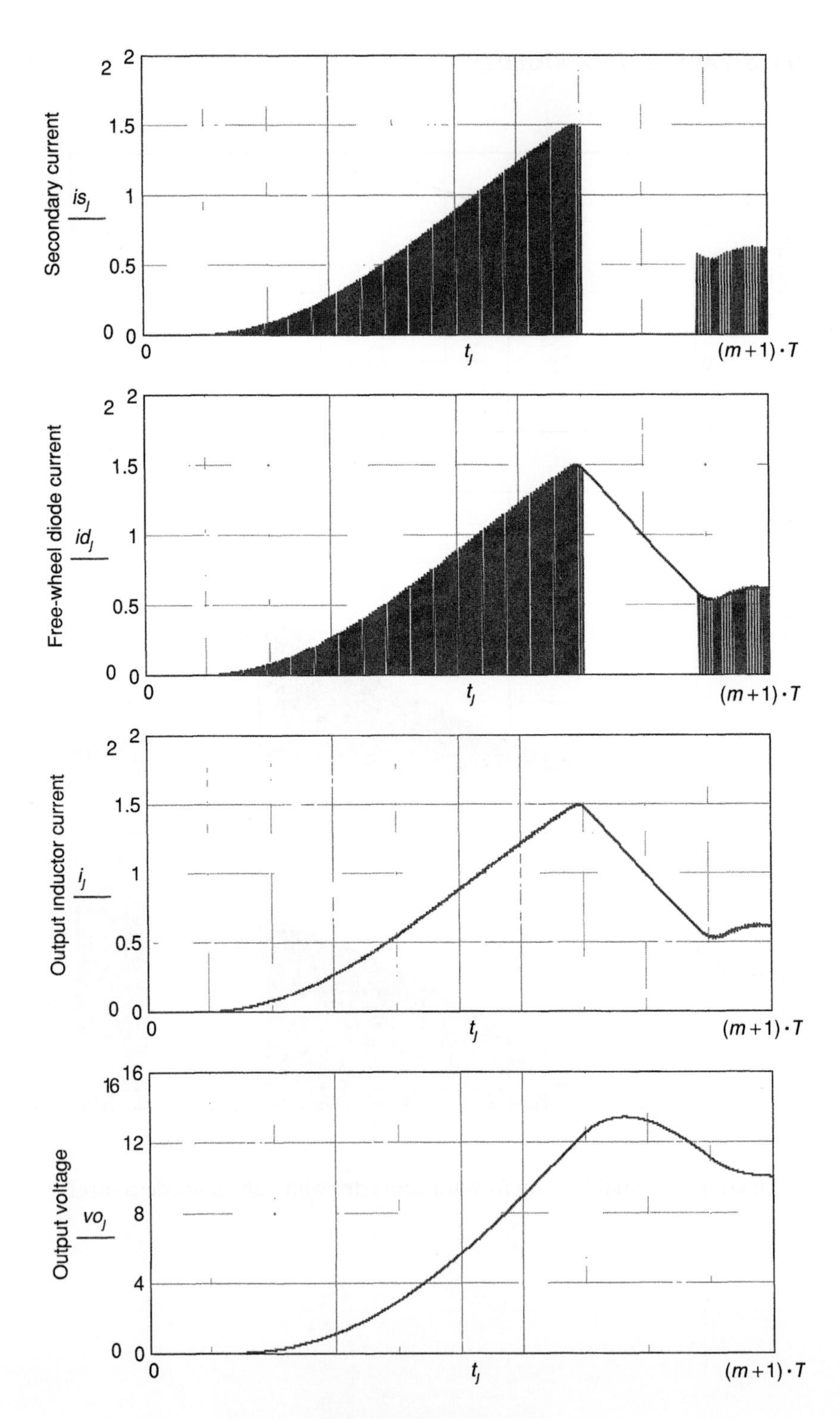

Figure 14.2: Continued

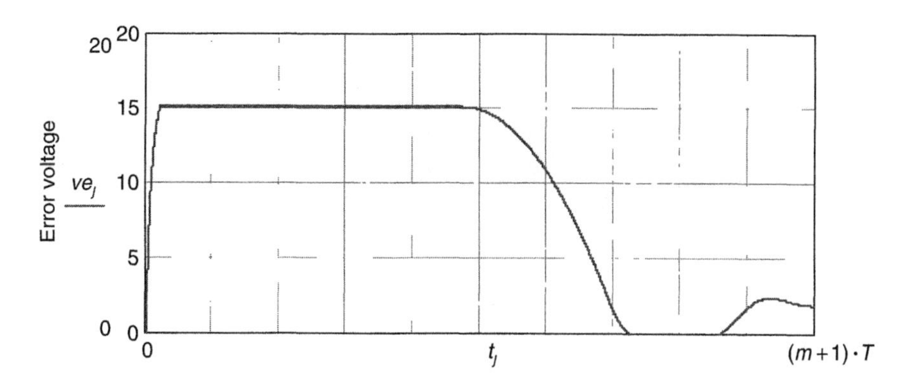

Figure 14.2: Continued

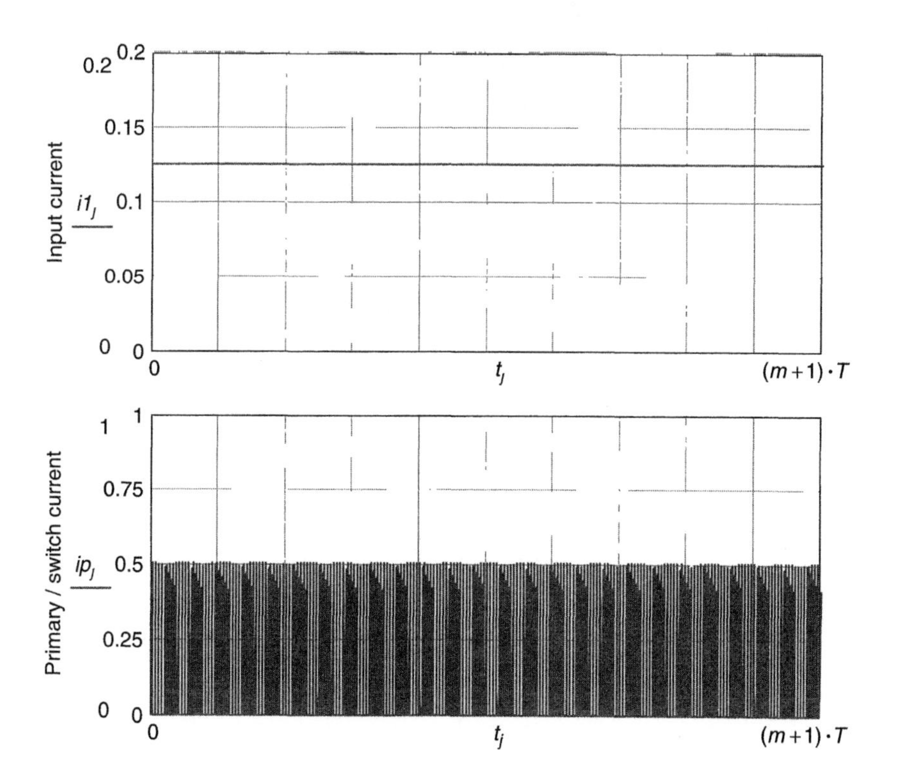

Figure 14.3: Steady-state forward converter with voltage-mode control

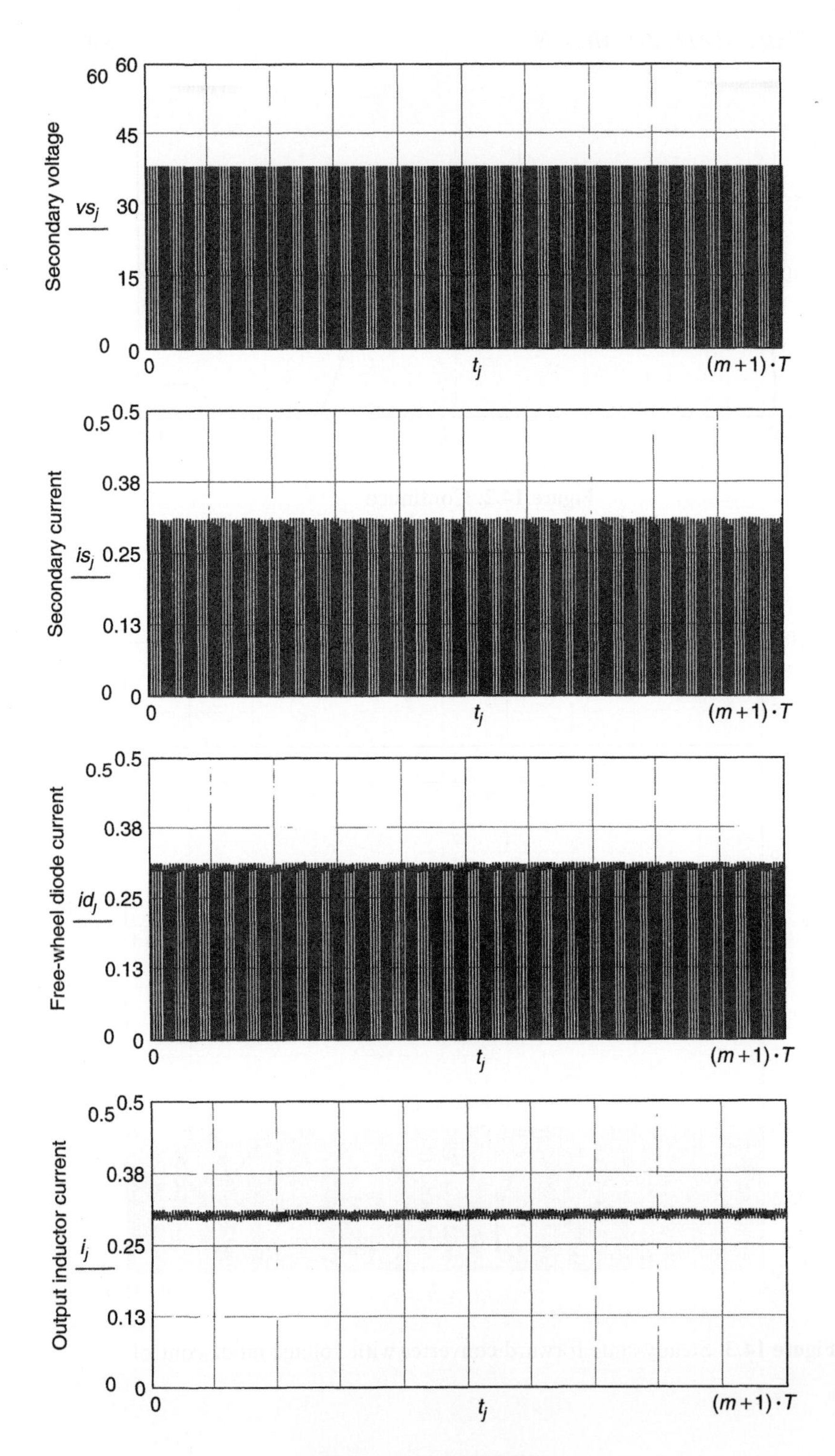

Figure 14.3: Continued

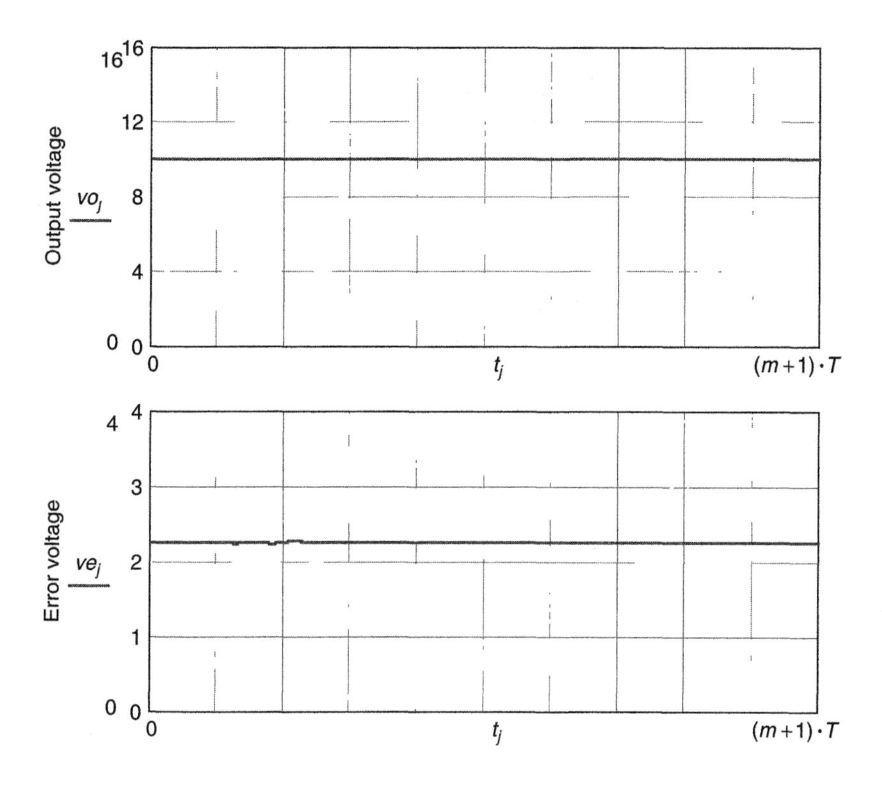

Figure 14.3: Continued

14.5 Load-Transient Forward Converter with Voltage-Mode Control

Figure 14.6 shows the converter performance when subjected to a step load increase, R_L, from 33 to 23.

Figure 14.7 shows the converter performance when subjected to a step load decrease, R_L, from 33 to 40.

A similar transient response can be performed by changing V_{in} from 28 volts to 30 volts or from 28 volts to 26 volts in step.

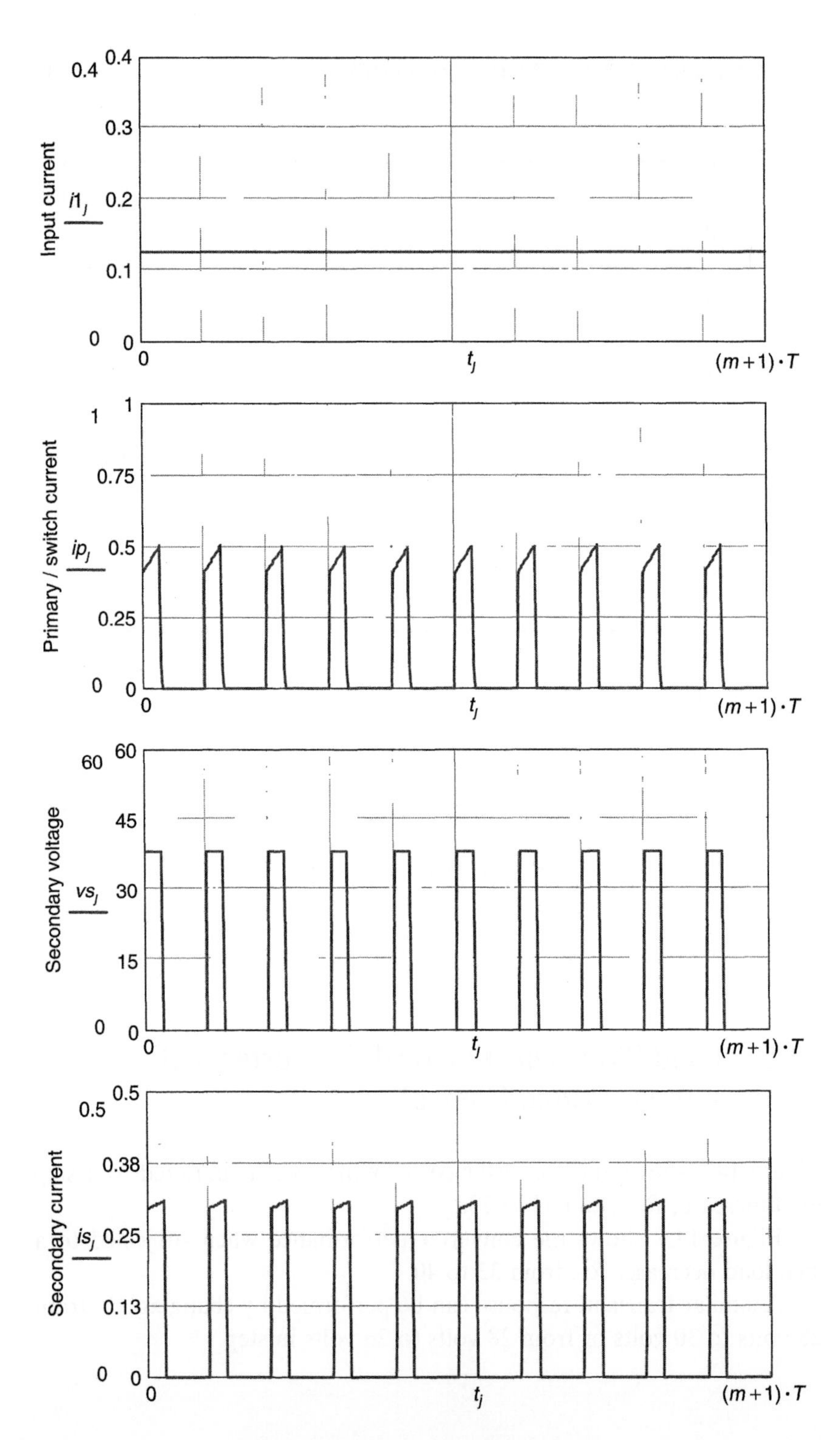

Figure 14.4: Steady state, zoomed in

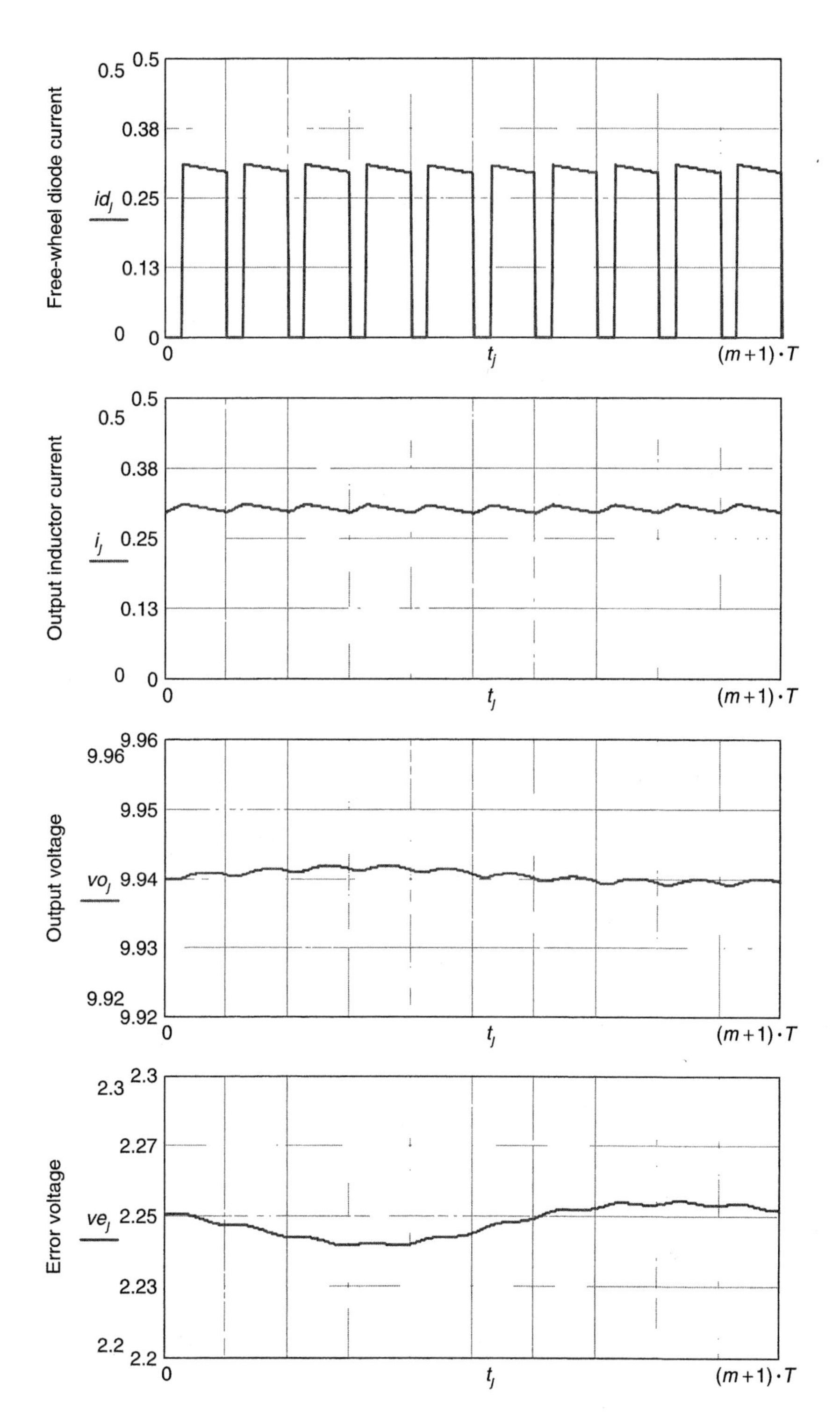

Figure 14.4: Continued

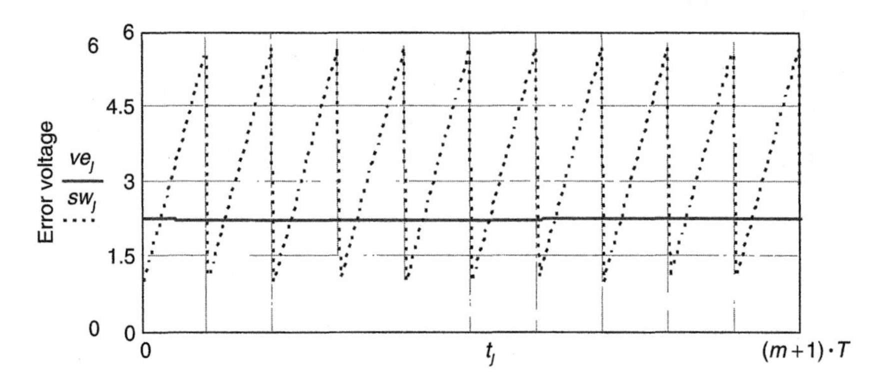

Figure 14.5: Voltage-mode pulse-width modulation

14.6 Dynamic Equations for a Forward Converter with Peak Current-Mode Control

The section presents the current-mode control. Figure 1.30 is expanded and component symbols are reassigned for consistency with the author's existing simulation file. This results in Figure 14.8. Since we already went through the step showing how the dynamic equations governing a converter are written, we will not repeat the step for this section.

However, an important feature of this circuit requires attention. As shown, the current feedback is compared with the effective error voltage, v_{eff}. When the current feedback intercepts the effective error, the comparator output swings to high from low. In practice, this low to high transition must be used somehow to terminate the power switch's conduction. Paradoxically, the conduction termination of the power switch quickly pulls the current feedback below the effective error. Thus, the comparator output also quickly changes state and swings to low. In other words, the comparator output exhibits a positive-going pulse of extremely short duration. This short pulse signals only the time of threshold crossing and is not suitable for commanding the power switch. As an effective pulse-width modulator, the short pulse must be converted into an appropriate on/off control with the right duration. This is done by a reset set flip-flop (RSFF) (Figure 14.9).

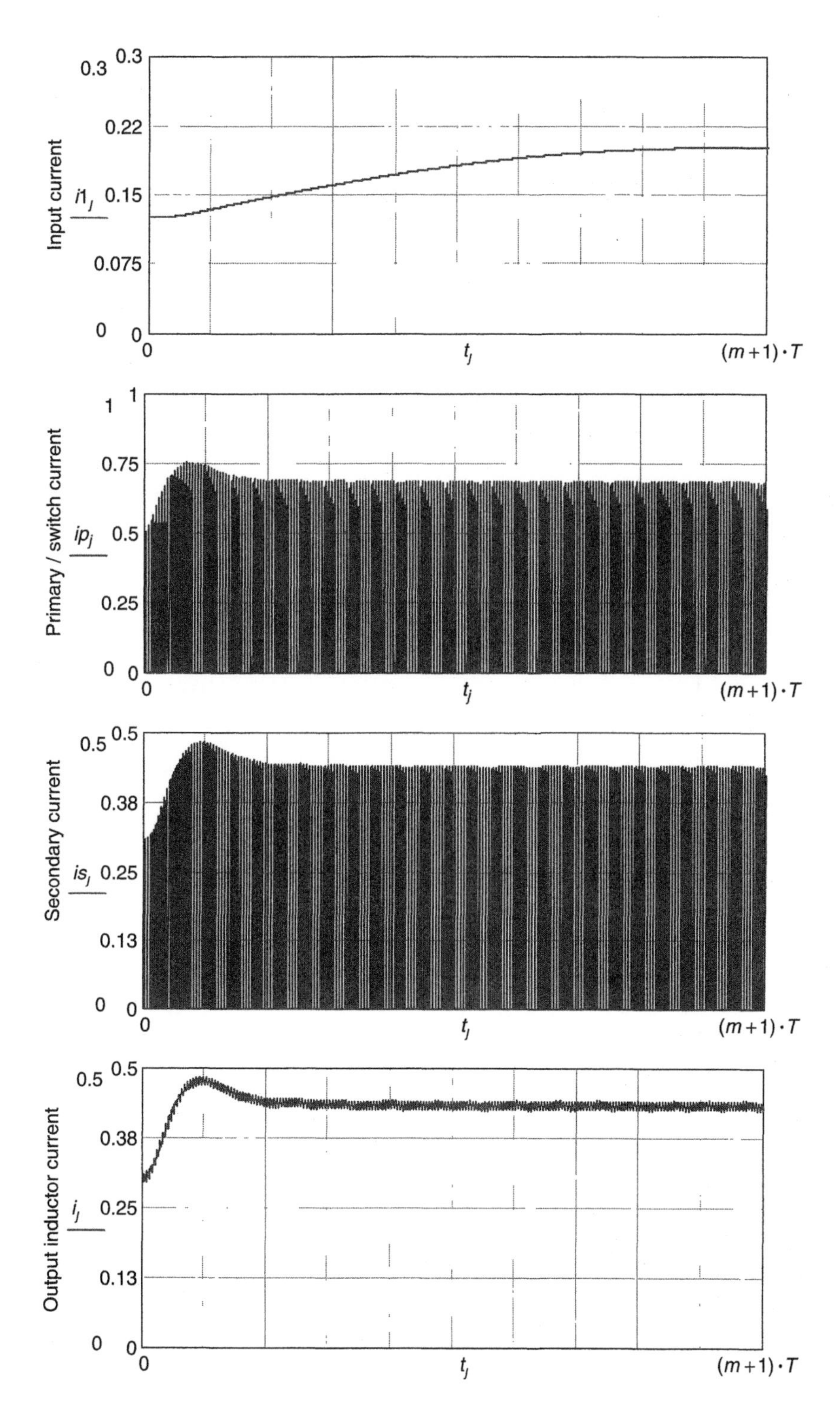

Figure 14.6: Transient load increase

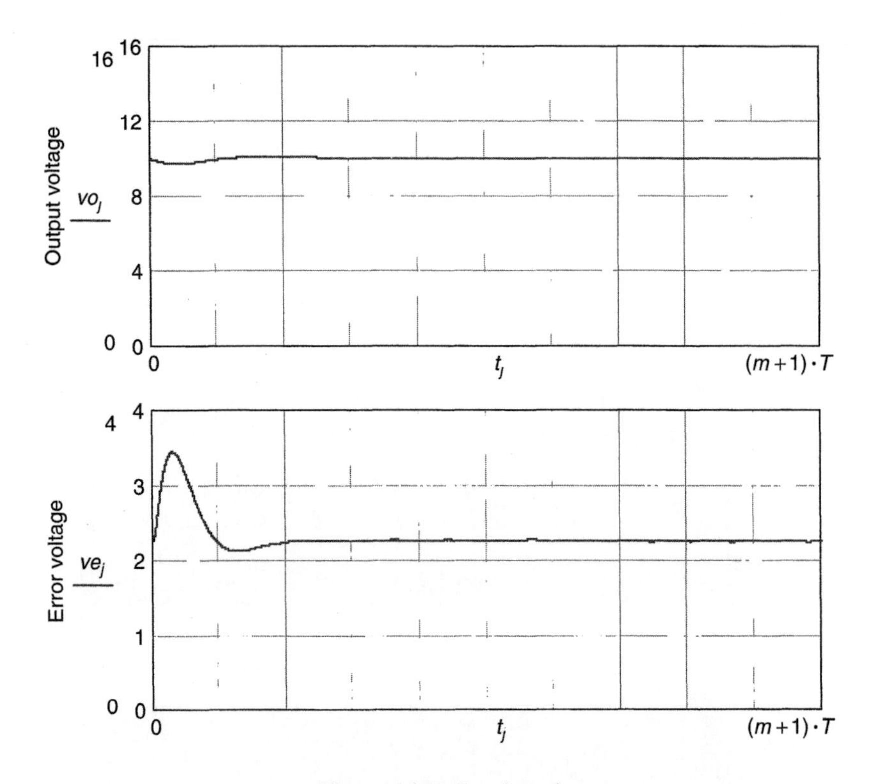

Figure 14.6: Continued

The RSFF accepts two inputs: R, the error-crossing intercept, and S, an external-pulse clock train. Mathematically, the RSFF's Q output can be expressed as

$$Q_{j+1} = \overline{R_j \cup (\overline{Q_j \cup S_j})} \tag{14.14}$$

We give here the discrete seeded iteration equations in their entirety without further procedural writing. The first equation is the expanded form of (14.14). It shall also be noted that, in the first equation, the effective error voltage is limited to larger than zero but less than 1 volt.

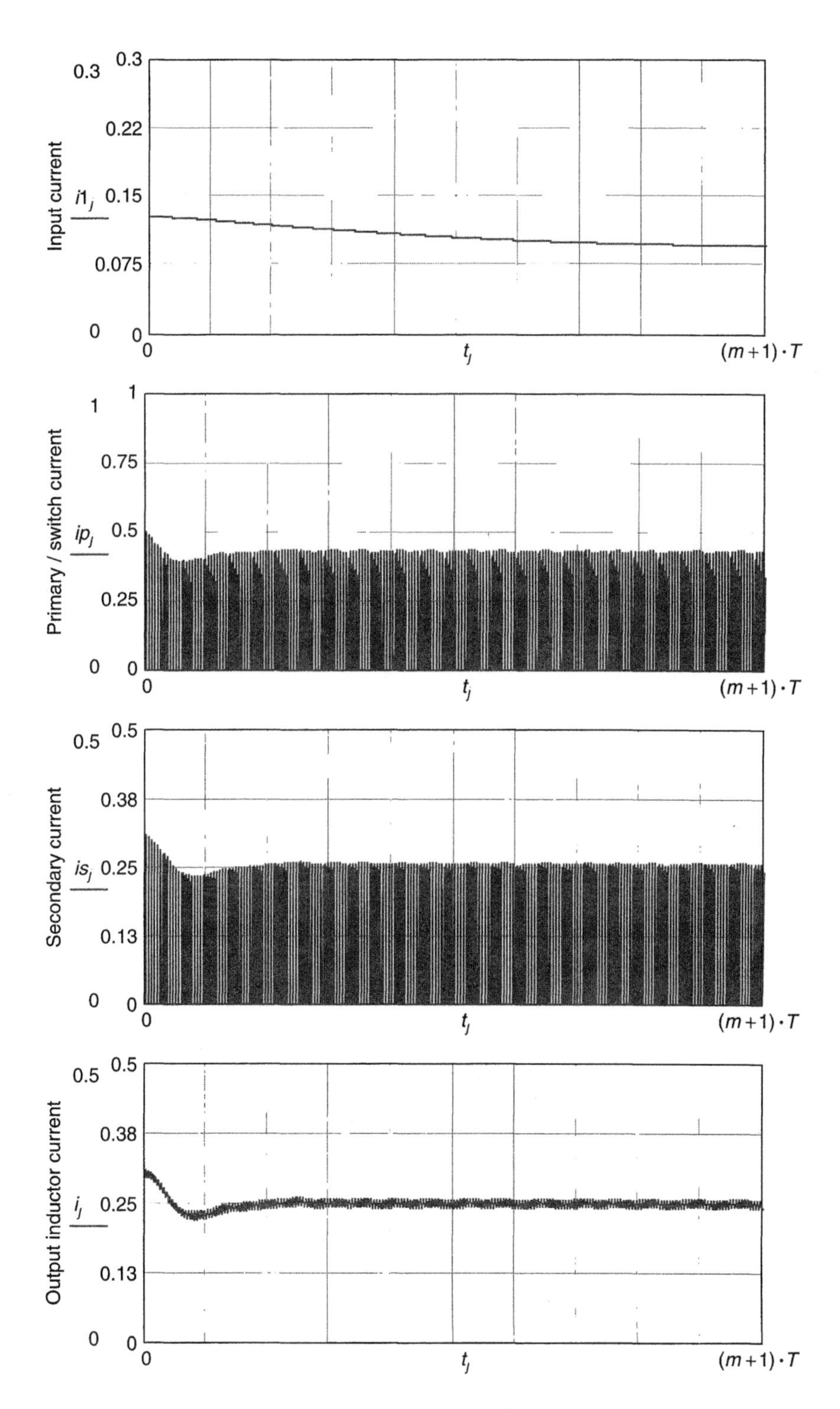

Figure 14.7: Transient load decrease

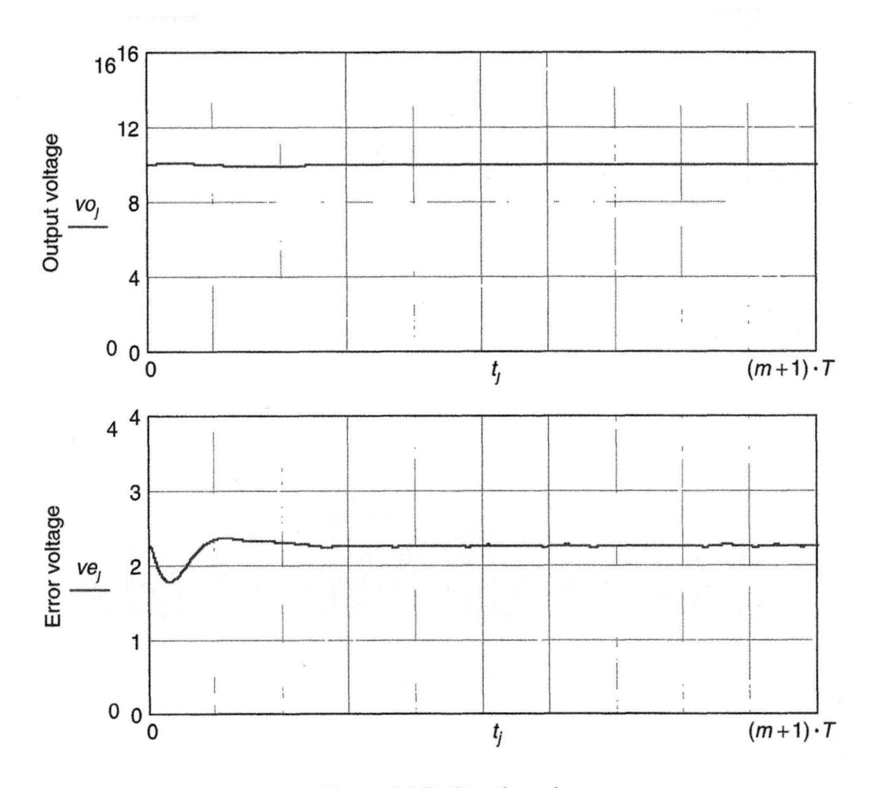

Figure 14.7: Continued

14.7 Simulation, Forward Converter with Peak Current-Mode Control

All simulations are performed with the following component values and operating parameters: $N_1 = 19$, $N_2 = 12$, $L_m = 1150\,\text{nH}$, $r_w = 0.1$, $R_{on} = 0.2$, $L_1 = 102\,\mu\text{H}$, $r_{L1} = 0.005$, $V_{in} = 28$, $L_2 = 385\,\mu\text{H}$, $r_{L2} = 0.077$, $R_L = 5.3/2.5$, $C_9 = 200\,\mu\text{F}$, $r_{C1} = 0.05$, $V_D = 0.5$, $V_{ref} = 2.5$, $R_{35} = 9.76\,\text{K}$, $R_{70} = 11.1\,\text{K}$, $R_{sen} = 36.5$, $\text{ni} = 100$, $R_{12} = 2\,\text{K}$, $R_{45} = 24.3\,\text{K}$, $C_{37} = 560\,\text{pF}$, $C_{29} = 2200\,\text{pF}$, $R_{43} = 10\,\text{K}$, $V_{be} = 0.7$, $R_{20} = 5.11\,\text{K}$, $R_{33} = 2.94\,\text{K}$, $h_{fe} = 150$, $R_{19} = 5.11\,\text{K}$, $R_{31} = 6.65\,\text{K}$, $R_{39} = 33.2\,\text{K}$, $C_{39} = 560\,\text{pF}$, $C_7 = 60\,\mu\text{F}$, $r_{C2} = 2.61$, and $C_8 = 60\,\mu\text{F}$.

The initial turn-on is conducted with a zero starting state for 200 switching cycles (Figure 14.10) at 125 KHz. After 400 cycles, it reaches

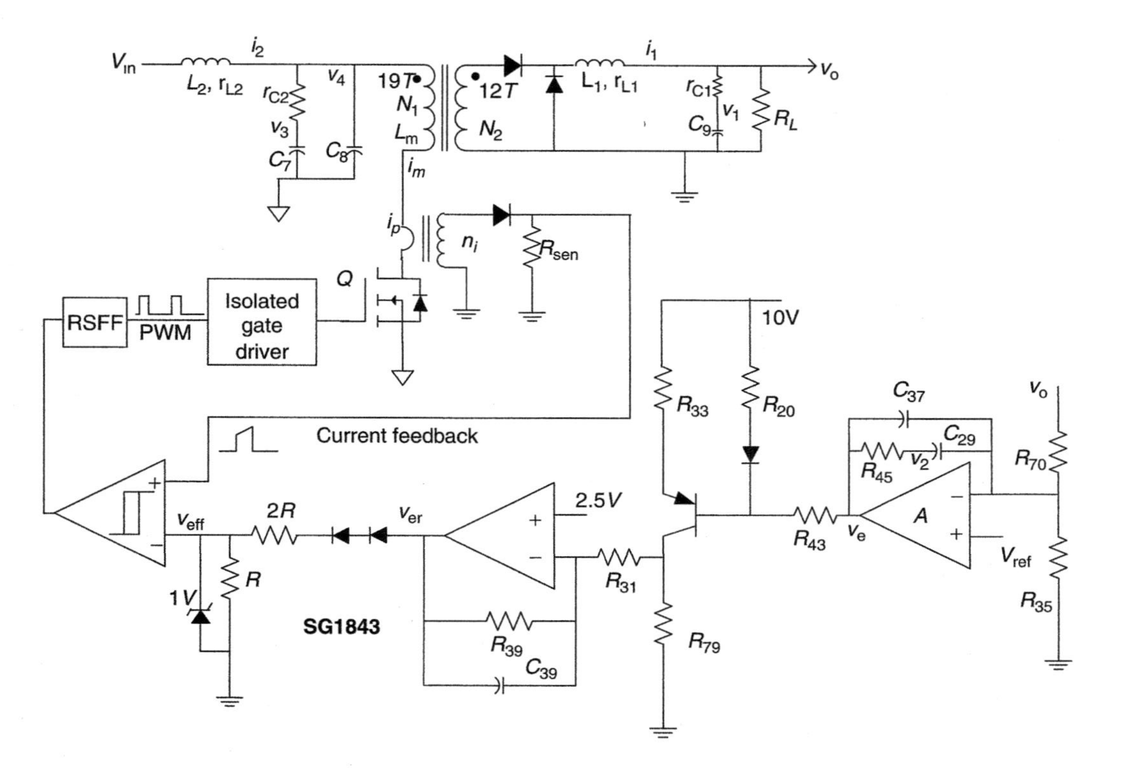

Figure 14.8: Isolated forward converter with peak current-mode control

$$
\begin{bmatrix} q_{j+1} \\ i1_{j+1} \\ v1_{j+1} \\ v2_{j+1} \\ ve_{j+1} \\ ver_{j+1} \\ i2_{j+1} \\ v3_{j+1} \\ v4_{j+1} \\ im_{j+1} \end{bmatrix} :=
\begin{bmatrix}
-\left[\left[\text{if}\left[\frac{Rsen}{ni}\cdot\left(im_j+\frac{N2}{N1}\cdot i1_j\cdot q_j\right)\geq \text{if}\left[(ver_j-2\cdot VD)\cdot\frac{1}{3}\geq 1,1,\text{if}\left[(ver_j-2\cdot VD)\cdot\frac{1}{3}\right.\right.\right.\right.\right. \\
\left.\left.\left.\left.\left.<0,0,(ver_j-2\cdot VD)\cdot\tfrac{1}{3}\right]\right],1,0\right]\right]\vee-[q_j\vee(if(clk_j=1,1,0))]\right] \\
i1_j+\delta t\cdot\left[-\left(\frac{rL1+Rp}{L1}\right)\cdot i1_j-\frac{k}{L1}\cdot v1_j+\frac{1}{L1}\cdot \text{if}\left[q_j=0,-VD,\frac{N2\left[v4_j-(rw+R_{\text{on}})\cdot\left(im_j+\frac{N2}{N1}\cdot i1_j\cdot q_j\right)\right]}{N1}-VD\right]\right] \\
v1_j+\delta t\cdot\left(\frac{Rp}{rc1\cdot C9}\cdot i1_j+\frac{k-1}{rc1\cdot C9}\cdot v1_j\right) \\
v2_j+\delta t\cdot\left[\frac{-kf\cdot Rp}{(R12C37)}\cdot i1_j+\frac{-kf\cdot k}{R12C37}\cdot v1_j-\left[\frac{1}{(R45C37)}+\frac{1}{R45C29}\right]\cdot v2_j+\left[\frac{1}{(R12C37)}+\frac{1}{(R45C37)}+\frac{1}{R45C29}\right]\cdot V_{\text{ref}}\right] \\
\text{if}\left[ve_j>5,5,\text{if}\left[ve_j<0,0,ve_j+\delta t\cdot\left[\frac{-kf\cdot Rp}{(R12C37)}\cdot i1_j+\frac{-kf\cdot k}{R12C37}\cdot v1_j-\frac{1}{(R45C37)}\cdot v2_j+\left[\frac{1}{(R12C37)}+\frac{1}{(R45C37)}\right]\cdot V_{\text{ref}}\right]\right]\right] \\
\text{if}\left[ver_j>10,10,\text{if}\left[ver_j<0,0,ver_j+\delta t\cdot\left[\frac{-GT}{R31C39}\cdot ve_j-\frac{1}{R39C39}\cdot ver_j+\left[\frac{1}{(R31)}+\frac{1}{(R39)}\right]\cdot\frac{1}{C39}\cdot V_{\text{ref}}-\frac{1}{R31C39}\cdot V_{\text{bias}}\right]\right]\right] \\
i2_j+\delta t\cdot\left(\frac{-rL2}{L2}\cdot i2_j-\frac{1}{L2}\cdot v4_j+\frac{1}{L2}\cdot Vin\right) \\
v3_j+\delta t\cdot\left(\frac{-1}{rc2\cdot C7}\cdot v3_j+\frac{1}{rc2\cdot C7}\cdot v4_j\right) \\
v4_j+\delta t\cdot\left[\frac{1}{C8}\cdot i2_j+\frac{1}{rc2\cdot C8}\cdot v3_j-\frac{1}{rc2\cdot C8}\cdot v4_j-\frac{1}{C8}\cdot\left(im_j+\frac{N2}{N1}\cdot i1_j\cdot q_j\right)\right] \\
\text{if}\left[q_j=0,0,im_j+\frac{\delta t}{Lm}\cdot\left[v4_j-(rw+R_{\text{on}})\cdot\left(im_j+\frac{N2}{N1}\cdot i1_j\cdot q_j\right)\right]\right]
\end{bmatrix}
$$

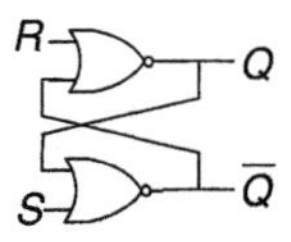

Figure 14.9: RSFF

the steady state (Figure 14.11). Figure 14.12 gives steady state zoomed in for 10 cycles. When subjected to a step load increase from 2.5 A to 2.6 A, the converter behaves very well (Figure 14.13). The same can also be said for a step load decrease from 2.5 A to 2.3 A (Figure 14.14). Both load transient responses settle in about 30 cycles.

14.8 State Transition Technique: Accelerated Steady State

In Sections 1.14, 3.7, and 5.7, we used state transition techniques and the concept of the continuity of state repeatedly to obtain the steady-state outputs (responses) of several simple circuits driven by periodic sources. Given its powerful utility, there is no reason the approach should be limited to dealing merely with simple circuits.

We also noted that simulation, either with commercially available software tools or custom-made programs given in the past several sections, must pass the transient phase. Not only does the transient phase invite trouble, as stated in the introduction of this chapter, it also wastes precious simulation time.

We now show that the state transition methods and the concept of the continuity of state can be just as effective in dealing with a complete converter, while bypassing the transient phase in the process.

Referring to Figure 14.15, which is based on Figure 14.8 with the input filter added, we assign nine state variables: i_1, the output filter choke current; v_1, the output filter capacitor voltage; v_2, the error amplifier feedback capacitor voltage; v_n, the inverting node voltage of the error amplifier; v_{n2}, the inverting node voltage of the PWM built-in error amplifier; i_2, the input filter choke current; v_3, the input filter capacitor voltage; v_4, the primary winding voltage; and i_m, the primary magnetizing

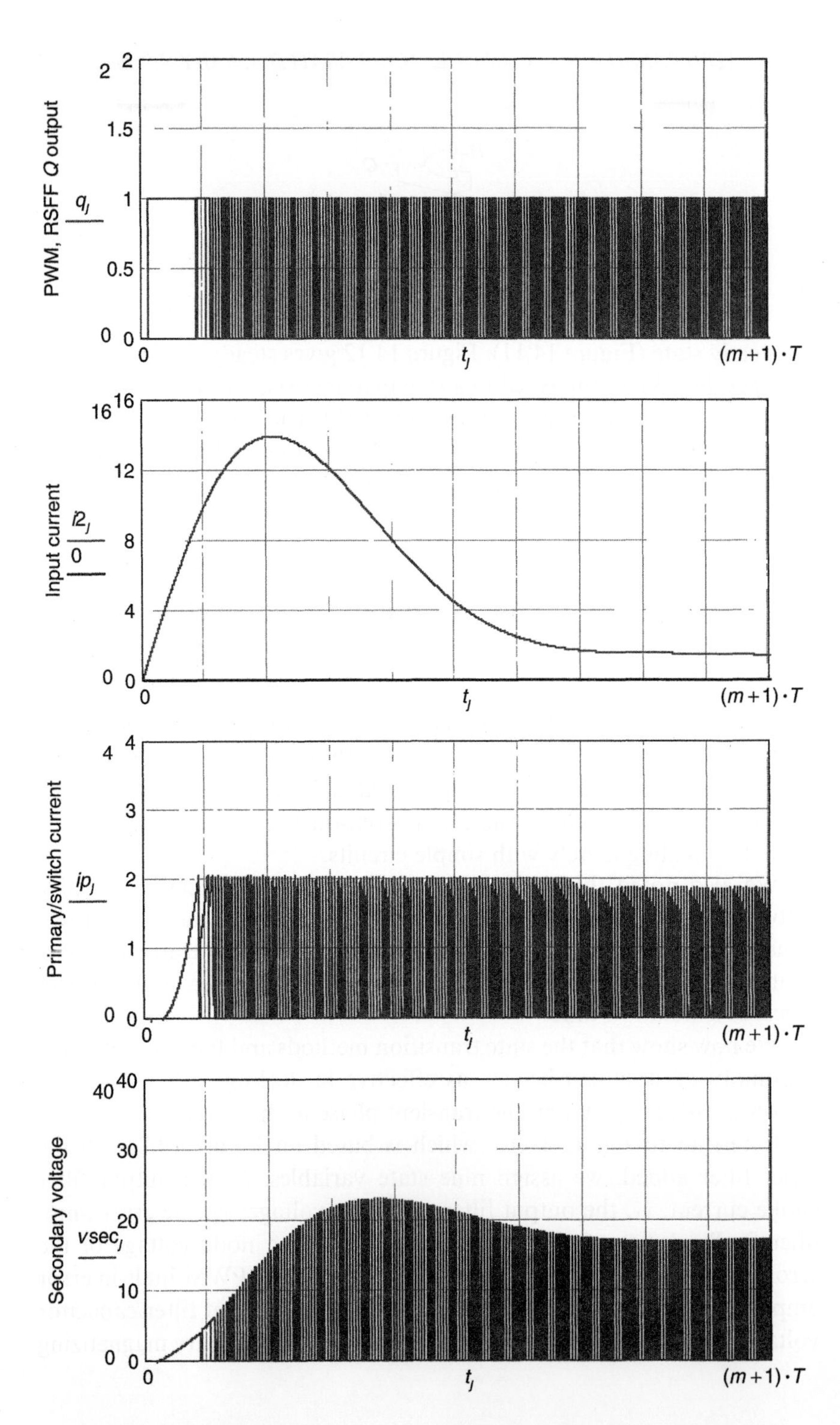

Figure 14.10: Initial turn-on forward converter with peak current-mode

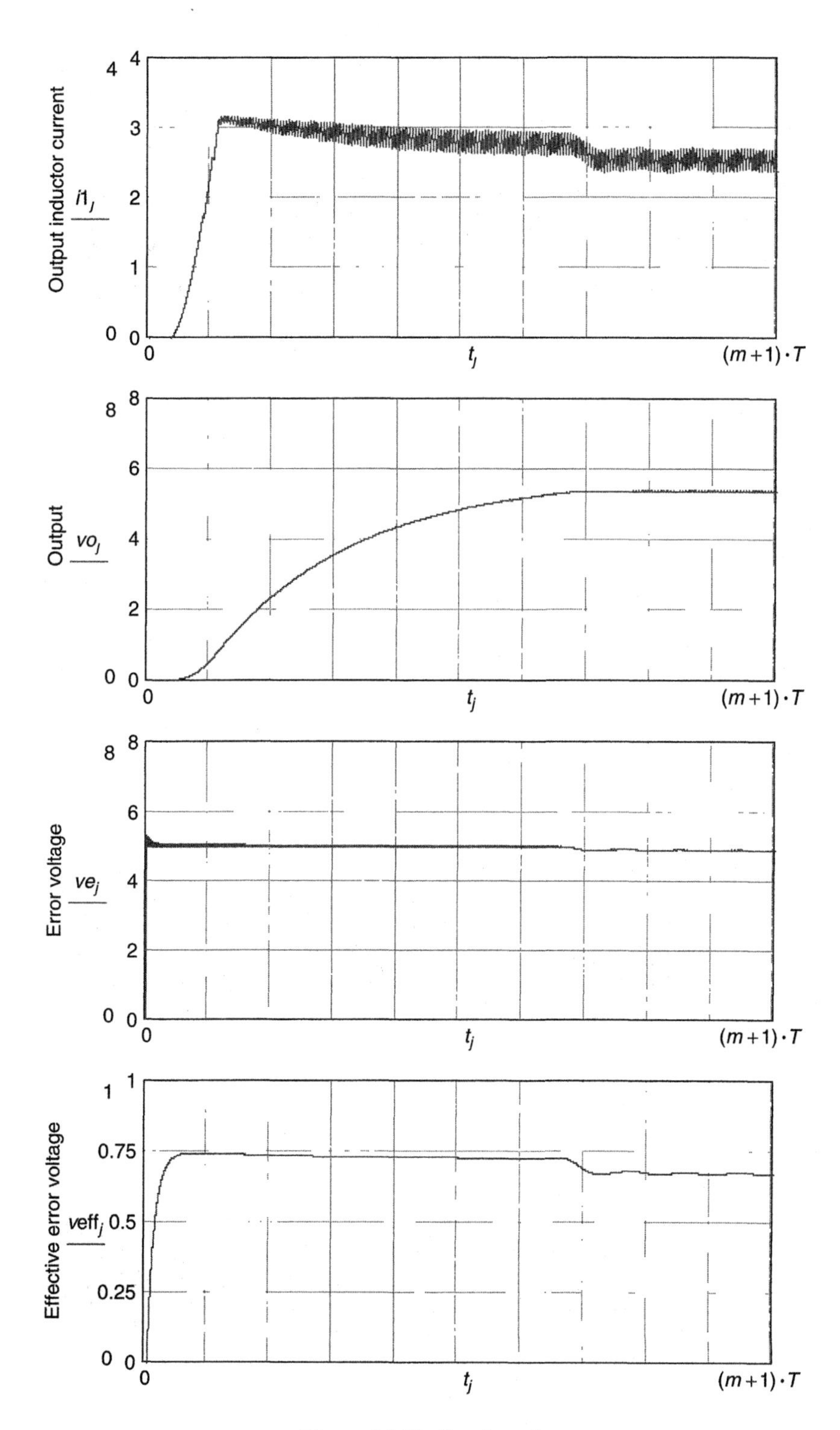

Figure 14.10: Continued

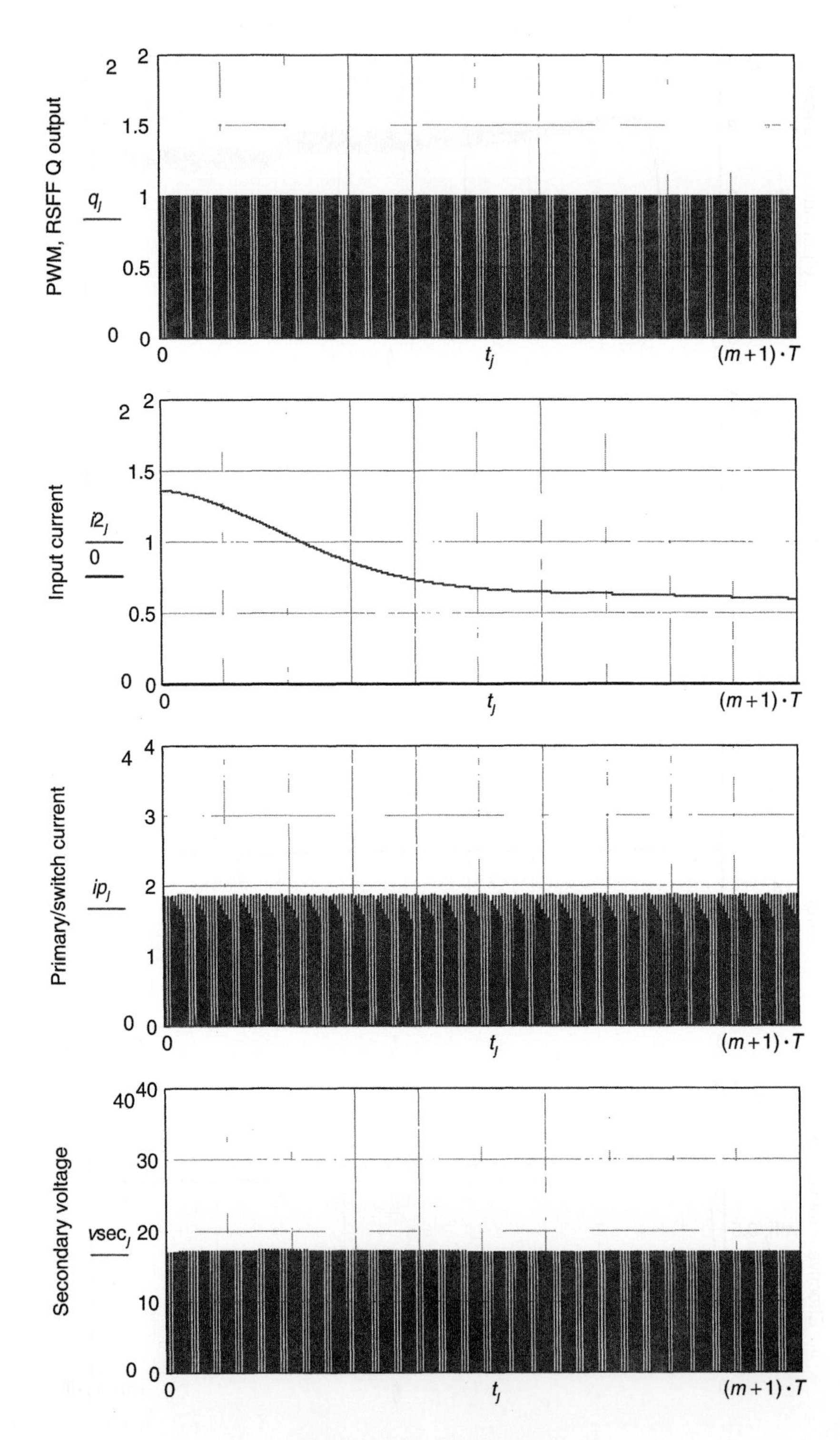

Figure 14.11: Steady state

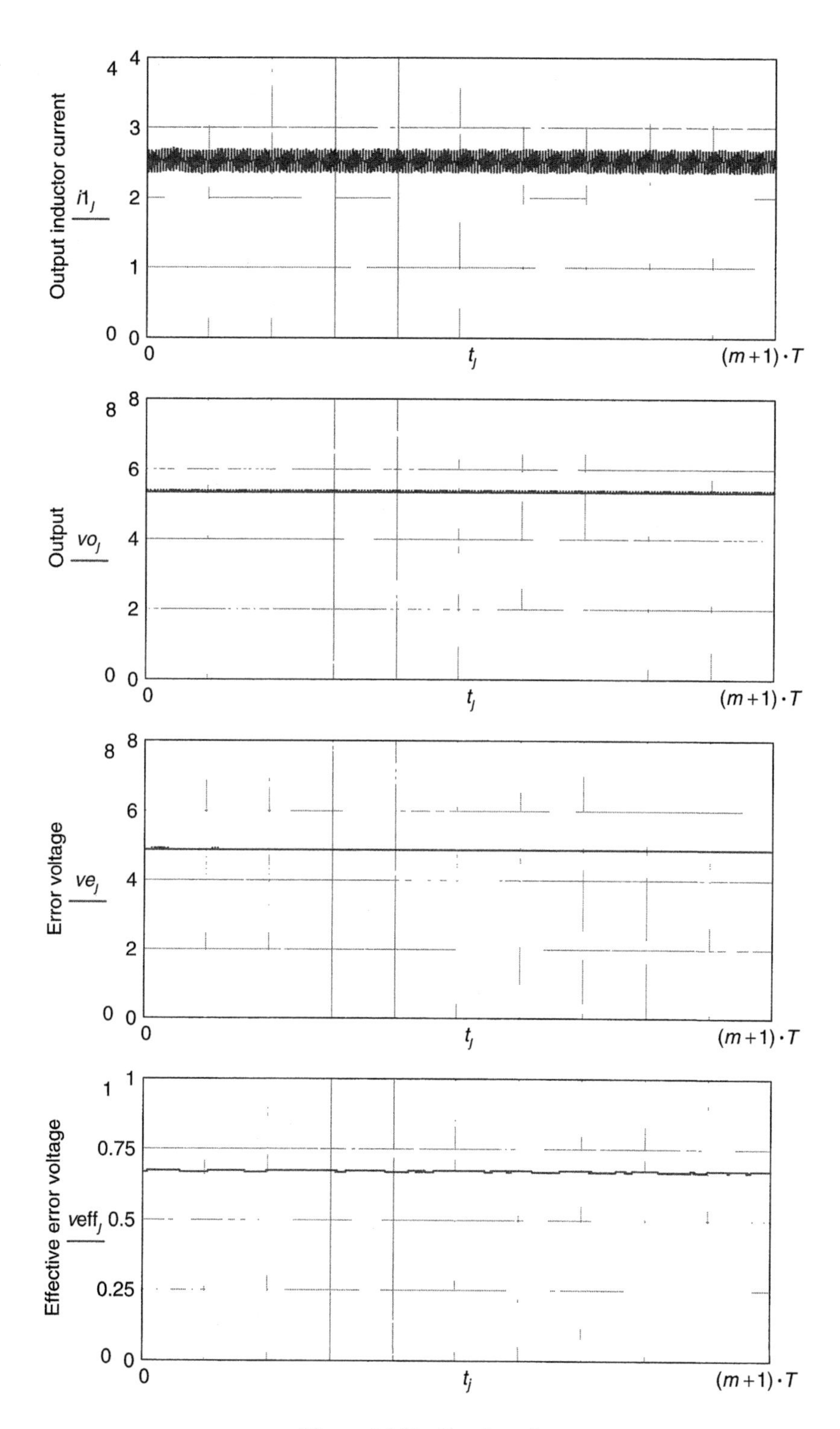

Figure 14.11: Continued

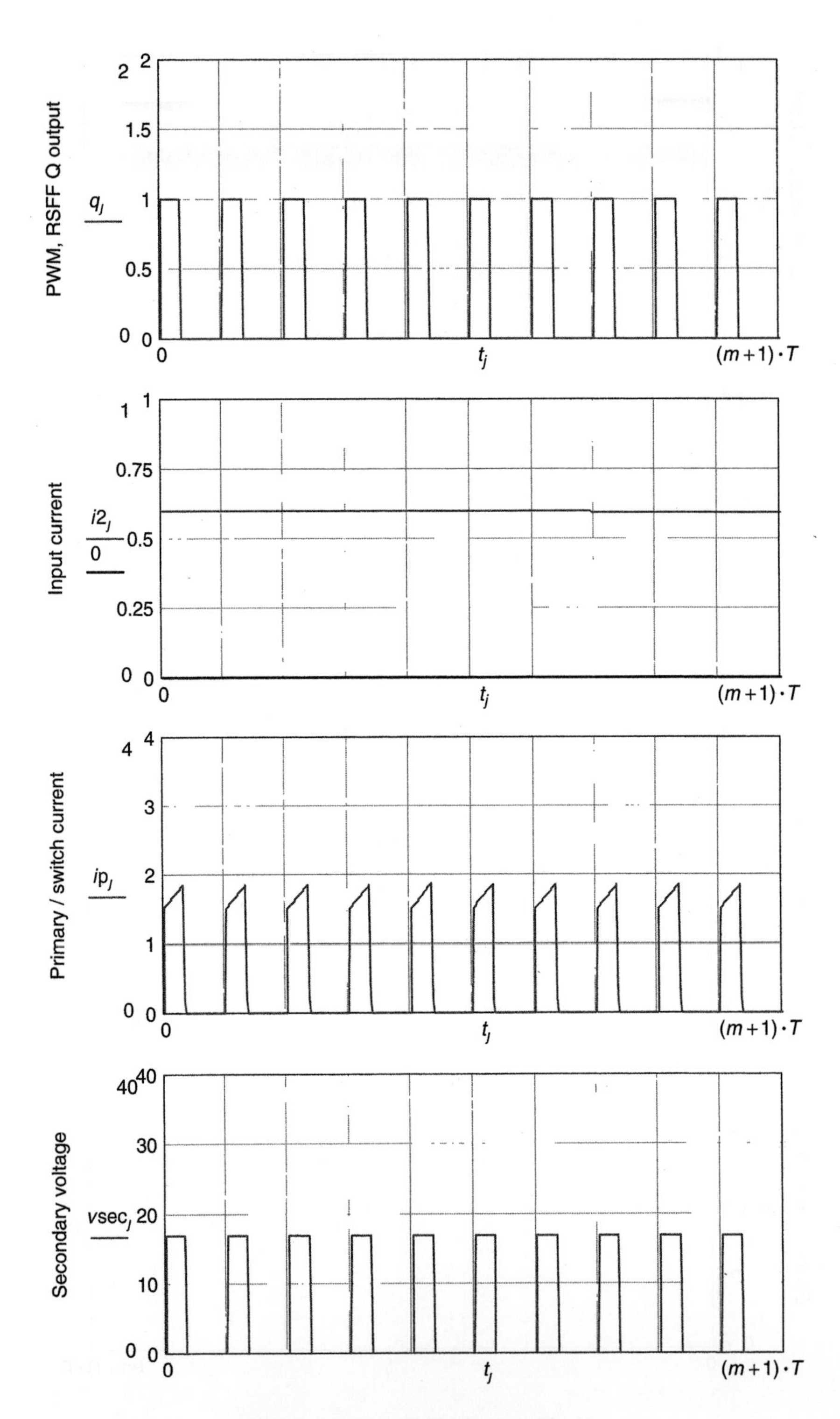

Figure 14.12: Steady state, zoomed in

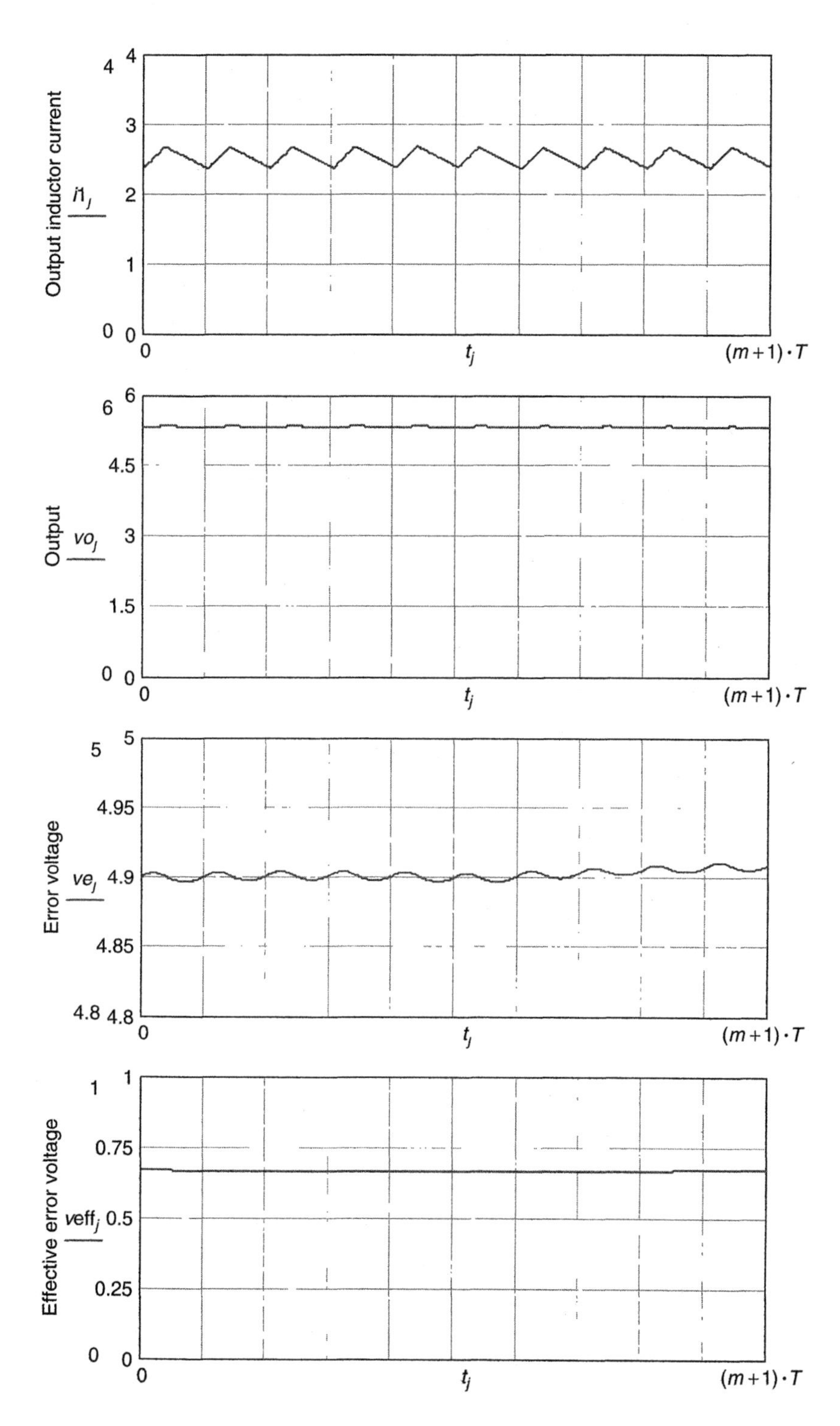

Figure 14.12: Continued

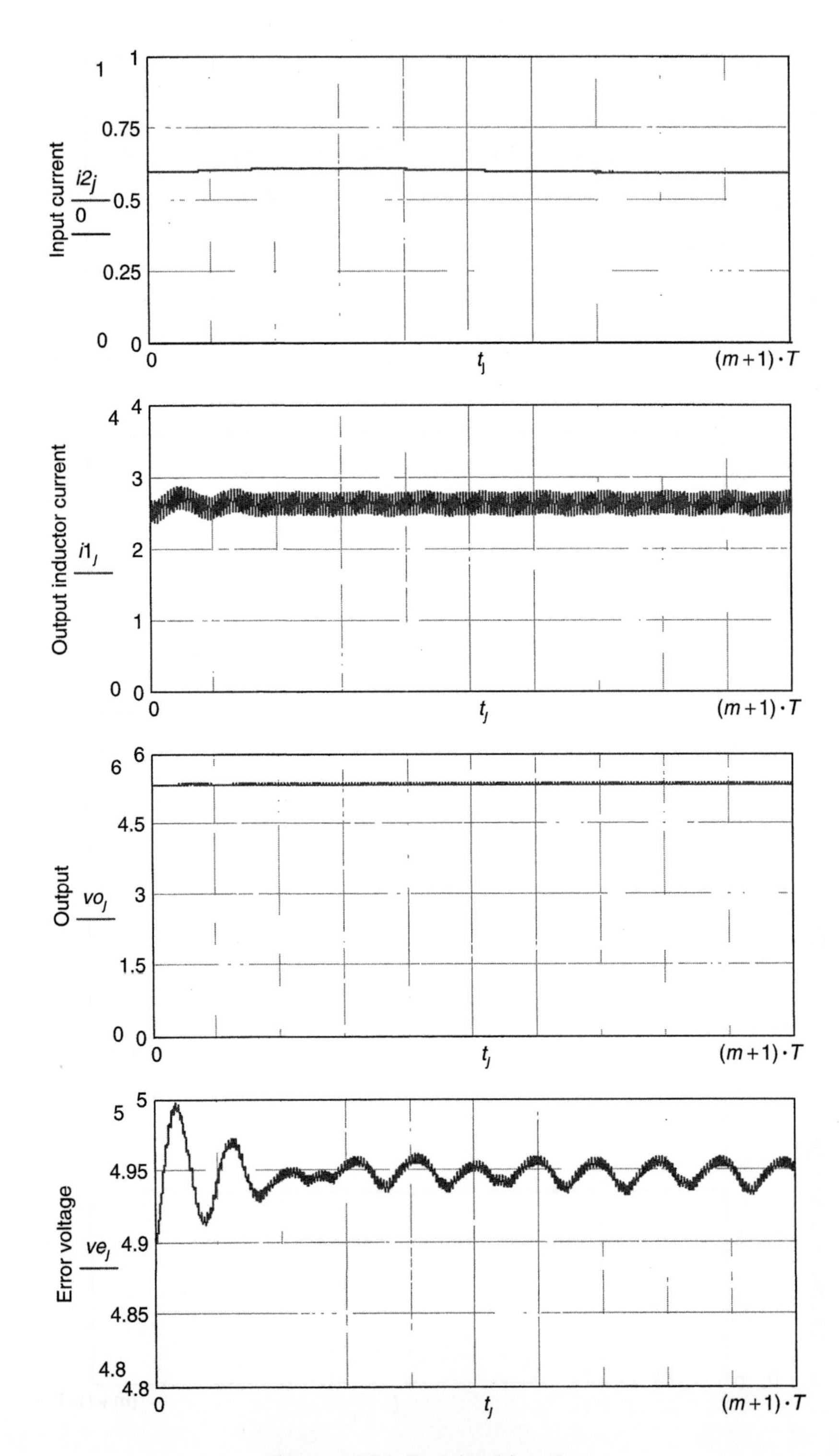

Figure 14.13: Step-load increase

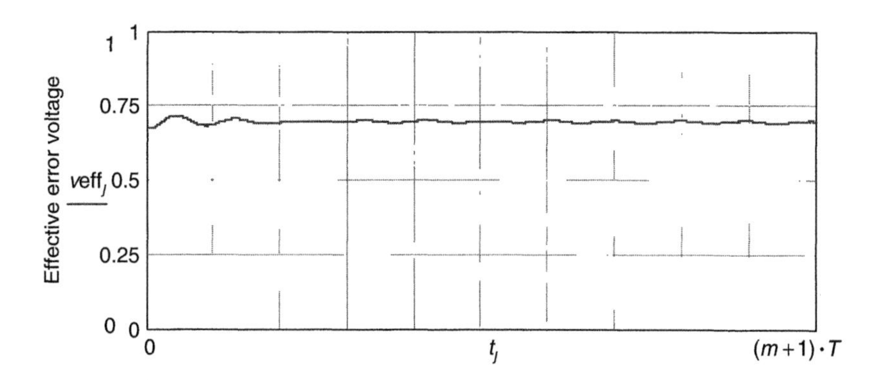

Figure 14.13: Continued

current. Without going through the derivations of the state equation one by one for all nine variables, we simply give the state matrix corresponding to switch, Q, on and off.

During switch-on, the state transition equation in differential form is $dx/dt = A_1x + B$; the source matrix and source are consolidated in a single matrix B. The state column matrix (vector) is

$$x = \begin{bmatrix} i_1 & v_1 & v_2 & v_n & v_{n2} & i_2 & v_3 & v_4 & i_m \end{bmatrix}^T$$

$$A_1 = \begin{bmatrix} al_{11} & al_{12} & 0 & 0 & 0 & 0 & 0 & al_{18} & al_{19} \\ al_{21} & al_{22} & 0 & 0 & 0 & 0 & 0 & 0 & 0 \\ al_{31} & al_{32} & al_{33} & al_{34} & 0 & 0 & 0 & 0 & 0 \\ al_{41} & al_{42} & al_{43} & al_{44} & 0 & 0 & 0 & 0 & 0 \\ 0 & 0 & 0 & al_{54} & al_{55} & 0 & 0 & 0 & 0 \\ 0 & 0 & 0 & 0 & 0 & al_{66} & 0 & al_{68} & 0 \\ 0 & 0 & 0 & 0 & 0 & 0 & al_{77} & al_{78} & 0 \\ al_{81} & 0 & 0 & 0 & 0 & al_{86} & al_{87} & al_{88} & al_{89} \\ al_{91} & 0 & 0 & 0 & 0 & 0 & 0 & al_{98} & al_{99} \end{bmatrix}$$

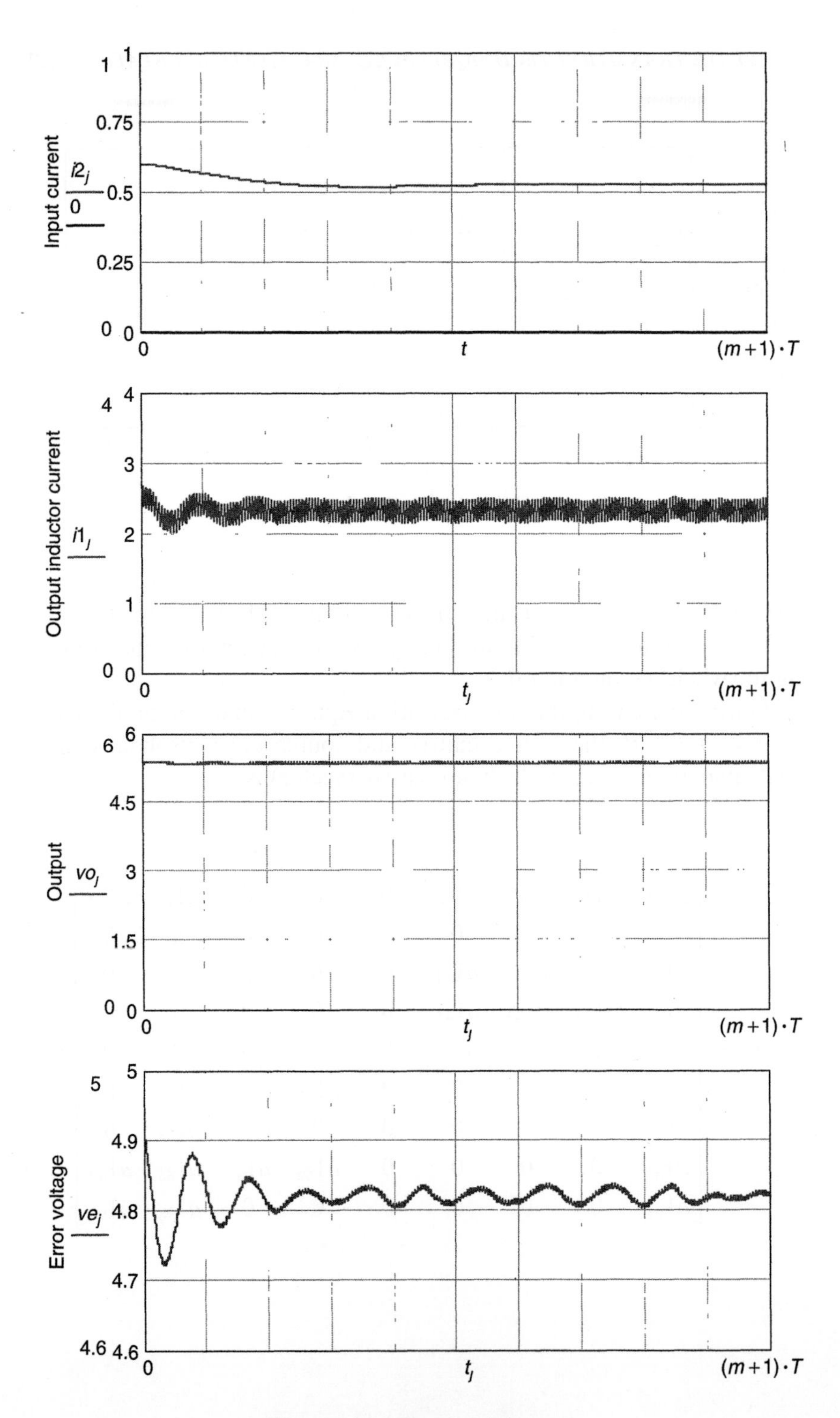

Figure 14.14: Step-load decrease

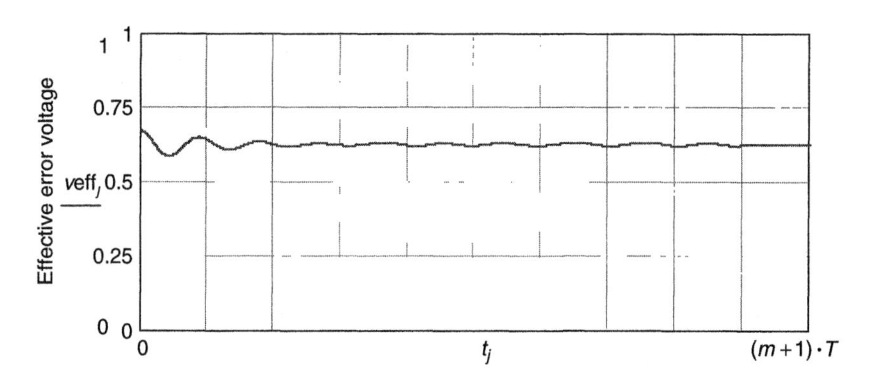

Figure 14.14: Continued

where

$$al_{11} = -\frac{r_{L1} + R_p + \left(\frac{N_2}{N_1}\right)^2 (r_w + R_{on})}{L_1}, \quad al_{12} = -\frac{k}{L_1}, \quad al_{18} = \frac{N_2}{L_1 N_1},$$

$$al_{19} = -\frac{N_2(r_w + R_{on})}{L_1 N_1}, \quad al_{21} = \frac{R_p}{r_{C1} C_9}, \quad al_{22} = \frac{k-1}{r_{C1} C_9},$$

$$al_{31} = \frac{-A \bullet k_f \bullet R_p}{(1+A) R_{12} C_{37}}, \quad al_{32} = \frac{-A \bullet k_f \bullet k}{(1+A) R_{12} C_{37}},$$

$$al_{33} = \frac{-A}{(1+A) R_{45} C_{37}} - \frac{1}{R_{45} C_{29}},$$

$$al_{34} = \frac{A}{(1+A) C_{37}} \left(\frac{1}{R_{12}} + \frac{1}{R_{45}} + \frac{1+A}{R_x}\right) + \frac{1}{R_{45} C_{29}},$$

$$al_{41} = \frac{k_f \bullet R_p}{(1+A) R_{12} C_{37}}, \quad al_{42} = \frac{k_f \bullet k}{(1+A) R_{12} C_{37}},$$

$$al_{43} = \frac{1}{(1+A) R_{45} C_{37}}, \quad al_{44} = -\frac{1}{(1+A) C_{37}} \left(\frac{1}{R_{12}} + \frac{1}{R_{45}} + \frac{1+A}{R_x}\right)$$

$$a1_{54} = \frac{h_{fe}\dfrac{R_{31}R_{19}}{R_{31}+R_{19}}}{(1+A)R_{31}C_{39}}K_{b2}, \quad a1_{55} = \frac{\dfrac{R_{19}}{(R_{19}+R_{31})R_{31}} - \dfrac{1}{R_{31}} - \dfrac{1+A}{R_{39}}}{(1+A)C_{39}},$$

$$a1_{66} = -\frac{r_{L2}}{L_2}, \quad a1_{68} = -\frac{1}{L_2}, \quad a1_{77} = -\frac{1}{r_{C2}C_7}, \quad a1_{78} = \frac{1}{r_{C2}C_7},$$

$$a1_{81} = \frac{-N_2}{N_1 C_8}, \quad a1_{86} = \frac{1}{C_8}, \quad a1_{87} = \frac{1}{r_{C2}C_8}, \quad a1_{88} = -\frac{1}{r_{C2}C_8},$$

$$a1_{89} = \frac{-1}{C_8}, \quad a1_{91} = \frac{-N_2(r_w + R_{\text{on}})}{N_1 L_m}, \quad a1_{98} = \frac{1}{L_m},$$

$$a1_{99} = \frac{-(r_w + R_{\text{on}})}{L_m},$$

$$B = \begin{bmatrix} -\dfrac{V_D}{L_1} \\ 0 \\ \dfrac{-A^2 V_{\text{ref}}}{(1+A)R_x C_{37}} \\ \dfrac{A \bullet V_{\text{ref}}}{(1+A)R_x C_{37}} \\ \dfrac{h_{fe}\dfrac{R_{31}R_{19}}{R_{31}+R_{19}}}{(1+A)R_{31}C_{39}}K_{b1} + \dfrac{A \bullet V_{\text{ref}}}{(1+A)R_{39}C_{39}} \\ \dfrac{V_{\text{in}}}{L_2} \\ 0 \\ 0 \\ 0 \end{bmatrix},$$

$$k_f = \frac{R_{35}}{R_{35} + R70}, \quad R_p = \frac{r_{C1}R_L}{r_{C1} + R_L}, \quad k = \frac{R_L}{r_{C1} + R_L},$$

$$K_{b2} = \frac{A \bullet R_{20}}{R_{20}R_{43} + (1 + h_{fe})R_{33}(R_{20} + R_{43})},$$

$$K_{b1} = \frac{-R_{43}(10 - V_D) + (R_{20} + R_{43})(10 - V_{be}) - A \bullet R_{20} \bullet V_{\text{ref}}}{R_{20}R_{43} + (1 + h_{fe})R_{33}(R_{20} + R_{43})}$$

During switch-off, the state transition equation in differential form is $dx/dt = A_2 x + B$:

$$A_2 = \begin{bmatrix} a2_{11} & a2_{12} & 0 & 0 & 0 & 0 & 0 & 0 & 0 \\ a2_{21} & a2_{22} & 0 & 0 & 0 & 0 & 0 & 0 & 0 \\ a2_{31} & a2_{32} & a2_{33} & a2_{34} & 0 & 0 & 0 & 0 & 0 \\ a2_{41} & a2_{42} & a2_{43} & a2_{44} & 0 & 0 & 0 & 0 & 0 \\ 0 & 0 & 0 & a2_{54} & a2_{55} & 0 & 0 & 0 & 0 \\ 0 & 0 & 0 & 0 & 0 & a2_{66} & 0 & a2_{68} & 0 \\ 0 & 0 & 0 & 0 & 0 & 0 & a2_{77} & a2_{78} & 0 \\ 0 & 0 & 0 & 0 & 0 & a2_{86} & a2_{87} & a2_{88} & a2_{89} \\ 0 & 0 & 0 & 0 & 0 & 0 & 0 & a2_{98} & a2_{99} \end{bmatrix}$$

where all but two ($a2_{11}$ and $a2_{99}$) nonzero elements of A_2 equal the corresponding nonzero elements of A_1. Here, $a2_{11} = -(r_{L1} + R_p)/L_1$, $a2_{99} = -(r_w + R_{\text{off}})/L_m$, because of switch-off.

Based on the linear system theory, the solution for $dx/dt = Ax + B$ has the form

$$x(t) = e^{At} \bullet X(0) + \int_0^t e^{A(t-\tau)} \bullet B \bullet d\tau$$

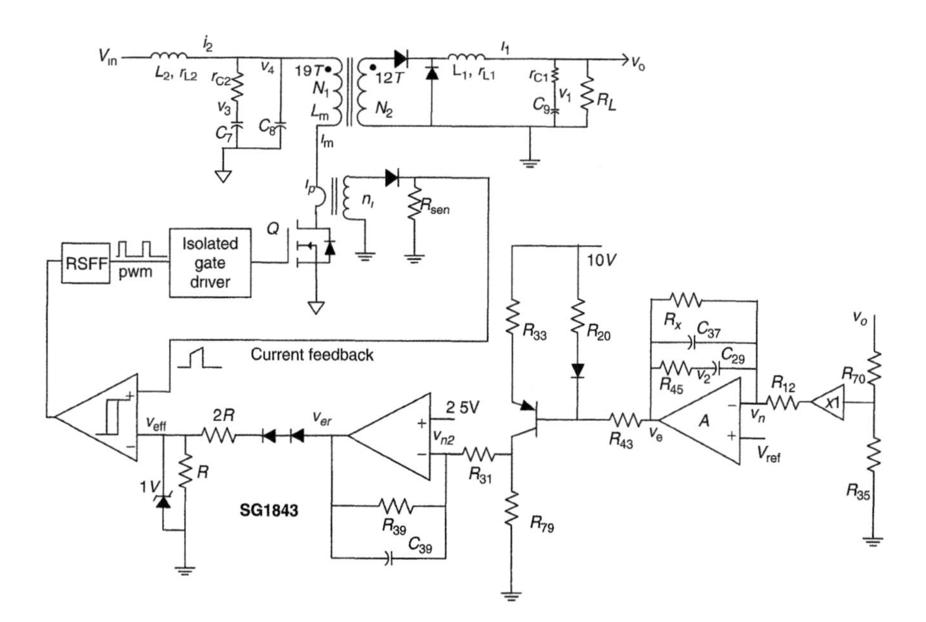

Figure 14.15: Isolated forward converter with peak current-mode control

given a starting state $X(0)$. This form can be further simplified if the driving source, B, is constant over the time interval of interest. If that is the case, we get

$$x(t) = e^{At} \bullet X(0) - A^{-1} \bullet (I - e^{At}) \bullet B$$

Therefore, during the switch-on time interval, we have

$$x_1(t) = e^{A_1 t} \bullet X_1 - A_1^{-1} \bullet (I - e^{A_1 t}) \bullet B$$

where a yet unknown starting state X_1 is invoked. And, for the switch-off time interval, we have

$$x_2(t) = e^{A_2(t-D \bullet T)} \bullet X_2 - A_2^{-1} \bullet [I - e^{A_2(t-D \bullet T)}] \bullet B$$

where the other unknown starting state, X_2, is assumed and a time delay, $D \bullet T$, is accounted for. (Duty cycle D is predetermined using the closed-loop technique presented in section 1.10.)

Next, at the switching transition boundaries, $t = D \bullet T$ and $t = I$, the continuity of state under steady-state operation demands

$$e^{A_1 D \bullet T} \bullet X_1 - A_1^{-1} \bullet (I - e^{A_1 D \bullet T}) \bullet B = X_2$$
$$e^{A_2(1-D)T} \bullet X_2 - A_2^{-1} \bullet [I - e^{A_2(1-D)T}] \bullet B = X_1$$

In other words, the two unknown starting states, X_1 and X_2, can be solved:

$$X_1 = (I - e^{A_4} e^{A_3})^{-1} [- e^{A_4} A_1^{-1} (I - e^{A_3}) B - A_2^{-1} (I - e^{A_4}) B]$$
$$X_2 = e^{A_3} \bullet X_1 - A_1^{-1} \bullet (I - e^{A_3}) \bullet B$$

where $A_3 = A_1 D \bullet T$, $A_4 = A_2(1 - D)T$. Readers should refer to linear algebra books for computing matrix exponentials $\exp(A_3)$ and $\exp(A_4)$ using eigenvalues, eigenvectors, and matrix diagonalization.

Next, we are supposed to evaluate time-domain matrix exponentials $\exp(A_3 t)$ and $\exp(A_4 t)$. However, the complexity involved in dealing with a 9×9 matrix exponential in the time domain is extremely discouraging. We invoke MATLAB and give the complete program listing and computation results in Appendix E.

Chapter 15

Power Quality and Integrity

Nowadays, laptop and desktop personal computers (PC) operate at 1–2 GHz. Not long ago, computer clocks running at 700 MHz were considered super. Clearly, it is safe to bet that the PC operating speed (clock rate) will keep going up with no limit in sight. After all, in theory, the frequency spectrum is unlimited, if infinite energy is available and controllable instantaneously. Given finite energy, what can be done to elevate the electronic clock speed?

First, let us look at the prevailing form of the electronic clock: a rectangular, cyclically varying waveform with two distinctive voltages: one low, V_L, the other high, V_H (Figure 15.1). We can see from the figure that two quantities, the voltage swing Δv and the time span Δt, play the ultimate role in defining clock speed.

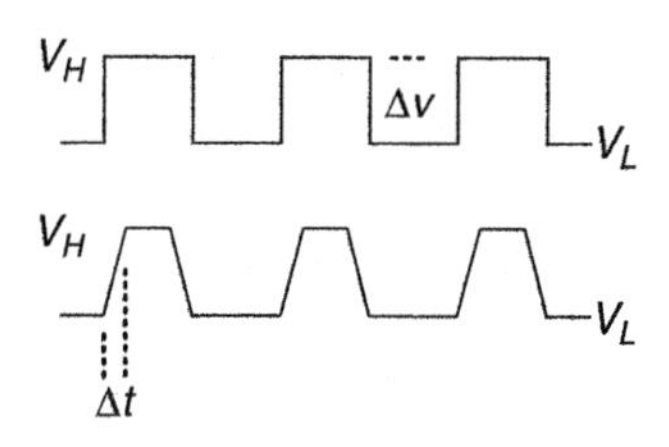

Figure 15.1: A clock signal

At this juncture, we examine the reason why slanted edges, either up or down, are recruited for depicting the clock waveform instead of the ideal straight edges.

It is understood that every circuit node has some capacitance in reference to the power return, attributed either to the capacitor as designed or device junctions and stray parasitics. The output node of a clock source must swing across a specific threshold voltage to signify a state (logic) change. It is also understood that the voltage swing across a capacitor is the result of current shuffling in and out within a given time. The explanation, of course, is just restating the working of a capacitor: $C\Delta v = i \cdot \Delta t$. However, the equation's simple appearance hides a lot of messages the capacitors are telling us (Appendix F). For our current discussion, we focus on our main objective—understanding the capacitive effects on clock speed.

Given a capacitance, there are basically two ways to speed up a clock. One way is to reduce the voltage swing required. The other is to make more current available. Obviously, a larger delta voltage requires longer time and more charge transfer. In contrast, a smaller voltage swing demands less time and charge. In other words, redefining a lower threshold is the primary means for high speed. The secondary means is the drive current. That is exactly the rationale why 5-V TTL (transistor transistor logic) had been largely replaced by 2.2–2.8-V logic. Moreover, as the demand on number crunching capability increases and the computer processing algorithm becomes more sophisticated, more and more devices (gates) are crammed into smaller and smaller areas in integrated circuits driven at higher and higher clock rates. The latter, which results in a very high step-load current, together with a lower working voltage, creates headaches for another subsystem of a computer—the power supply.

In this chapter, we look at those electric DC power issues generated by the advance of technology, electronic computers in particular. We examine the power converter output regulations, steady-state output ripples, output behavior under transient modes, transient behavior associated with bandwidth, and many others that all influence in a subtle way the quality of power.

15.1 Tolerance of Components, Devices, and Operating Conditions

Power supply output(s), single or multiple, is (are) generally specified as a nominal value at a nominal load and input. However, in actual operation, the supply output almost never meets the specification dead on. Many factors contribute to the deviation of actual output level from the specification. Among them, component tolerance, circuit layout, and stray parasitics of the interconnection are the three that stand out. Of the three, the impact of circuit layout (artwork) is not easily quantified and is omitted from the discussion. The effect of interconnection parasitics can be considered equivalent to adding elements to components. We therefore evaluate only the component effects on regulation.

At least three subfactors constitute the total component tolerance. Due to material variability and differences in processing sequence and environment, all electronic components show initial tolerance. For instance, in the data sheets resistor manufacturers provide, we see the term *tolerance* and ranges from 0.02% to 5% in irregular increments. What the term means is, given a specified, desired (nominal) value, the actual value is enveloped by an upper bound equal to nominal times (1 + tolerance/100) and a lower bound equal to nominal times (1 − tolerance/100). For instance, if a 2.2-Kohm 1% resistor is selected for a circuit, the actual value for that resistor picked from a production batch is expected to reside somewhere between $2.2 \cdot 10^3 \cdot (1 + 0.01)$ and $2.2 \cdot 10^3 \cdot (1 - 0.01)$.

Once installed in the circuit and in operation, each component also generates heat in addition to the heat produced by other components. Heat is the source of temperature rise. When subjected to temperature changes, component values also change. These temperature-induced changes in component value are named the *temperature coefficient* (TC) on the component data sheet and specified in units of parts per million, ppm/°C or %/°C. That is, given a delta temperature, ΔT, the component value is expected to change (TC) $\cdot$ ΔT ppm, or %.

Components are also understood to change value if they have not been used and are stored in a protective enclosure, due to the aging of the material. Quantification of material aging is impractical. It is generally given a so-called end-of-life degradation figure (EOL), which tends to be more a guess than a statistical confidence.

In all, over the life of a component, its value can in theory vary between two extremes, (nominal)(1 + initial tolerance/100 +TC • ΔT+ EOL/100) and (nominal)(1 + initial tolerance / 100 −TC • ΔT− EOL/100).

So far, we seem to confine the component tolerance discussion to passive components (*R*, *L*, and *C*). However, if we examine the closed-loop equation for a converter, for instance, (1.99), we immediately realize that there are many other operating conditions (switching frequency, load), device parameters (rectifier forward drop, transistor gain, op-amp offset), and voltage variables (reference voltage, input voltage) that play some role in setting the actual output voltage. Their effects, of course, must also be accounted for. That is what we will do next.

15.2 DC Output Regulation and Worst-Case Analysis

Equation (1.99) for a forward converter and many other similar equations for other topologies show that numerous factors, directly or indirectly, have an impact on the actual output. The question is how those factors produce the effects and by how much. The answer lies in the concept of output sensitivity against the variable of interest. For instance (refer to Figure 1.30), the output feedback ratio is set by R_1 and R_2. Both resistors constitute a voltage divider that samples the output and compares it to a precision reference voltage, V_{ref}. We ask how much the output deviates from the ideal specification if real discrete resistors are used. Well, we can find out by asking first, How much does the output change if R_2 changes by 1 ohm?

This question actually is equivalent to a partial derivative of, for instance, (1.99) against R_2; that is, the output sensitivity S_{R2} against R_2 in unit volts/ohm:

$$S_{R2} = \frac{\partial V}{\partial R_2} \tag{15.1}$$

We also know from the previous section that R_2 is expected to have uncertainties in value; that is,

$$\Delta R_2 = R_2(\% \text{ initial tol.}/100 + \text{TC} \cdot \Delta T + \text{EOL}\%/100) \tag{15.2}$$

Therefore, the expected output delta due to R_2 change is

$$\Delta V_{R2} = S_{R2} \cdot \Delta R_2 \tag{15.3}$$

Similar calculations are performed for all the factors involved. When the output can be expressed in an explicit function, such as (1.99), calculation of sensitivity is quite straightforward. If the closed-loop output is expressed in multiple implicit functions, the technique of Jacobian determinant can be employed.

By so doing, one may be tempted to expect that the worst-case output deviation is

$$\Delta V = \sum_k S_k \cdot \Delta k \tag{15.4}$$

The form is correct, but it misses one very important point. Depending on the selected variable, the corresponding sensitivity figure may have either a positive or a negative sign. A positive sensitivity simply means that the output deviates in increasing sense when the variable of interest increases by one unit. A negative sensitivity, of course, means that the output decreases when the variable increases by one unit. Therefore, the reader must use this equation with care. However, the equation still leaves one discussion open. One may question the argument of worst-case component values deviating in opposite directions—some increase, others decrease, simultaneously. After all, a circuit board, and consequently its components, experiences almost the same environment. There is little reason to expect some components to experience a temperature increase while others encounter a decrease. To alleviate this philosophical controversy, we rename (15.4) the extreme case analysis (ECA) if the sensitivity sign is also properly taken care of. Here, we also introduce the root squared statistics (RSS) that neutralize the direction debate:

$$\Delta V_{\text{RSS}} = \sqrt{\sum_k (S_k \cdot \Delta k)^2} \tag{15.5}$$

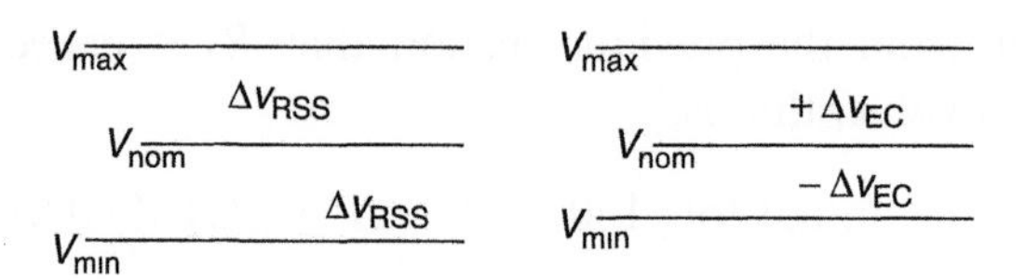

Figure 15.2: Output distribution

Once either (15.4) or (15.5) is done, the actual output level then has the distribution of Figure 15.2.

In some cases, $|+\Delta_{EC}|$ does not equal $|-\Delta_{EC}|$.

15.3 Supply Output Ripple and Noise

For switch-mode power supplies, Figure 15.2 is misleading, since all outputs in a time domain are not ripple and noise free. For example, Figure 1.41 (bottom trace) shows the actual output for a 28-V-in, 5.3-V-out DC–DC converter. In other words, a typical output of a switch-mode power converter exhibits the ripple shown in Figure 15.3.

Not only are Δv_1, Δv_2, and Δv_3 not equal, their cyclical time-domain ripple shapes are not identical because components interact differently and the duty cycle changes under different operating conditions. Figure 1.41 also shows spikes ringing at transition edges. Those high-frequency behaviors are attributed to parasitic elements associated with transformer winding, transformer core, rectifier recovery, and so on. They are almost impossible to quantify or analyze; hence, we set them aside.

It is very desirable to be able to estimate the ripple magnitude at the fundamental switching frequency, however. One way is by simulation.

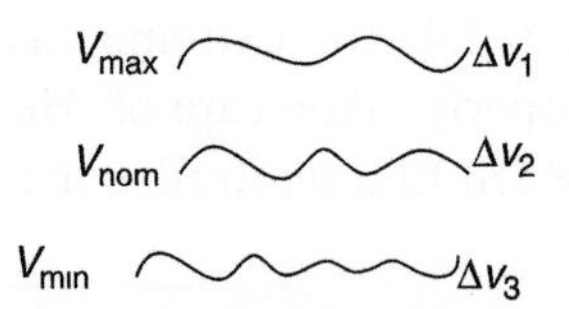

Figure 15.3: Typical output ripple under various operating conditions

For this approach, two techniques exist. One is a time-domain cycle-by-cycle model. The other is an average model. Because of the theory on which the model is built, the average model does not give the output ripple at the switching frequency, so it is excluded from the discussion at this point. The cycle-by-cycle model suffers from long simulation time and convergence issues that plague even the best simulator. Therefore, both techniques are unable to meet the objective.

Luckily, a third technique is believed to fulfill the goal. Actually, the technique was already shown in sections 1.12 and 1.14. The technique, coupled with the determination of the duty cycle given, for instance, in sections 1.1 and 1.2, can give the steady-state output ripple very quickly. Bypassing the turn-on transient equivalent to starting conventional simulation at a zero state, the third technique takes only one switching cycle time to zero in on the steady state. The key is in finding the starting states of the steady state, which are completely embedded in the circuit (system) matrix that represents alternating circuit topologies. By invoking the concept of the continuity of state, the systematic extension of the continuity of the capacitor voltage and inductor current, the starting states, such as (1.91), are explicitly obtained. Once the starting states are found, the steady-state expressions, such as (1.92), follow. In addition to the advantage of giving a closed-form solution, the technique also yields new results in one switching cycle, when components are given new values. In contrast, the conventional cycle-by-cycle simulation, even if it does not diverge and offers an end result, has to crawl through the transient phase before reaching a new steady state, because changes in component values are equivalent to a new operating environment. Clearly, this technique has a lot of untold edges over the other two. Best of all, it can identify the maximum ripple, given changing components, without wading through the mud of transients prone to divergence.

15.4 Supply Output Transient Responses

In the last section, Figure 15.3 shows the steady-state voltage of an output under constant input and constant load. The image can be misleading, given that modern computers rarely stay idle while awake.

A wakening microprocessor draws lots of current that can sink a power supply, or a refrigerator's compressor kicks in and momentarily depresses the AC line voltage, which in turn, also sinks the power supply. In the first case, it is quite straightforward to explain the mechanism leading to the transient behavior (Figure 15.4) when subjecting a power supply to a step load, up or down.

Basically, all outputs have nonzero output (source) impedance. Higher loads (currents), as a result, generate higher internal drops, which, of course, means lower effective output, while lower loads cause the effective output to rise.

In contrast, an input-induced transient is harder to explain. Signs of output sensitivity against input voltage (signs of $\partial v_o/\partial v_{\text{in}}$) may give the sense of how the output moves when step input takes place. A rising or falling input may set off either an overshoot or an undershoot. From Figure 15.4, we see two quantities—output deviation Δv and settling time Δt—that characterize the transient responses. The value of Δv is basically, but not exactly, inversely proportional to the control-loop gain, while that of Δt is also roughly inversely proportional to the control-loop bandwidth (BW). The larger the gain around a control loop, the smaller the output deviation Δv. The wider the bandwidth of a control loop, the shorter the settling time. The control-loop gain for a real system, unfortunately, cannot be kept high across a wide bandwidth. The reality is that a minute inductance behaves like a large resistance at very high frequency and blocks current flow, while a minute capacitance behaves as a short at very high frequency and diverts current flow. Given the parasitic elements everywhere in a real system, the control-loop gain is expected to diminish at higher and higher frequencies. This fact of

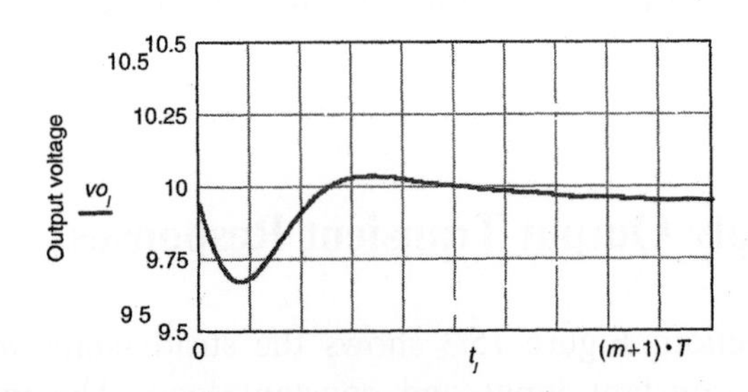

Figure 15.4: Typical output transient under a step load

nature is expressed as a combination of both the gain and the bandwidth: the (gain)(bandwidth) product; assuming a single-pole roll-off (–20dB/decade). We examine this more in the following sections.

15.5 The Concepts of Frequency and Harmonic Content

Webster's dictionary defines *frequency* as "the number of repetitions of a periodic process in a unit of time." The definition, strictly speaking, is incorrect, for in a sense, it is correct for only half of the cases. Yes, the repetition of a periodic process gives the sense of frequency: the number of recurrence. But frequency means much more than just counting repetition. In order not to give the impression of nitpicking, we examine the dimension of the frequency measure.

As we know, the unit of frequency is Hertz (1/second). As a matter of fact, it can be generalized as the "inverse of time." But, this definition still misses something. The key is in the "1/time"; that is, the inverse of time elapsed. Any process that evolves over time has frequency. For instance, the slanted edges of Figure 15.1 have a frequency content and so does a step function. As long as state changes take place over an observation time interval (Figure 15.5) frequency is involved. Signals with an identical alternation or repetition number, but with different rising or falling edges, possess different frequency contents. Figure 15.6 gives a full-bridge DC–AC converter. By appropriate on/off control of the Q_1/Q_4 pair and the Q_2/Q_3 pair, voltage V_{ab} is known to exhibit an ideal waveform, given in Figure 15.7.

The harmonic content of the Figure 15.7 waveform can be obtained by several approaches. The first approach shifts the time zero to the right

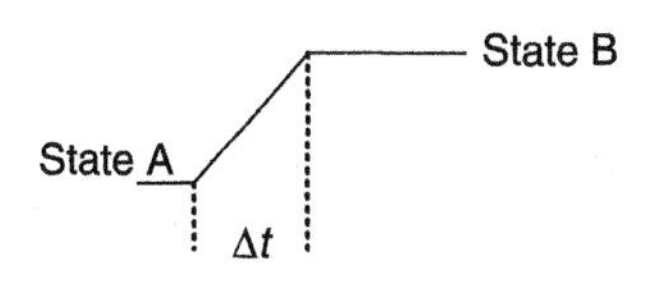

Figure 15.5: Definition of frequency

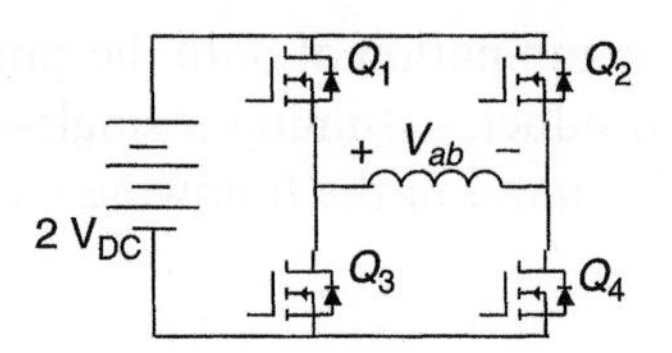

Figure 15.6: Full-bridge DC–AC converter

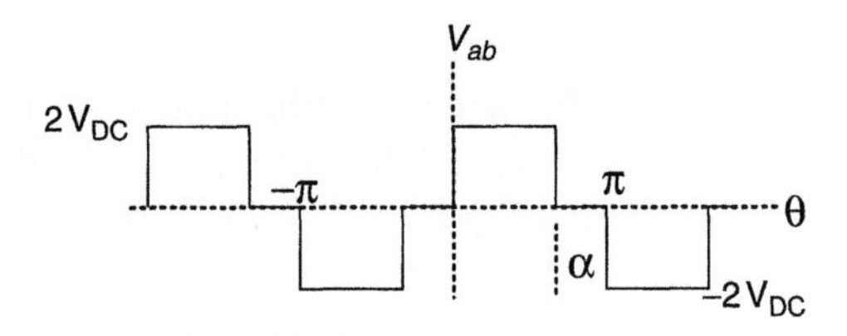

Figure 15.7: Ideal, full-bridge converter output

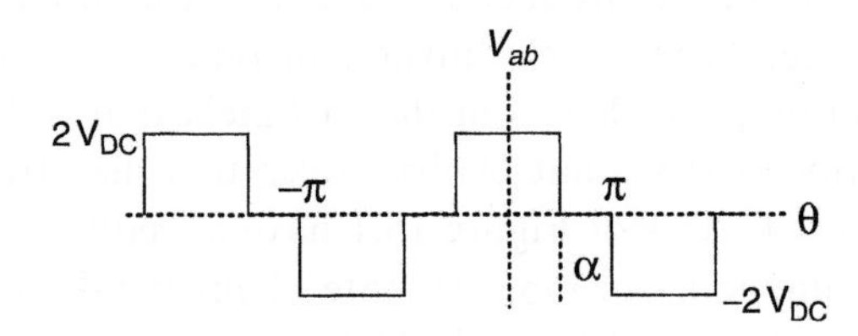

Figure 15.8: Full-bridge converter output, time-zero right shifted

and yields Figure 15.8. The time-zero right-shifted waveform contains only cosine series, $\sum a_n \cos(n\theta)$. The coefficient of cosine series can be given as

$$a_n = \begin{cases} +\dfrac{8\mathrm{V_{DC}}}{n \bullet \pi} \cos\dfrac{n \bullet \alpha}{2}, & n = 1,5,9,\ldots \\ -\dfrac{8\mathrm{V_{DC}}}{n \bullet \pi} \cos\dfrac{n \bullet \alpha}{2}, & n = 3,7,11,\ldots \end{cases}$$

$$= \frac{8\mathrm{V_{DC}}}{n \bullet \pi} \sin\left[n\left(\frac{\pi - \alpha}{2}\right)\right], \quad n = 1,3,5,7,\ldots \tag{15.6}$$

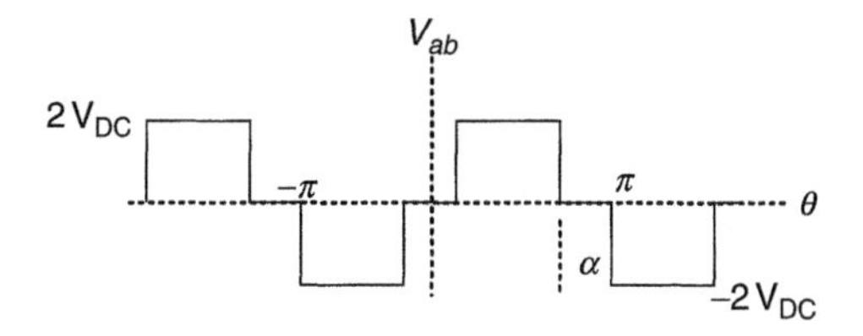

Figure 15.9: Full-bridge converter output, time-zero left shifted

The time zero of Figure 15.7 can also be shifted to the left and yields Figure 15.9. The time-zero left-shifted wave contains only sine series $\sum b_n \sin(n\theta)$. The coefficient of sine series can be proven to be

$$b_n = +\frac{8V_{DC}}{n \bullet \pi} \cos \frac{n \bullet \alpha}{2}, \quad n = 1,3,5,7,\ldots \tag{15.7}$$

The third approach keeps the time zero as it is in Figure 15.7. The Fourier series for the waveform then contains both cosine and sine terms, $\sum [a_n \cos(n\theta) + b_n \sin(n\theta)]$:

$$\begin{aligned} a_n &= \frac{4V_{DC}}{n \bullet \pi} \sin n\alpha, \quad n = 1,3,5,7,\ldots \\ b_n &= \frac{4V_{DC}}{n \bullet \pi} (\cos n\alpha + 1), \quad n = 1,3,5,7,\ldots \end{aligned} \tag{15.8}$$

We, of course, can also use complex-form Fourier series. The coefficient is simply $(a_n - j \bullet b_n)/2$. The interesting, and very important, point is that the magnitude of coefficients obtained by various means is identical. This is to be expected because the Fourier content (magnitude) of a periodic signal must be invariant when a snapshot of that signal is taken. In other words, taking a snapshot for a steady periodic signal at any moment does not alter its harmonic contents.

Now, let us alter Figure 15.7 slightly, by cutting in some realistic signal properties: finite rise and fall times at transition edges. Figure 15.10 gives such a more realistic signal with time zero already moved to simplify the extraction of sine series coefficients.

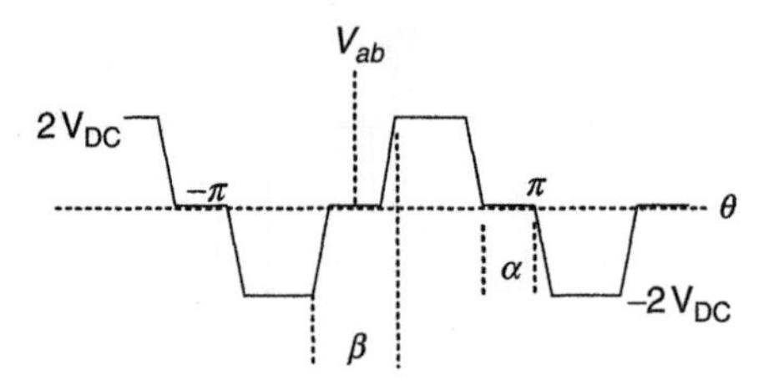

Figure 15.10: Practical, full-bridge converter output

The sine series coefficients can be proven to be

$$b_n = \frac{16V_{DC}}{n^2 \bullet \pi \bullet (\beta - \alpha)} \left(\sin \frac{n \bullet \beta}{2} - \sin \frac{n \bullet \alpha}{2} \right), \quad n = 1,3,5,7,\ldots \tag{15.9}$$

To compare the bandwidth of the two waveforms (Figures 15.7 and 15.10), we reconstruct the two waveforms using a finite number of terms of Fourier coefficients and compare the reconstruction. It turns out that, on the basis of reproducing an equal overshoot, fewer terms (Figure 15.11) are needed to reconstruct Figure 15.10 than Figure 15.7.

Another view is that, to create a reconstruction that resembles the source waveform, many more terms are needed for Figure 15.7 than for Figure 15.10, as shown in Figure 15.12.

By now, it is quite clear from the discussion that short-term transition dynamics actually have even more significance than longer-term repetition alone in determining the frequency content of a periodic process.

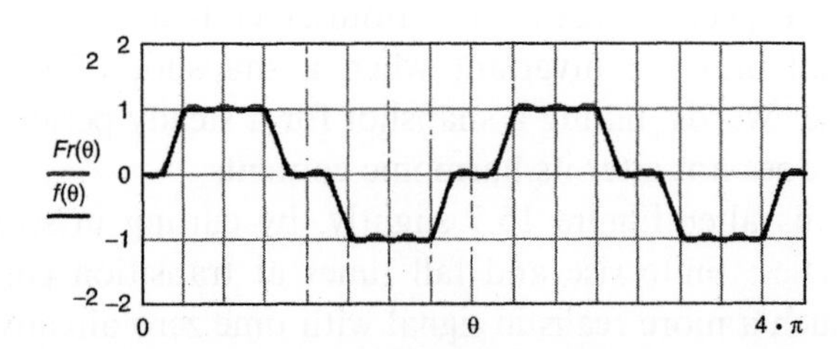

Figure 15.11: Nine-term reconstruction, $Fr(\theta)$, of Figure 15.10

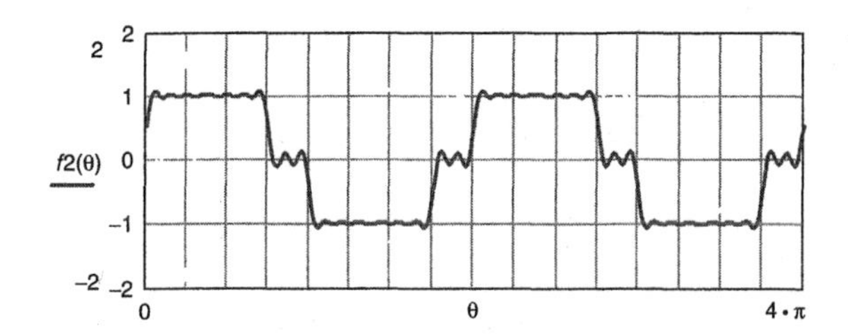

Figure 15.12: Nineteen-term reconstruction of Figure 15.7

15.6 Control-Loop Bandwidth

It was made amply clear in the last section that the time rate at which a process changes state, or how a signal rises or falls, is critically linked to the bandwidth, or spectrum, of the process. If the process represents the load current on (for example, the 5-V supply for a microprocessor or the voltage input to a DC–DC converter providing 5 V), the converter loop must have a response speed equivalent to the bandwidth of the perturbing process. Otherwise, the 5-V supply suffers.

One may ask how the bandwidth in the unit of frequency is tied to the process transition in the unit of time. We examine the output of a simple, single-pole system subjected to a step input (Figure 15.13(a)). The output/input transfer function is given as

$$H(s) = \frac{1}{R \cdot C \cdot s + 1} \tag{15.10}$$

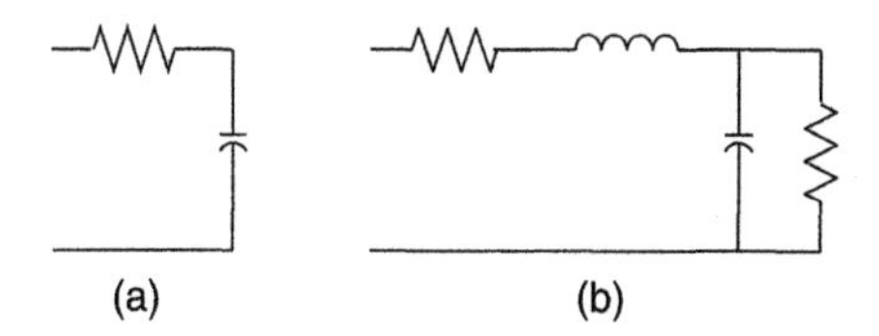

Figure 15.13: First-order and second-order systems

The function can be placed in another form as

$$H(f) = \frac{1}{j\left(\frac{f}{1/(2\pi R \cdot C)}\right)\ldots + 1} = \frac{1}{j\left(\frac{f}{f_{BW}}\right) + 1} \tag{15.11}$$

where $f_{BW} = 1/(2\pi RC)$ is the 3dB bandwidth.

It is also known that, given a step input A, the output is

$$v_o(t) = A(1 - e^{-t/RC}) \tag{15.12}$$

If the output rise time, t_r, is defined as the time needed for the output to rise from zero to 90% of the input step, both the bandwidth and the rise time are interrelated by

$$f_{BW} = \frac{0.366}{t_r} \tag{15.13}$$

The switch-mode power converters we are interested in, however, rarely behave as simple, single-pole systems. It is then completely reasonable to raise the system order by 1. We then also ask if a relationship similar to (15.13) exists for a second-order system. To answer the question, we first have to define the bandwidth of such a system (Figure 15.13(b)).

It turns out that the transfer function for Figure 15.13 has a typical shape, as shown in Figure 15.14. The figure immediately raises a funda-

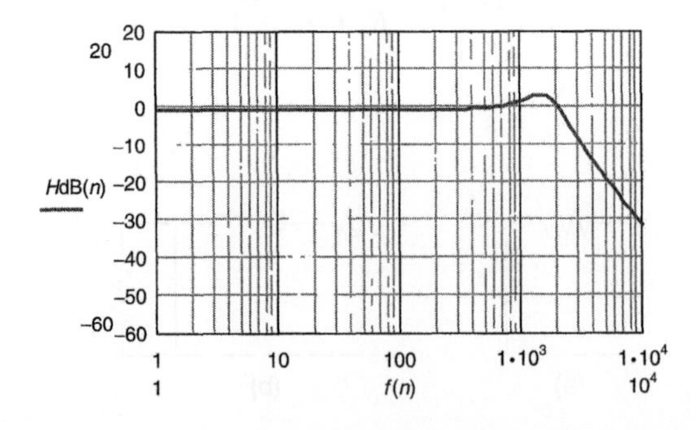

Figure 15.14: Transfer function of second-order system

mental question: What is the definition of *bandwidth*? It is not clear at what frequency the concept, or definition, of bandwidth can be applied. The best we can do is to pick the high-frequency point where the transfer function crosses zero decibel, $f_{0\text{-dB}}$. By so doing, we establish a criteria that is reasonably consistent among different high-order systems. However, we are not out of the woods yet.

Unlike the first-order system that possesses a smooth, exponential rise when subjected to a step input, the second-order system response takes on many shapes, depending on system components and operating conditions. One possible shape is oscillatory (Figure 15.15). Given such a step response, the definition of rise time is also blurred. The best we can do is base the definition on the rising envelope (Figure 15.15). The envelope is given by

$$e(t) = A\left(1 - \frac{\omega_o}{\omega} e^{-\zeta\omega_o t}\right) \tag{15.14}$$

while the 0dB crossing frequency is given by

$$f_{0-\text{dB}} = \frac{\sqrt{\omega^2 - \zeta^2\omega_o^2 + \sqrt{-4\omega^2\zeta^2\omega_o^2 + \omega_o^4}}}{2\pi} \tag{15.15}$$

If we use the same definition of rise time as the time it takes for the exponential to rise (in this case, the envelope, from 0 to 90%), the rise time for a second-order system is

$$t_{r-2} = -\frac{1}{\zeta\omega_o} \ln\left(\frac{\omega}{10 \bullet \omega_o}\right) \tag{15.16}$$

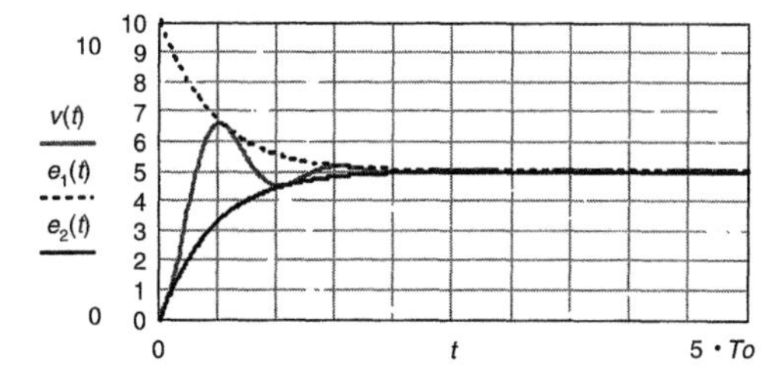

Figure 15.15: Underdamped step response of a second-order system

Unfortunately, both equations, in the given forms, do not allow a simple, symbolic relationship resembling (15.13) that connects rise time and bandwidth for a first-order system. But, we are not at a loss, since a numerical example gives us some indication as to the difference between first-order and second-order systems. For instance, given $L = 100\,\mu\text{H}$, $C = 100\,\mu\text{F}$, $R = 5$, and $A = 5$, the bandwidth, $f_{0-\text{dB}}$, is related to the rise time, t_{r-2}, as

$$f_{0-\text{dB}} = \frac{1.32}{t_{r-2}} \tag{15.17}$$

This equality tells us in simple terms that a second-order system needs a much wider bandwidth (that is, the ability to respond to higher frequencies) than a first-order system reacting to a step disturbance.

15.7 Step Response Test

By now, a picture begins to emerge of the close link between the hidden bandwidth and the observable step response. We also show, in section 15.5, that the periodic process with sharp edge transitions contains very high-frequency components. It is therefore quite appropriate to expect that a step stimulus is the best test source, since its numerous harmonic contents can elicit system responses at numerous frequencies simultaneously in place of multiple single-frequency tests. However, a single step produces a single response that is so short-lived it makes data acquisition difficult. Obviously, a stream of repetitive steps eases the problem. This is the rationale of using a periodic, rectangular drive as the perturbation source for testing.

With the test source resolved, we give several possible step responses with an ever-increasing degree of complexity in addition to Figures 15.4 and 15.15. Figure 15.16 shows the presence of higher harmonic frequency. Its corresponding mathematical form is

$$v(t) = A\left\{1 - \frac{\omega_o}{\omega} e^{-\zeta\omega_o t}[\sin(\omega t + \alpha) + \sin(\omega_n t + \beta)]\right\} \tag{15.18}$$

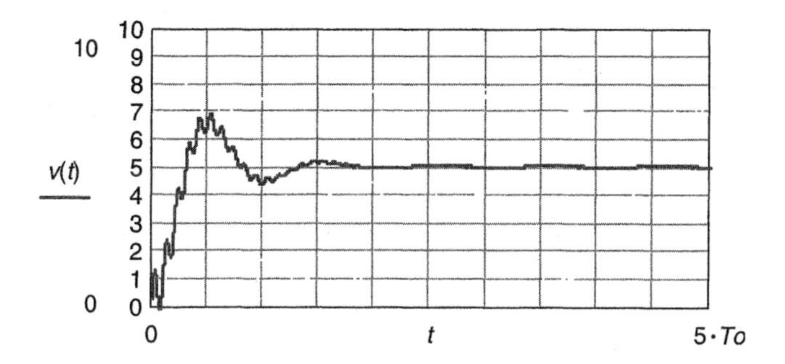

Figure 15.16: Step response containing additive high frequency

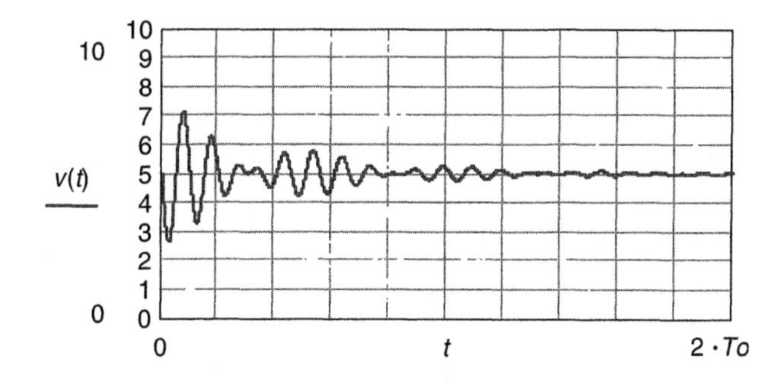

Figure 15.17: Step response containing multiplicative high frequency

As shown, the effects of high frequency are additive. More contributions from other higher frequencies can also show up if the system allows. Figure 15.17 shows the presence of higher harmonic frequency in multiplicative form:

$$v(t) = A\left(1 - \frac{\omega_o}{\omega} e^{-\zeta\omega_o t}[\sin(\omega t + \alpha)\sin(\omega_n t + \beta)]\right) \tag{15.19}$$

15.8 Bandwidth and Stability

In Figure 15.15, a damped oscillation at one transition edge was shown to dissipate completely prior to the initiation of the next transition. It is reasonable to expect that, under certain circumstances, the oscillation lasts longer than the time duration between two consecutive transitions. In such a case, the repetitive step response appears as in Figure 15.18 or 15.19.

Just what do all these mean? They are the exhibition of system stability in the time domain. They reveal the stability margin that was discussed extensively in previous chapters. However, the exact correlation between such transient time-domain behavior and loop-gain (stability) characteristics is not obvious. We know from the classical control system theory that the most prominent manifestation associated with control-loop instability is a prolonged oscillatory system response when perturbed. System instability is further understood to be embedded in the characteristics of the control-loop frequency response. In other words, system transient behaviors are intimately tied to control-loop frequency responses. Without exceeding the scope of this writing, we accept the well-established conclusion that a zero-phase margin (phase curve) at

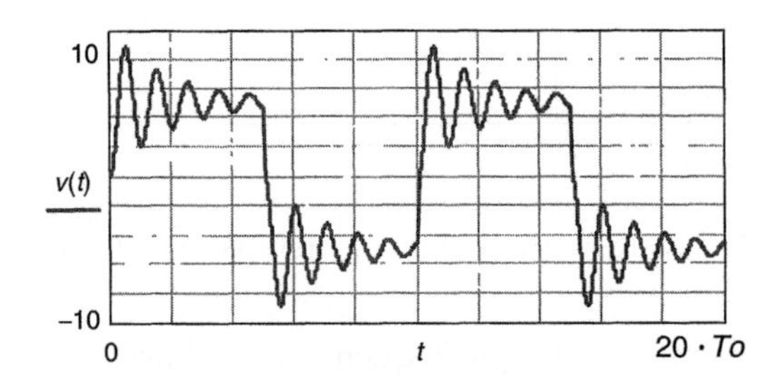

Figure 15.18: Underdamped response

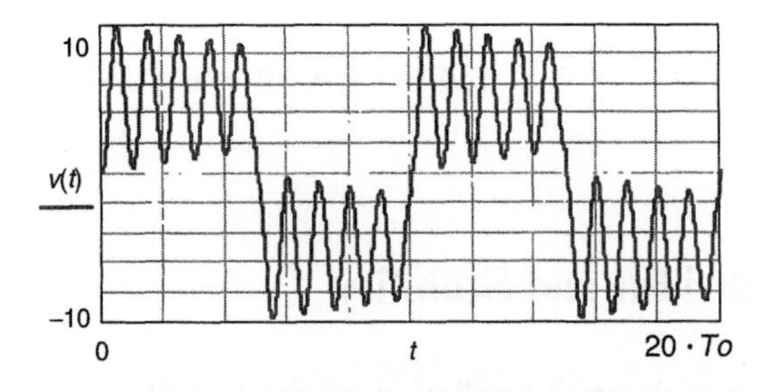

Figure 15.19: Undamped response

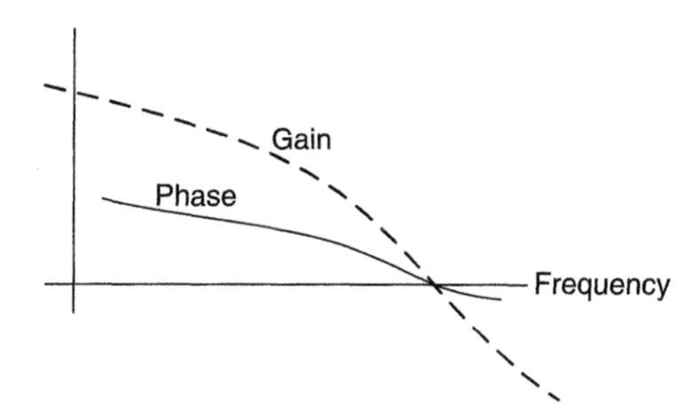

Figure 15.20: Unstable system frequency response

the 0dB crossing (gain curve) for a small-signal frequency response (Bode plot) signifies instability (Figure 15.20).

The time-domain response associated with Figure 15.20 is a sustained oscillation at the frequency crossing 0dB and zero phase. It is therefore logical to ask what is the same between Figures 15.21(a) and 15.21(b).

Naturally, the answer is that both are unstable. The system corresponding to Figure 15.21(b) gives an oscillation at a higher frequency. Similarly, Figures 15.22(a) and 15.22(b) give different transient responses, in that the smaller phase margin contributes to a longer settling time (Figure 15.23), while Figures 15.24(a) and 15.24(b) give different transient responses, in that the damped oscillation of Figure 15.24(b) occurs at a higher frequency.

Next, we consider the implications of the gain margin. Figures 15.25(a) and 15.25(b) give two systems with different gain margins. It is understood that the smaller the gain margin, the longer the decay time.

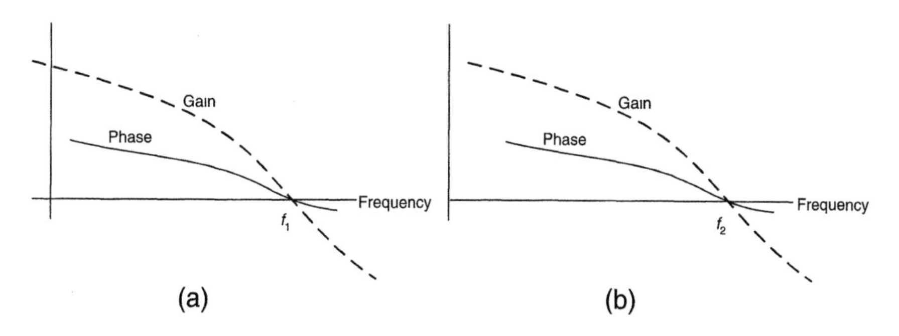

Figure 15.21: Two unstable systems, different frequencies, $f_1 < f_2$

So far, we give a most probable response profile corresponding to a given Bode plot in forward reasoning. However, we are not sure, given a time response, if the reverse pinpoints a corresponding Bode plot.

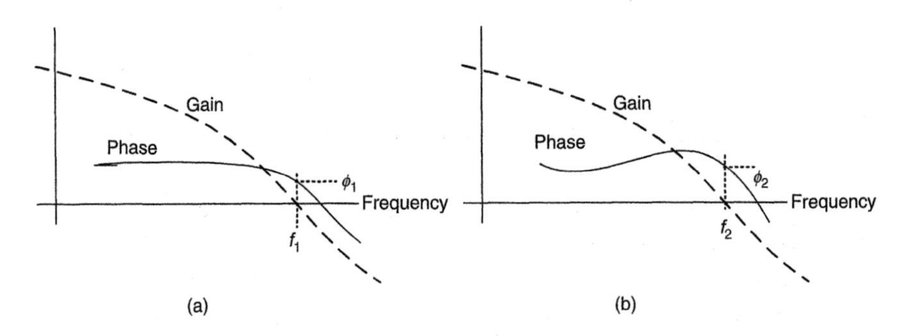

Figure 15.22: Two stable systems with different phase margins, $f_1 = f_2$, $\Phi_1 < \Phi_2$

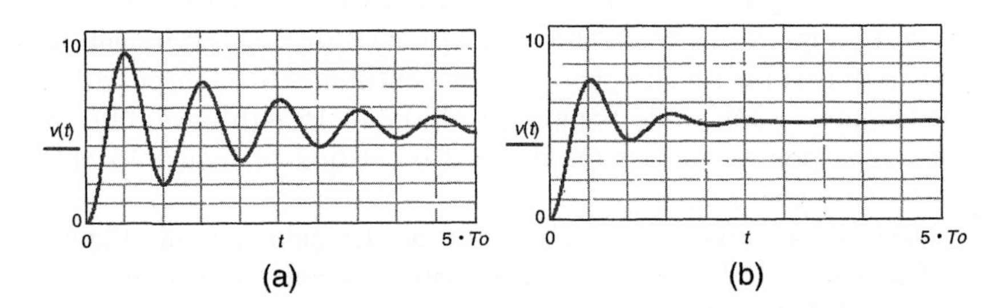

Figure 15.23: Transient response: (a) Figure 15.22(a), (b) Figure 15.22(b)

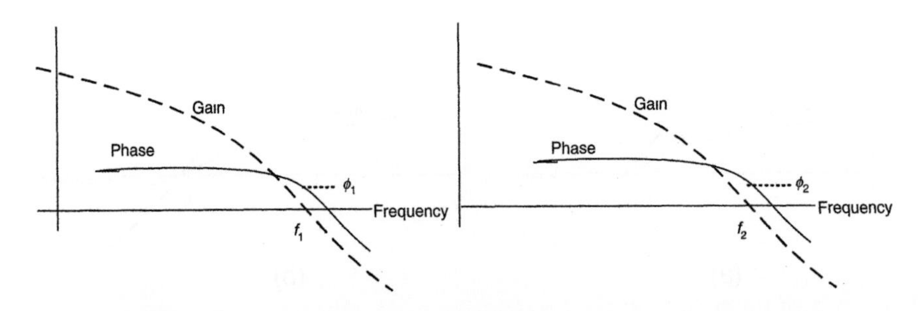

Figure 15.24: Two stable systems with the same phase margin, $\Phi_1 = \Phi_2$

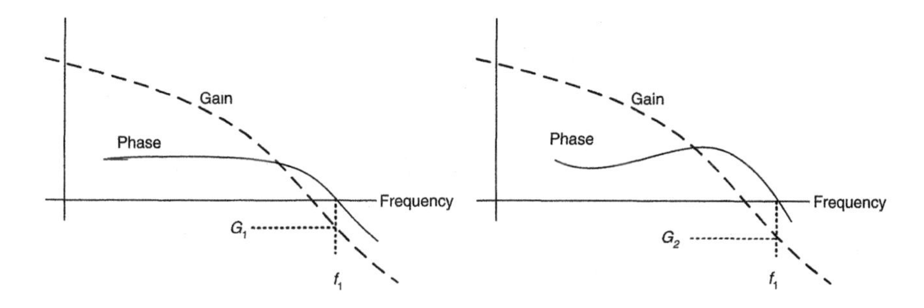

Figure 15.25: Two stable systems with different gain margins, $G_1 < G_2$

15.9 Electromagnetic Harmonic Emissions

By the late 1990s, the trend of personal computers penetrating every sector of business was very clear. Riding along with it, the high-efficiency switching-power supplies found their unchallenged foothold in the PC revolution, albeit serving in the background as unsung heroes. The picture is not all rosy, however. Sitting next to many other pieces of EMI (electromagnetic interference)-sensitive equipment, the omnipresent switch-mode power converters create a highly electromagnetically polluted environment in the office or factory space and on the utility grid. The source of pollution can be confidently traced to the on/off action of power converter switches and the resulting AC currents flowing in the switch and other power-processing components (rectifiers and inductors, in particular).

In section 15.3 we gave our attention to the output ripple voltage alone. We now focus our attention on the dynamic currents residing in the power train of every switch-mode converter. With this in mind, the first thing that comes to mind is the input filter loaded by a pulsating current attributed to the turning on and off of the main power switch, for instance Q of Figure 1.30. The input filter contains an inductor in series with the input source. However, the series inductor cannot totally block the pulsating current.

Considering the stringent requirements various international standards demand, designers of power converters need to estimate, accurately if possible, one requirement: the conducted emission injected onto the input bus by the pulsating current. In dealing with such a study, most engineers gravitate toward simulation. The approach, however, has the

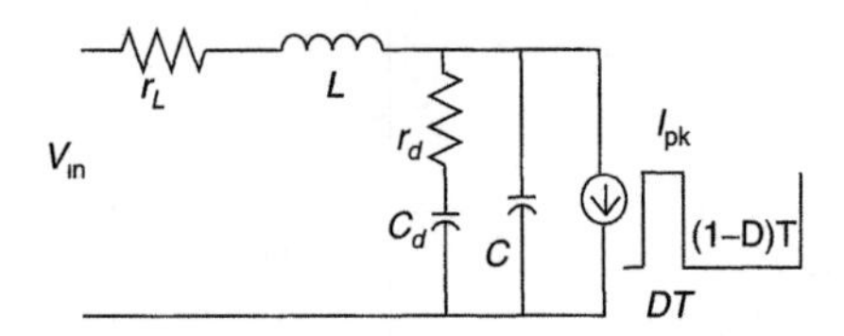

Figure 15.26: Input filter with pulsating current load

shortcoming discussed in Chapter 14. We instead address this specific topic employing our old friend the accelerated steady state (refer to Figure 15.26). Again, two differential equation sets, $dx/dt = A \bullet x + B \bullet u_1$ and $dx/dt = A \bullet x + B \bullet u_2$, are given, corresponding to two time segments, $0 < t < DT$ and $DT < t < T$, where

$$A = \begin{bmatrix} -\frac{r_L}{L} & 0 & -\frac{1}{L} \\ 0 & -\frac{1}{r_d C_d} & \frac{1}{r_d C_d} \\ \frac{1}{C} & \frac{1}{r_d C} & -\frac{1}{r_d C} \end{bmatrix}, \quad B = \begin{bmatrix} \frac{1}{L} & 0 \\ 0 & 0 \\ 0 & -\frac{1}{C} \end{bmatrix},$$

$$u_1 = \begin{bmatrix} V_{\text{in}} \\ I_{\text{pk}} \end{bmatrix}, \quad u_2 = \begin{bmatrix} V_{\text{in}} \\ 0 \end{bmatrix} \tag{15.20}$$

Using MATLAB and the concept of the continuity of states, Appendix G gives the input ripple current and the ripple voltage at the junction of source resistance and input choke. The harmonic content of the ripple voltage is also analyzed. The additional step yields a noise spectrum that forms the basis of evaluation and approval for meeting various EMI requirements and standards.

15.10 Power Quality

So far, we put in a great deal of effort to have an end product that is robust in performance and reliable in quality, which is a totally separate issue in its own right and not covered at all in this writing. So, what is the measure of performance? Obviously, every user with neither understanding of the workings of modern switch-mode power converters nor an idea about the cost factors involved in making them will spell out a wish list as shown in the following for a DC–DC converter:

Input line DC voltage range.
Output DC voltage range.
Output DC load range.
Output DC voltage-load regulation.
Output DC voltage-line regulation.
Output-load transient response (overshoot, undershoot, settling time).
Output-line transient response (overshoot, undershoot, settling time).
Output-overvoltage threshold.
Output-overload threshold.

The word *wish* may sound offensive. But, many of these performance figures are indirectly mutually exclusive. One may be made acceptable at the expense of others. Furthermore, the list does not include cost considerations, which in general is not at the disposal of the designer. Size and weight should also be part of the equation. And how about cooling, the operating environment, electromagnetic interference, safety, reliability, and other weird requirements? The point is that, once those considerations are factored in, a proper specification for a power converter is no longer a short list but a document containing many pages. Still, electrical performance tops the list and is what electrical engineers expect to optimize.

As discussed previously, steady-state output sensitivities against components can be numerically evaluated. It was also understood that steady-state output regulation is the cumulative result of component tolerances and operating conditions. Given the ability of computing sensitivity, the most sensitive components can be identified and their sensitivities (volts/ohm, volts/unit change) quantified. At a higher cost, steady-state regulation requirements can be met by tightening the tolerance of those offending components with high sensitivity; for example, V_{ref} and feedback resistors.

Next we tackle the time-varying aspect of the list. It was understood from previous studies that high-speed response, or short settling time, requires a wide bandwidth. Wide bandwidth, in turn, requires pole–zero locations at high frequency as part of the closed-loop gain.

High-frequency pole–zero locations are translated into small capacitor values that set those pole–zero frequencies. The small capacitor values run against the requirements of small overshoot under transient perturbation. Stated differently, to support a high-speed transient load (high *di/dt*) without large excursions off the regulation band, additional output capacitors are generally called for. But large capacitor values degrade bandwidth. Clearly, the transient performance of power converters does not enjoy the advantage regulation possesses, in which performance improvement is easily achievable by way of tightening component tolerances alone. Compromises among conflicting requirements are therefore called for.

Worse yet, sensitivity figures of bandwidth, phase margin, gain margin, and overshoot against components are not readily available. Obtaining symbolic solutions for those performance parameters is prohibitively impossible. In theory, the frequency at which loop gain's 0dB crossing must first be found and expressed in explicit, closed form. With that, the phase margin at that frequency also must be identified in closed form. Phase margin sensitivity can then be evaluated. A similar process is carried out for the gain margin. The difficulty, however, lies at the juncture of finding 0dB crossing frequency and expressing it in symbolic form. Without it, all bets are off. That being said, it is possible to do the following given a loop gain $T(s)$. Theoretically speaking, the 0dB crossing frequency, f_{BW}, can be obtained numerically:

$$|T(j2\pi \cdot f_{BW})| - 1 = 0 \tag{15.21}$$

With f_{BW} identified, the phase margin is given by $\angle T(j2\pi \cdot f_{BW}, \ldots, R_x)$ and its sensitivity against R_x is $\partial \angle T(j2\pi f_{BW}, \ldots, R_x)/\partial R_x$. Similarly, the gain margin and sensitivity can be obtained by first finding the zero-phase frequency, f_0, with the imaginary part of the loop gain set to zero:

$$\mathrm{Im}[T(j2\pi \cdot f_0)] = 0 \tag{15.22}$$

The gain margin, $|T(j2\pi \cdot f_0, \ldots, R_x)|$, and sensitivity, $\partial \angle T(j2\pi f_0, \ldots, R_x)/\partial R_x$, follows. However, another difficulty arises. Whatever improvement is made for the phase margin based on phase margin sensitivity figures may work against the gain margin, or vice versa. In essence, the uncertainty is recognized. But, we are awkwardly equipped to provide an answer. This unpleasant situation spills over to our inability to control overshoot. Not only are we unable to formulate the closed-loop transient overshoot in symbolic form, we are also unable to correlate the response with bandwidth and stability.

In summary, we understand what constitutes high quality for power converter performance under normal and abnormal operations, but we are awfully short of means, even if cost is not a concern. At least we understand at a qualitative level how things work. This is fertile ground where a lot of innovations are waiting to be made.

Appendix A

Additional Filtering for Forward Converter Current Sensing

The equivalent circuit for the current sensor with additional filter (R_f and C_f) is depicted in Figure 1.30. For the steady-state analysis, the driving current source is placed in $(a + bt)$ form by rewriting equation (1.50). It is understood that

$$a = \bullet \frac{1}{n_i} \frac{N_s}{N_p} \left(I_o - \frac{(V_o + V_D)(1 - D)}{2L \bullet f_s} \right) \tag{A.1}$$

$$b = \frac{1}{n_i} \left(\frac{N_s}{N_p} \frac{\frac{N_s}{N_p} V_{\text{in}} - V_D - V_o}{L} + \frac{V_{\text{in}}}{L_p} \right) \tag{A.2}$$

Given an unknown initial voltage, V_Φ, the output voltage in Laplace transformation form is

$$V(s) = \frac{V_\Phi}{s + \frac{1}{\tau}} + k \bullet a \bullet \frac{1}{s\left(s + \frac{1}{\tau}\right)} + k \bullet b \bullet \frac{1}{s^2\left(s + \frac{1}{\tau}\right)} \tag{A.3}$$

where

$$\tau = (R_s + R_f)C_f, \quad k = \frac{R_s}{(R_s + R_f)C_f}$$

During the on-time $D \cdot T_s$, the output voltage in time domain is given as

$$v_1(t) = [V_{\Phi 1} \cdot e^{-t/\tau} + k \cdot a \cdot f_1(t) + k \cdot b \cdot f_2(t)][u(t) - u(t - D \cdot T_s)] \quad \text{(A.4)}$$

while during the off-time it is given as

$$v_2(t) = \left[V_{\Phi 2} \cdot e^{-\frac{t - D \cdot T_s}{\tau}} + k \cdot a \cdot f_1(t - D \cdot T_s) + k \cdot b \cdot f_2(t - D \cdot T_s) \right] [u(t - D \cdot T_s) - u(t - T_s)] \quad \text{(A.5)}$$

where $f_1(t)$ and $f_2(t)$ are the inverse transforms of their corresponding transfer functions in (A.3).

The continuity of states at the time-domain transition boundaries requires

$$V_{\Phi 1} \cdot e^{\frac{D \cdot T_s}{\tau}} + k \cdot a \cdot f_1(D \cdot T_s) + k \cdot b \cdot f_2(D \cdot T_s) = V_{\Phi 2} \quad \text{(A.6)}$$

$$V_{\Phi 2} \cdot e^{\frac{T_s - D \cdot T_s}{\tau}} + k \cdot a \cdot f_1(T_s - D \cdot T_s) + k \cdot b \cdot f_2(T_s - D \cdot T_s) = V_{\Phi 1} \quad \text{(A.7)}$$

Both unknowns, $V_{\Phi 1}$ and $V_{\Phi 2}$, can be solved from (A.6) and (A.7) and expressed in terms of V_{in}, V_o, D, and other components. The steady-state duty cycle is then determined by

$$V_{\Phi 1} \cdot e^{\frac{D \cdot T_s}{\tau}} + k \cdot a \cdot f_1(D \cdot T_s) + k \cdot b \cdot f_2(D \cdot T_s) = v_{\text{er}} \quad \text{(A.8)}$$

We define an implicit function

$$w(V_{\text{in}}, V_o, D, v_{\text{er}}) = V_{\Phi 1} \cdot e^{\frac{D \cdot T_s}{\tau}} + k \cdot a \cdot f_1(D \cdot T_s) + k \cdot b \cdot f_2(D \cdot T_s) - v_{\text{er}} = 0 \quad \text{(A.9)}$$

Appendix B

MathCAD Listing, Steady-State Output for Figure 1.42

DC–DC converter output filter, Steady-state output, Laplace transform approach:

$Np := 19$

$Ns := 12$

$Vbus := 28$

Expected output voltage:

$Vo := 5.3$

Load:

$Io := 2.5$

$$RL := \frac{Vo}{Io}$$

Output filter:

$L := 102 \cdot 10^{-6}$

$C := 100 \cdot 10^{-6}$

$rL := 0.005$

$rC := 0.01$

$$Rp := \frac{rC \cdot RL}{rC + RL}$$

$$k := \frac{RL}{rC + RL}$$

Rectifier drop

$VD := 0.5$

Auxilary equation

$vo(t) = Rp \cdot i(t) + k \cdot v(t)$

Expected switching duty cycle

$$Du := \frac{Vo + VD + rL \cdot Io}{\left(\frac{Ns}{Np}\right) \cdot Vbus}$$

Switching frequency

$fs := 125 \cdot 10^3$

$$Ts := \frac{1}{fs}$$

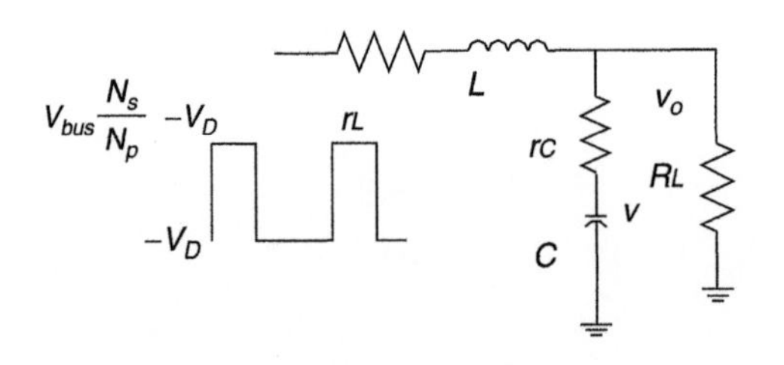

$$D(s) := s^2 + \left[\frac{-1}{(rC \cdot C)} \cdot k + \frac{1}{(rC \cdot C)} + \frac{1}{L} \cdot Rp + \frac{1}{L} \cdot rL\right] \cdot s + \frac{1}{L} \cdot \frac{rL}{(rC \cdot C)} - \frac{1}{L} \cdot \frac{rL}{(rC \cdot C)} \cdot k + \frac{1}{L} \cdot \frac{Rp}{(rC \cdot C)}$$

$$F1(s) := \frac{s + \frac{1-k}{rC \cdot C}}{D(s)}$$

$$F2(s) := -\frac{\frac{k}{L}}{D(s)}$$

$$F3(s) := \frac{\left(\frac{Ns}{Np} \cdot Vbus - VD\right) \cdot \frac{1}{L} \cdot \left(s + \frac{1-k}{rC \cdot C}\right)}{s \cdot D(s)}$$

$$G1(s) := \frac{\frac{Rp}{rC \cdot C}}{D(s)}$$

$$G2(s) := \frac{s + \frac{rL + Rp}{L}}{D(s)}$$

$$G3(s) := \frac{\left(\frac{Ns}{Np} \cdot Vbus - VD\right) \cdot \frac{1}{L} \cdot \frac{Rp}{rC \cdot C}}{s \cdot D(s)}$$

$$F4(s) := \frac{-VD \cdot \frac{1}{L} \cdot \left(s + \frac{1-k}{rC \cdot C}\right)}{s \cdot D(s)}$$

$$G4(s) := \frac{-VD \cdot \frac{1}{L} \cdot \frac{Rp}{rC \cdot C}}{s \cdot D(s)}$$

$f1(t) := invlaplace(F1(s), s, t)\ float, 3 \rightarrow$

$f2(t) := invlaplace(F2(s), s, t)\ float, 3 \rightarrow$

$f3(t) := invlaplace(F3(s), s, t)\ float, 3 \rightarrow$

$f4(t) := invlaplace(F4(s), s, t)\ float, 3 \rightarrow$

$g1(t) := invlaplace(G1(s), s, t)\ float, 3 \rightarrow$

$g2(t) := invlaplace(G2(s), s, t)\ float, 3 \rightarrow$

$g3(t) := invlaplace(G3(s), s, t)\ float, 3 \rightarrow$

$g4(t) := invlaplace(G4(s), s, t)\ float, 3 \rightarrow$

$I := identity\ (2)$

$$A1 := \begin{pmatrix} f1(Du \cdot Ts) & f2(Du \cdot Ts) \\ g1(Du \cdot Ts) & g2(Du \cdot Ts) \end{pmatrix}$$

$$B1 := \begin{pmatrix} f3(Du \cdot Ts) \\ g3(Du \cdot Ts) \end{pmatrix}$$

$$A2 := \begin{bmatrix} f1[(1 - Du) \cdot Ts] & f2[(1 - Du) \cdot Ts] \\ g1[(1 - Du) \cdot Ts] & g2[(1 - Du) \cdot Ts] \end{bmatrix}$$

$$B2 := \begin{bmatrix} f4[(1 - Du) \cdot Ts] \\ g4[(1 - Du) \cdot Ts] \end{bmatrix}$$

On starting:

$$X1 := (I - A2 \cdot A1)^{-1} \cdot (A2 \cdot B1 + B2)$$

Off starting:

$$X2 := A1 \cdot X1 + B1$$

Inductor current:

$$\mathrm{i(t)} := \left[X1^T \cdot \begin{pmatrix} f1(t) \\ f2(t) \end{pmatrix} + f3(t)\right] \cdot (\Phi(t) - \Phi(t - Du \cdot Ts)) \ldots$$
$$+ \left[X2^T \cdot \begin{pmatrix} f1(t - Du \cdot Ts) \\ f2(t - Du \cdot Ts) \end{pmatrix} + f4(t - Du \cdot Ts)\right]$$
$$\cdot (\Phi(t - Du \cdot Ts) - \Phi(t - Ts))$$

Output voltage:

$$\mathrm{v(t)} := \left[(X1)^T \cdot \begin{pmatrix} g1(t) \\ g2(t) \end{pmatrix} + g3(t)\right] \cdot (\Phi(t) - \Phi(t - Du \cdot Ts)) \ldots$$
$$+ \left[(X2)^T \cdot \begin{pmatrix} g1(t - Du \cdot Ts) \\ g2(t - Du \cdot Ts) \end{pmatrix} + g4(t - Du \cdot Ts)\right]$$
$$\cdot (\Phi(t - Du \cdot Ts) - \Phi(t - Ts))$$

$$\mathrm{vo(t) := Rp \cdot i(t) + k \cdot v(t)}$$

$$\mathrm{v(t)} := \sum_{\mathrm{n=0}}^{4} \mathrm{vo(t - n \cdot Ts)_0}$$

$$\mathrm{iL(t)} := \sum_{\mathrm{n=0}}^{4} \mathrm{i(t - n \cdot Ts)_0}$$

$$\mathrm{t := 0, 0.001 \cdot Ts..5 \cdot Ts}$$

Appendix C

MATLAB Listing, Steady-State Output for Figure 1.42

```
Np=19;
Ns=12;
L=102*10^ - 6;
rL=0.005;
rc=0.01;
C=100*10^ - 6;
Vbus=28;
VD=0.5;
Vo=5.3;
Io=2.5;
RL=Vo/Io;
fs=125*10^3;
T=1/fs;
k=RL/(RL+rc);
```

```
Rp=(rc*RL)/(rc+RL);
Du=(Vo+VD+Io*rL)/((Ns/Np)*Vbus);
vg1=(Ns/Np)*Vbus-VD;
vg2=-VD;
a11=-(rL+Rp)/L;
a12=-k/L;
a21=Rp/(rc*C);
a22=-(1-k)/(rc*C);
A=[ a11 a12; a21 a22 ];
B=[1/L; 0];
A1=A*Du*T;
A2=A*(1-Du)*T;
I=eye(2,2);
X1=inv(I-expm(A2)*expm(A1))*(-expm(A2)*inv(A)*
    (I-expm(A1))*B*vg1-inv(A)*(I-expm(A2))*B*vg2)
X2=expm(A1)*X1-inv(A)*(I-expm(A1))*B*vg1
Y1=[];
for t=0:T/100:Du*T;
Y1=[Y1 expm(A*t)*X1-inv(A)*(I-expm(A*t))*B*vg1];
end
Y2=[];
for t=Du*T:T/100:T;
Y2=[Y2expm(A*(t-Du*T))*X2-inv(A)*(I-expm(A*
    (t-Du*T)))*B*vg2];
end
```

```
Time=0:T/100:T;
Current=[Y1(1,:) Y2(1,:)];
Voltage=[Y1(2,:) Y2(2,:)];
Output=Current*Rp+k*Voltage;
figure(1)
plot(Time,Current);
figure(2)
plot(Time,Output);
```

Appendix D

MathCAD Listing, Steady-State Current-Sensing Output

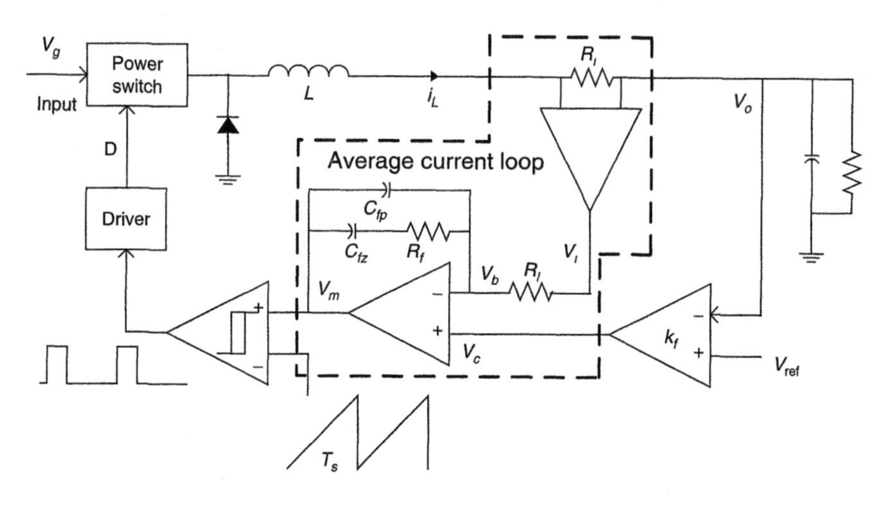

$K := 10^3$

$u := 10^{-6}$

$m := 10^{-3}$

$n := 10^{-9}$

$p := 10^{-12}$

$Fs := 50 \cdot K$

$Vg := 14$

$Vo := 5$

$L := 37.5 \cdot u$

$C := 380 \cdot u$

$Rc := 20 \cdot m$

$Ri := 0.1$

$R1 := 2.2 \cdot K$

$Rf := 30.5 \cdot K$

$Cfz := 5.8 \cdot n$

$Cfp := 220 \cdot p$

$T := \frac{1}{Fs}$

$d := \frac{Vo}{Vg}$

Load resistance:

$R := 1$

Control voltage:

$Vc := 0.5$

$\delta i := \frac{Vg - Vo}{L} \cdot d \cdot T$

$$\tau 1 := Cfp \cdot \frac{R1 \cdot Rf}{R1 + Rf}$$

$$IL := \frac{Vo}{R}$$

$$Ib := IL - \frac{\delta i}{2}$$

$$\tau 2 := Rf \cdot Cfp$$

$$\tau 3 := R1 \cdot Cfp$$

$$\tau 4 := Rf \cdot Cfz$$

$$Sn := \frac{V_g - V_o}{L}$$

Let

$$ton := d \cdot T$$

$$tof := (1 - d) \cdot T$$

Current slope alone:

$$\omega p := \frac{1}{\tau 4} + \frac{1}{\tau 2}$$

$$\omega z := \frac{1}{\tau 4}$$

$$\omega x := \frac{1}{\tau 1} + \frac{1}{\tau 4}$$

$$\omega y := \frac{1}{\tau 3}$$

$$\omega h := \frac{1}{\tau 2}$$

$$a1 := \frac{V_g - V_o}{L \cdot Ib}$$

$$B1 := \frac{a1}{\omega p}$$

$$A1 := \frac{1}{\omega p} \cdot (1 - B1)$$

$$D1 := -A1$$

$$b1 := \frac{Ri \cdot Ib}{\tau 3} \cdot \left(A1 + B1 \cdot ton + D1 \cdot e^{-\omega p \cdot ton}\right) - \frac{\omega x}{\omega p} \cdot \left(1 - e^{-\omega p \cdot ton}\right) \cdot Vc$$

$$a11 := e^{-\omega p \cdot ton}$$

$$a12 := -1$$

$$a2 := \frac{-Vo}{L \cdot (Ib + \delta i)}$$

$$B2 := \frac{a2}{\omega p}$$

$$A2 := \frac{1}{\omega p} \cdot (1 - B2)$$

$$D2 := -A2$$

$$b2 := \frac{Ri \cdot (Ib + \delta i)}{\tau 3} \cdot \left(A2 + B2 \cdot tof + D2 \cdot e^{-\omega p \cdot tof}\right) - \frac{\omega x}{\omega p} \cdot \left(1 - e^{-\omega p \cdot tof}\right) \cdot Vc$$

$$a21 := -1$$

$$a22 := e^{-\omega p \cdot tof}$$

$$vbx := \frac{\begin{vmatrix} b1 & a12 \\ b2 & a22 \end{vmatrix}}{\begin{vmatrix} a11 & a12 \\ a21 & a22 \end{vmatrix}}$$

$$vby := \frac{\begin{vmatrix} a11 & b1 \\ a21 & b2 \end{vmatrix}}{\begin{vmatrix} a11 & a12 \\ a21 & a22 \end{vmatrix}}$$

$$vb1(t) := vbx \cdot e^{-\omega p \cdot t} + Vc \cdot \frac{\omega x}{\omega p} \cdot \left(1 - e^{-\omega p \cdot t}\right) - \frac{Ri \cdot Ib}{\tau 3} \cdot \left(A1 + B1 \cdot t + D1 \cdot e^{-\omega p \cdot t}\right)$$

$$vb2(t) := vbye^{-\omega p(t-ton)} + Vc \cdot \frac{\omega x}{\omega p} \cdot \left[1 - e^{-\omega p \cdot (t-ton)}\right] - \frac{Ri(Ib + \delta)}{\tau 3} \cdot \left[A2 + B2(t - ton) + D2e^{-\omega p \cdot (t-ton)}\right]$$

Voltage at the Rf–Cfz node:

$$vb(t) := vb1(t) \cdot (\Phi(t) - \Phi(t - ton)) + vb2(t) \cdot (\Phi(t - ton) - \Phi(t - T))$$

$$t := 0,\ 0.002 \cdot T..T$$

$$Idc := \frac{1}{Rf \cdot T} \cdot \left[\int_0^T (Vc - vb(t))dt\right]$$

The compensator output:

$$vm(t) := vb(t) - \frac{1}{Cfz} \cdot \int_0^t \left(\frac{Vc - vb(\tau)}{Rf} - Idc\right) d\tau$$

The source function during ton:

$$v1(t) := Ri \cdot (Ib + Sn \cdot t)$$

during tof:

$$v2(t) := Ri \cdot \left[Ib + \delta i - \frac{Vo}{L} \cdot (t - d \cdot T)\right]$$

$$vi(t) := v1(t) \cdot (\Phi(t) - \Phi(t - d \cdot T)) + v2(t) \cdot (\Phi(t - d \cdot T) - \Phi(t - T))$$

Slope of vm(t) at t = ton according to the preceding calculation:

$$\mathrm{Sm(t)} := \frac{\mathrm{Vc}}{\tau 1} - \frac{\mathrm{v1(t)}}{\tau 3} - \frac{\mathrm{vb(t)}}{\tau 2}$$

$$|\mathrm{Sm(ton)}| = 1.714 \times 10^5$$

Slope of vm(ton) according to (2) of paper:

$$\mathrm{Sn} \cdot \mathrm{Ri} = 2.4 \times 10^4$$

$$\omega \mathrm{i} := \frac{1}{\mathrm{R1} \cdot (\mathrm{Cfp} + \mathrm{Cfz})}$$

$$\mathrm{Sn}' := \omega \mathrm{i} \cdot \mathrm{Sn} \cdot \left[\mathrm{d} \cdot \mathrm{T} + \left(\frac{1}{\omega \mathrm{z}} - \frac{1}{\omega \mathrm{p}}\right) \cdot \left(1 - \mathrm{e}^{-\omega \mathrm{p} \cdot \mathrm{d} \cdot \mathrm{T}}\right)\right] \cdot \mathrm{Ri}$$

$$\mathrm{Sn}' = 2.195 \times 10^5$$

$$\frac{\mathrm{Sn}'}{\mathrm{Sn} \cdot \mathrm{Ri}} = 9.145$$

$$\frac{|\mathrm{Sm(ton)}|}{\mathrm{Sn} \cdot \mathrm{Ri}} = 7.143$$

Finding external ramp waveform:

$$\mathrm{Se} := 0.15 \cdot 10^6$$

$$\mathrm{vx} := 1$$

Given

$$\mathrm{Se} \cdot \mathrm{ton} = \mathrm{vm(ton)}$$

$$\mathrm{se} := \mathrm{Find(Se)}$$

$$\mathrm{se} = 6.647 \times 10^4$$

$$\mathrm{ve(t)} := (se \cdot t) \cdot (\Phi(t) - \Phi(t - 0.99T)) + \left[se \cdot 0.99T - \frac{se \cdot 0.99T}{0.01T} \cdot (t - .99T)\right] \cdot (\Phi(t - .99T) - \Phi(t - T))$$

Appendix E

MATLAB Listing, Converter Simulation

```
N1=19;
N2=12;
Lm=1150*N1^2*10^-9;
rw=0.1;
Ron=0.2;
Roff=10^7;
L1=102*10^-6;
rL1=0.005;
Vg=28;
L2=385*10^-6;
rL2=0.077;
RL=5;
fs=125*10^3;
T=1/fs;
```

```
C9=100*10^-6;
rc1=0.01;
VD=0.5;
Vref=2.5;
A=10^5;
R35=9.76*10^3;
R70=11.11*10^3;
kf=R35/(R35+R70);
Rp=(rc1*RL)/(rc1+RL);
k=RL/(rc1+RL);
R12=2*10^3;
R45=24.3*10^3;
C37=560*10^-12;
C29=2200*10^-12;
R43=10*10^3;
Vbe=0.7;
R20=5.11*10^3;
R33=2.94*10^3;
hfe=150;
R19=5.11*10^3;
R31=6.65*10^3;
R39=33.2*10^3;
C39=560*10^-12;
C7=60*10^-6;
rc2=2.61;
```

```
C8=60*10^-6;
Rx=50*10^6;
Du=0.328;
a11=-(rL1+Rp)/L1-((rw + Ron)*(N2/N1)^2)/L1;
a12=-k/L1;
a18=N2/(L1*N1);
a19=-N2*(rw+Ron)/(L1*N1);
a21=Rp/(rc1*C9);
a22=(k-1)/(rc1*C9);
a31=-(A*kf*Rp)/((1+A)*R12*C37);
a32=-(A*kf*k)/((1+A)*R12*C37);
a33=-A/((1+A)*R45*C37)-1/(R45*C29);
a34=(A/((1+A)*C37))*(1/R12+1/R45+(1+A)/Rx)+1/(R45*C29);
a41=(kf*Rp)/((1+A)*R12*C37);
a42=(kf*k)/((1+A)*R12*C37);
a43=1/((1+A)*R45*C37);
a44=-(1/((1+A)*C37))*(1/R12+1/R45+(1+A)/Rx);
Kb2=(A*R20)/(R20*R43+(1+hfe)*R33*(R20+R43));
a54=((hfe*R31*R19/(R31+R19))/((1+A)*R31*C39))*Kb2;
a55=(R19/((R19+R31)*R31)-1/R31-(1+A)/R39)/ (1+A)*C39);
a66=-rL2/L2;
a68=-1/L2;
a77=-1/(rc2*C7);
a78=1/(rc2*C7);
a81=-N2/(N1*C8);
```

```
a86=1/C8;
a87=1/(rc2*C8);
a88=-a87;
a89=-a86;
a91=-N2*(rw+Ron)/(N1*Lm);
a98=1/Lm;
a99=-(rw+Ron)/Lm;
A1 = [ a11 a12 0 0 0 0 0 a18 a19; a21 a22 0 0 0 0 0 0 0; a31 a32 a33 a34 0 0
    0 0 0;a41 a42 a43 a44 0 0 0 0 0;0 0 0 a54 a55 0 0 0 0;0 0 0 0 0 a66 0
    a68 0;0 0 0 0 0 0 a77 a78 0;a81 0 0 0 0 a86 a87 a88 a89;a91 0 0 0 0 0 0
    a98 a99];
b1=-VD/L1;
b3=-A^2*Vref/((1+A)*Rx*C37);
b4=A*Vref/((1+A)*Rx*C37);
Kb1=(-R43*(10-VD)+(R20+R43)*(10-Vbe)-A*R20*Vref)/
    (R20*R43+(1+hfe)*R33*(R20+R43));
b5=((hfe*R31*R19/(R31+R19))/((1+A)*R31*C39))*Kb1+(A*Vref)/((1
   +A)*R39*C39);
b6=Vg/L2;
B=[b1;0;b3;b4;b5;b6;0;0;0];
d11=-(rL1+Rp)/L1;
d99=-(rw+Roff)/Lm;
A2 = [ d11 a12 0 0 0 0 0 0 0; a21 a22 0 0 0 0 0 0 0; a31 a32 a33 a34 0 0 0 0
    0;a41 a42 a43 a44 0 0 0 0 0;0 0 0 a54 a55 0 0 0 0;0 0 0 0 0 a66 0 a68
    0;0 0 0 0 0 0 a77 a78 0;0 0 0 0 0 a86 a87 a88 a89;0 0 0 0 0 0 0 a98
    d99];
A3=A1*Du*T;
A4=A2*(1-Du)*T;
```

```
I=eye(9,9);
X1=inv(I-expm(A4)*expm(A3))*(-expm(A4)*inv(A1)*(I-
    expm(A3))*B-inv(A2)*(I-expm(A4))*B)
X2=expm(A3)*X1-inv(A1)*(I-expm(A3))*B
Y1=[];
for t=0:T/100:Du*T;
Y1=[Y1 expm(A1*t)*X1-inv(A1)*(I-expm(A1*t))*B];
end
Y2=[];
for t=Du*T+T/100:T/100:T;
Y2=[Y2 expm(A2*(t-Du*T))*X2-inv(A2)*(I-expm(A2*
    (t-Du*T)))*B];
end
%output inductor current
Current=[Y1(1,:) Y2(1,:)];
iL=repmat(Current,[1 5]);
%ideal cap
Voltage=[Y1(2,:) Y2(2,:)];
%output voltage
Output=Rp*Current+k*Voltage;
vo=repmat(Output,[1 5]);
%transformer secondary current
%switch current
%C29 node
VC29=[Y1(3,:) Y2(3,:)];
v2=repmat(VC29,[1 5]);
```

```
%error amp inverting node
Vinvert=[Y1(4,:) Y2(4,:)];
vn=repmat(Vinvert,[1 5]);
%pwm inverting node
Vpwminv=[Y1(5,:) Y2(5,:)];
vn2=repmat(Vpwminv,[1 5]);
%input current
Iin=[Y1(6,:) Y2(6,:)];
i2=repmat(Iin,[1 5]);
%input filter cap one
VC7=[Y1(7,:) Y2(7,:)];
v3=repmat(VC7,[1 5]);
%primary winding voltage
Vpri=[Y1(8,:) Y2(8,:)];
v4=repmat(Vpri,[1 5]);
%mag current
imag=[Y1(9,:) Y2(9,:)];
im=repmat(imag,[1 5]);
%plotting
Time = 0:T/100:5*T-T/100;
figure(1)
plot(Time,iL);
figure(2)
plot(Time,vo);
figure(3)
```

```
plot(Time,v2);
figure(4)
plot(Time,vn);
figure(5)
plot(Time,vn2);
figure(6)
plot(Time,i2);
figure(7)
plot(Time,v3);
figure(8)
plot(Time,v4);
figure(9)
plot(Time,im);
```

Appendix F

Capacitor and Inductor

Capacitor

The mathematical equation governing the capacitor behavior is given by

$$i_c = C\frac{dv_c}{dt}$$

in differential form. It certainly can be placed in integral form:

$$v_C = v_0 + \frac{1}{C}\int_0^t i_C(\tau)d\tau$$

Both equations give quite a few important insights:

1. When $i_c = 0$, $dv_c/dt = 0$. The latter signifies analytically the existence of extreme values for variable v_c at the particular moment. In other words, v_{c_max} or v_{c_min} occurs at the zero crossing of the capacitor current; figure follows:

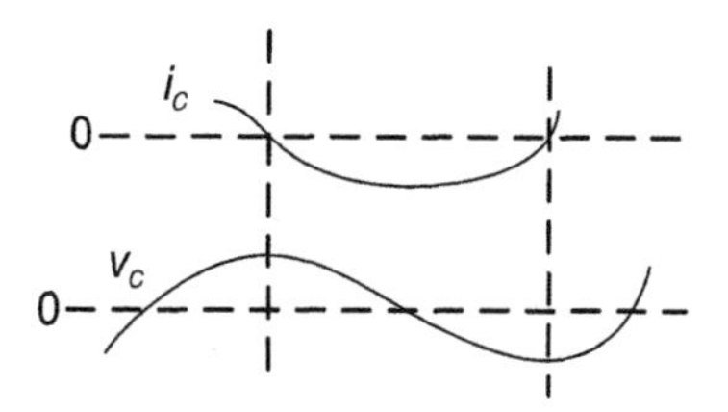

2. Being the integral of the other variable i_c (integrand), capacitor voltage lags behind the corresponding current.
3. The second fact also implies that the capacitor does not take in direct current for long. If it does, the terminal voltage exceeds its maximum rating and destroys its dielectric material.
4. Direct voltage does not contribute to the capacitor current. The capacitor blocks direct current.
5. The capacitor can be biased with a direct voltage.

Inductor

The mathematical equation governing the inductor behavior is given by

$$v_L = L\frac{di_L}{dt}$$

in differential form. It certainly can be placed in integral form:

$$i_L = i_0 + \frac{1}{L}\int_0^t v_L(\tau)d\tau$$

Both equations give quite a few important insights:

1. When $v_L = 0$, $di_L/dt = 0$. The latter signifies analytically the existence of extreme values for variable i_L at the particular moment. In other words, i_{L_max} or i_{L_min} occurs at the zero crossing of the inductor voltage; figure follows:

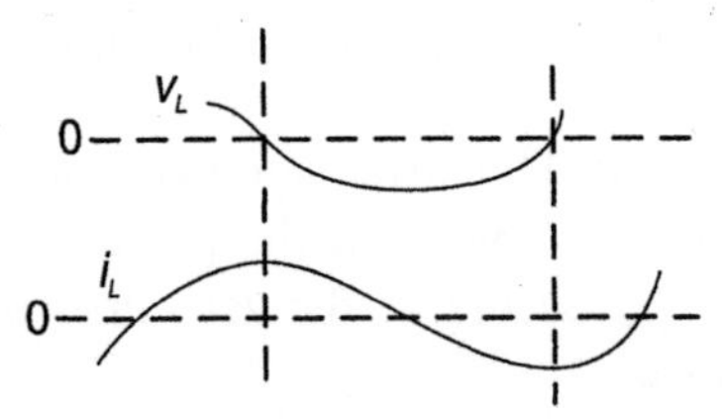

2. Being the integral of the other variable v_L (integrand), the inductor current lags behind the corresponding voltage.
3. The second fact also implies that the inductor does not take in direct voltage for long. If it does, the terminal current exceeds saturation.
4. Direct current does not contribute to the inductor voltage.
5. The inductor can be biased with a direct current.

Appendix G

MATLAB Listing for an Input Filter with a Pulsating Load

```
N1=19;
N2=12;
L=385*10^-6;
rL=0.005;
rd=2.6;
Cd=60*10^-6;
C=60*10^-6;
Vin=28;
Vo=5.3;
VD=0.5;
Io=2.5;
Ipk=(N2*Io)/N1;
fs=125*10^3;
T=1/fs;
```

```
Du=(Vo+VD+Io*rL)/((N2/N1)*Vin);
a11=-rL/L;
a12=0;
a13=-1/L;
a21=0;
a22=-1/(rd*Cd);
a23=-a22;
a31=1/C;
a32=1/(rd*C);
a33=-a32;
A=[ a11 a12 a13; a21 a22 a23; a31 a32 a33];
B=[1/L 0; 0 0; 0 -1/C];
A1=A*Du*T;
A2=A*(1-Du)*T;
I=eye(3,3);
u1=[Vin; Ipk];
u2=[Vin; 0];
X1=inv(expm(A*T)-I)*(expm(A2)*inv(A)*(I-expm(A1))*B*u1+
    inv(A)*(I-expm(A2))*B*u2)
X2=expm(A1)*X1-inv(A)*(I-expm(A1))*B*u1
Y1=[];
for t=0:T/100:Du*T;
Y1=[Y1 expm(A*t)*X1-inv(A)*(I-expm(A*t))*B*u1];
end
Y2=[];
for t=Du*T+T/100:T/100:T;
```

```
Y2=[Y2 expm(A*(t-Du*T))*X2-inv(A)*(I-expm(A*(t-Du*T)))*B*u2];
end
Current=[Y1(1,:) Y2(1,:)];
Voltage=[Y1(3,:) Y2(3,:)];
iL=repmat(Current,[1 5]);
vo=repmat(Voltage,[1 5]);
Time=0:T/100:5*T-T/100;
figure(1)
plot(Time,iL);
figure(2)
plot(Time,vo);
```

References

[1] R.D. Middlebrook and Slobodan Cuk. *Advances in Switched-Mode Power Conversion*. Pasadena, CA: TESLco.

[2] Keng C. Wu. *Pulse Width Modulated DC-DC Converters*. London: Chapman & Hall, 1997.

[3] Wei Tang and Fred C. Lee. "Small-Signal Modeling of Average Current-Mode Control." *IEEE Trans. on Power Electronics*. PE-8, no. 2 (April 1993): 112.

[4] Texas Instruments. "Phase Shifted, Zero Voltage Transition Design Considerations and the UC3875 PWM Controller." (Unitrode) Application Note, U-136A.

[5] Kwang-hwa Liu, Ramesh Orugnati, and Fred C. Lee. "Quasi-Resonant Converters—Topologies and Characteristics." *IEEE Trans. on Power Electronics* PE-2, no. 1 (January 1987): 62.

[6] Marian K. Kazimierczuk. "Class-E DC/DC Converters with a Capacitive Impedance Inverter." *IEEE Trans. on Industrial Electronics*. IE-36, no. 3 (August 1989): 425.

[7] Marian K. Kazimierczuk. "Class-E Amplifiers with an Inductive Impedance Inverter." *IEEE Trans. on Industrial Electronics*. IE-37, no. 2 (April 1990): 160.

[8] Nathan O. Sokal and Alan D. Sokal. "Class E-A New Class of High-Efficiency Tuned Single-Ended Switching Power Amplifiers." *IEEE Journal of Solid-State Circuits*. SC-10, no. 3 (June 1975).

[9] Jacek J. Jozwik and Marian K. Kazimierczuk. "Analysis and Design of Class-E^2 DC/DC." *IEEE Trans. on Industrial Electronics*. IE-37, no. 2 (April 1990): 173.

[10] Keng C. Wu. *Transistor Circuits for Spacecraft Power System.* Norwell, MA: Kluwer Academic Publishers, 2002.

[11] Ron-Jie Tu and Chern-Lin Chen. "A New Space-Vector-Modulated Control for a Unidirectional Three-Phase Switch-Mode Rectifier." *IEEE Trans. on Industrial Electronics.* IE-45, no. 2 (April 1998): 256.

[12] K. Mark Smith and Keyue Ma Smedley. "Lossless Passive Soft-Switching Methods for Inverters and Amplifiers." *IEEE Trans. on Power Electronics.* PE-15, no. 1 (January 2000): 164.

Index

Printed and bound by CPI Group (UK) Ltd, Croydon, CR0 4YY
08/05/2025
01864862-0001